AF517956

ENCYCLOPAEDIA OF ENVIRONMENTAL WASTE POLLUTION

Encyclopaedia of Environmental Waste Pollution

VOLUME I

G R CHHATWAL

ANMOL PUBLICATIONS PVT. LTD.
NEW DELHI-110 002

ANMOL PUBLICATIONS PVT. LTD.
4374/4B, Ansari Road, Daryaganj
New Delhi-110002

Encyclopaedia of Environmental Waste Pollution

First Edition 1997

ISBN 81-7488-439-4 (Set)

PRINTED IN INDIA

Published by J.L. Kumar for Anmol Publications Pvt. Ltd., New Delhi-110002 and Printed at Mehra Offset Press, Delhi.

Preface

Excessively consumer-oriented culture and human greed have done irreparable damages to the environment and ecology of this planet. Waste and its disposal is the gravest problem before the mankind. In the prevalent words pollution is the wrong amount of the wrong thing in the wrong place at the wrong time. Environmental pollution due to waste, whatever may its source be, has reached menacing dimensions. This present Encyclopaedia of Environmental Waste Pollution which is compiled and edited judiciously in two volumes is veritable mine of authoritative scientific information on the problem of waste pollution.

No doubt, the present encyclopaedia will prove most useful reference tool to the researchers, teachers, policy planners and activists in the field of environment. This will prove useful to the general readers as well.

Editor

Preface

Excessively consumer-oriented culture and human greed have done irreparable damages to the environment and ecology of this planet. Waste and its disposal is the gravest problem before the mankind. In the parlance of waste, pollution is the wrong amount of the wrong thing in the wrong place at the wrong time. Environmental pollution due to waste, whatever may its source be, has reached threatening dimensions. This present Encyclopaedia of Environmental Waste Pollution which is compiled and edited judiciously in two volumes contains remarkable amount of authentic and scientific information on the problem of waste pollution.

No doubt, the present encyclopaedia will prove a most useful reference tool to the researchers, teachers, policy planners and activists in the field of environment. This will prove useful to the general readers as well.

Editor

Contents

1

The Problem of Waste

Waste, refuse, rubbish, garbage and trash are all words used to describe something unpleasant—and something produced by man alone. The problem of waste, and waste disposal, is therefore entirely man-made. Before *Homo sapiens* appeared on earth, there were no real waste materials, although even our most distant ancestors soon saw to it that a waste problem was created.

The body constituents of most plants and animals are efficiently recycled after they die. The dead organism is rapidly broken down by other animals, by fungi and by bacteria so that its essential chemicals are liberated into the soil. There they serve as nutrients for growing plants, which in turn are consumed by herbivorous animals. Most of the nutrients are then quickly returned to the soil in the herbivore's excrement. The herbivores themselves also serve as food for carnivores, which similarly recycle nutrients in their excrement. Finally, the bodies of the carnivores decay after their death, and their substances are incorporated into the soil for further use by the plants.

In the wild, it is unusual for either dung or corpses to accumulate in objectionable amounts, because the organisms concerned with recycling are generally ready and waiting to perform their essential duties. The only exception is when alien introductions produce dung to which the native consumers are unadapted. Thus in Australia, the insects which dealt efficiently with kangaroo faeces could not break down cowpats which then produced swarms of objectionable flies. Today, dung beetles from the countries which were the homes of the cattle have been introduced into Australia to reduce this nuisance. On the other hand, we have often made the recycling of man-made wastes a slow, difficult and unpleasant process—the situation being greatly complicated by the sophistications of civilization.

Decay and recycling is the rule for most dead animal and plant materials, but under certain circumstances other processes take over. Our so called fossil fuels — coal, oil and natural gas — were produced many millions of years ago when the dead vegetation did not decompose. We can see the beginning of the processes which produced these energy-rich fuels in a modern peat bog. Here the acid conditions inhibit decay, and the plant material is preserved in the form of peat which, with compression and time, could produce coal. It is true that other substances, including animal matter, may also be preserved in peat bogs. Archaeologists have found wooden artefacts, textiles and even bodies (Tollund man was dug up in Denmark, and a similar corpse from a Cheshire bog in a remarkable state of preservation). However, these are special cases—breakdown and decay is the general rule.

The distinguished archaeologist Jacquetta Hawkes writes 'Waste disposal by human beings may be said to have begun when hunters of the earliest Stone Age tossed their gnawn bones over their shoulders.' She points out that Palaeolithic hunters 'were not cave-proud. They allowed bones and other animal debris, wood ash and waste flakes of flint knapping to accumulate under them until the cave might be reduced to half its height' So their methods of disposal were not very efficient, and most ancient settlements were eventually buried under their own debris.

Excrement, dung and urine, have always produced hygienic and aesthetic problems for man and, in fact, for other animals. Birds are said not to 'foul their own nests' and may take precautions not to do so. Thus the young songbirds in the nest may produce their excrement in sacs which are tidily removed from the nests by the parents and dropped some distance away. Many mammals, including rabbits, have areas away from their dens which serve as 'latrines', where the ground first shows signs of great fertility and then, with continued use, of pollution from overfertilization, which kills the vegetation.

Nomadic people were, and often still are, unhygienic, moving their camp sites when they become too unpleasant. However, the ancient Israelites were aware of the problem. In Deuteronomy, Chapter 23, verses 12 and 13 we read: 'Thou shalt have a place without the camp, whither thou shalt go forth abroad, And thou shalt have a paddle upon thy weapon; and it shall be, when thou wilt ease thyself abroad; thou shalt dig therewith, and shalt turn back and cover that which cometh from thee. That is the Authorised Version. In this case I prefer the James Moffatt translation, first published in Scotland in 1924, which puts it rather more concisely; 'Also, you must have a latrine outside the camp and go outside to it; you must carry a spade among your weapons, and when you relieve yourself outside, you must dig a hole with it, to cover up your filth'.

Jacquetta Hawkes comments on London as follows: 'The city of London may be regarded as a modest form of tell (i.e. an artificial mound of accumulated remains of ancient settlements) with the streets of the Roman town lying some 20 feet below our own. It provides a convincing proof of the greater civic efficiency of the Romans over Londoners of the Middle Ages. At both periods pits were dug and filled with waste, but while in Roman times these held ash and other dry materials, those of a thousand years or so later were used for organic domestic refuse that must have added to the notorious stink of medieval London' Similarly, reports of excavations of Anglo-Saxon York show that much of the rubbish often remained within the house, where the floors were covered with material which had 'many resemblances to that of herbivorous dung' , and which supported a similar fauna of scavenging insects.

The wastes produced by our unhygienic ancestors, and the materials found in their 'kitchen middens' have enabled archaeologists to produce a picture of their life-style. In the same way modern, investigative journalists, often in defiance of law, examine the contents of the dustbins of the famous or the notorious, and by counting champagne bottles or caviar jars, produce articles describing the conspicuous eating habits of their victims.

Until a hundred years ago, waste disposal and sewage treatment in London and other cities was very ineffective, although many of the great houses and the monasteries had reasonably salubrious privies. In the cities most refuse was allowed to rot in the street, where it encouraged vermin of various types, including rats, mice and crows. Until the end of the 18th century, the Red Kite (***Milvus milvus***) was common in London, scavenging on the refuse. The presence of this species, now a rare denizen of the more remote parts of Wales, was resented by many of the inhabitants. They were probably wrong in their attitude, for kites got rid of stinking and decaying food, and probably discouraged rats, which were potentially more dangerous carriers of disease. Kites became scarce in the 19th century when they ran short of food in the cities, and were harried by gamekeepers in the country. In 1900 only 12 pairs remained in Wales; the bird was extinct throughout the rest of Britain. The Red Kite is now strictly preserved, and its population has increased to over 40 breeding pairs. One reason for this increase may be that they have returned to the refuse tips of Wales (just as their relatives do in Africa and Australia), and this extra food supply may be as useful to them as it is to the hordes of gulls which also frequent the refuse disposal sites.

When humans came to live in towns and cities, there was usually some attempt to reduce pollution of the whole environment from excrement, although this was not always successful. Contemporary reports suggest that even such grand places as the Palace of Versailles stank rather badly Attempts were made to alleviate the smell by the use of pomanders, and by spreading sweet herbs on the floor. Royalty and the aristocracy were constantly travelling around the country. These peregrinatory habits were partly to escape from the ordure and other noxious waste materials which the court always generated.

Until a couple of centuries ago, when there was no main drainage, chamber pots in Edinburgh were emptied from upper windows into the street; hence the warning cry of 'Gardy loo' (Scotch-French for 'beware of the water') so that passers-by could try to avoid being soaked with the contents. One of the earliest types of public lavatory, featured in a series of prints of London's street life, was a bucket carried by a woman, into which passers-by who were caught short could defecate covered by the woman's cloak (with only their head appearing rather like a young 'joey' in a kangaroo's pouch).

However, in the towns at any rate, much of the excrement was voided into some sort of latrine. James Greig has described the contents of a 15th century barrel latrine in Worcester, where the concentrated faecal matter was well preserved. Greig found traces of 20 different plants, mainly fruits and vegetables, together with a rich fauna of human intestinal parasites. Night soil from earth closets was known to be a rich fertilizer, and was generally collected (together with material left about the streets) and used in market gardens, producing excellent, nutritious if unhygienic and parasite-infested vegetables. The market gardens of Hertfordshire, for instance, were famous for their fertility, which was based on the use of London's night soil. It is likely that all members of the population, from the richest to the paupers, carried a selection of round worms and tapeworms as well as many bacterial and protozooal parasites.

Actually, human dung is still employed as manure in many countries. Its use in China is particularly well developed, and here the material, perhaps because it is often partially sterilized by the heat when it is composted, appears to avoid some of the infestation from intestinal parasites common in other countries. Until recently, less effort was made to prevent pollution from urine than from faeces, and in Britain we still find notices in quiet passages in our cities urging people to 'Commit no Nuisance' Where these notices are ignored, the plants, in particular algae and bryophytes, which grow on the walls, are likely to be affected. At first they will benefit from the rich source of plant nutrients, but will soon suffer from a surfeit of riches.

The introduction of the water closet provided the opportunity for improving the situation. This convenience is generally believed to have been invented by Sir John Harington, godson of Queen Elizabeth, in the 16th century. It is said that the Queen installed a model in Windsor Castle, and that she kept a copy of Sir John's masterpiece The *Metamorphosis of Ajax* (i.e. a jakes, that is a privy), which describes his invention, as reading matter in the appropriate room. However, like so many British inventions, this one was not widely exploited at home, and WCs do not seem to have been at all common until the appropriately named Mr. Crapper went into business in London in the early 19th century. Flushing water closets were then widely introduced, with a great improvement in urban conditions; but as the contents of the WCs were discharged untreated into the Thames, the effect on the tidal stretches of the river was horrendous. At times Parliament had to suspend sitting because of the stench. During the summer the windows were covered with cloths soaked in disinfectant. The river became‘ ‘dead’ and all fish and most other forms of aquatic life were eliminated. 1856 became known as ‘the year of the great stink’

The situation improved from the end of the 19th century onward. The capital’s raw sewage was diverted from the London stretch of the Thames, and discharged further down the river (where, of course, damage continued to occur,but since the area was much less densely populated, fewer people, and no Members of Parliament, complained). Then various methods of sewage treatment were installed, not only in London but in all towns further up the river. By the 1960s fish were found to have returned even to what had been the most polluted reaches. The importance of ‘sewage farms’ and other forms of treatment are discussed later in this book.

Today we have a major industry disposing of our wastes as is described later in this book. Domestic refuse is regularly collected from practically every dwelling in Britain and in other developed countries. This has considerably reduced the food supply of urban vermin, although foxes may still pillage dustbins. In Britain landfill is much the most important form of refuse disposal. This is, in effect, a development of the refuse pits of Roman and medieval London. These waste disposal sites may develop considerable ecological interest, as was shown by Arnold Darlington in 1969 in his pioneer work *The Ecology of Refuse Tips*. The management of such sites obviously has a considerable effect on the animals and plants found on them.

To most of us, the 20,000,000 tonnes of domestic waste — sometimes called ‘municipal waste’ — constitute the main problem in Britain today. This is the material we put out of our houses every week in dustbins or plastic sacks, the collection of which adds substantially to our rates. It is what so disfigures many of our streets, where leaking sacks may be ripped open by dogs in search of bones, and where cardboard cartons add to the litter. However, as is indicated in the eleventh report of the Royal Commission on Environmental Pollution (Cmd 9675, 1985) it constitutes only about four per cent of the total waste of some 482,000,000 tonnes generated in Britain. Agriculture makes a contribution of 200,000,000 tonnes, mineral working of 135,000,000 tonnes and industry of 100,000,000 tonnes.

All these different types of waste produce different ecological problems. Recently there has been much concern about so-called ‘toxic wastes’ although many substances so classified are not particularly poisonous, and those that are may be present in very small quantities. Perhaps the public is most worried by the problem of the safe disposal of radioactive waste from nuclear power stations. The problems of disposing of these different wastes are discussed in later chapters of this book. Surprising results are sometimes obtained. Those wastes which might be expected to have the greatest environmental effects, such as radioactive materials, often do little damage, and the sites containing the most toxic substances may develop the most interesting flora.

In addition to the solid and liquid wastes produced directly by humans and by industry, we have many gaseous wastes given out mainly as a byproduct of the liberation of energy from fossil fuels. When these gases are in high enough concentrations, they can do serious damage to living organisms and to whole ecosystems. As well as these direct effects, they may produce important results in areas for distant from the site of emission, as the gases may suffer chemical changes as they pass through the atmosphere. This results in the phenomenon commonly called ‘acid rain’. As will be seen in Chapters 5 and 6, this is a much more complicated subject than is generally realized. Furthermore, these waste gases, and other emission, from industrial processes and domestic apparatus, may have global effects, possibly affecting climate and the amount of ultraviolet radiation reaching the earth's surface.

Waste heat, particularly from electric generating stations, affects many rivers, lakes and even parts of the sea, when water is abstracted and used for cooling purposes. This heat may have important consequences for fresh water and for marine flora and fauna.

There is public concern about the potential danger from radioactive waste material from nuclear power stations, and also from possible 'leaks' of radiation from nuclear power stations, themselves. The ecological effects of all such radiation have often been considerably exaggerated, but nevertheless this is clearly a subject which merits our attention.

The purpose of this book is to describe the effects of all important types of waste upon the countryside and the environment in which we live. I have discussed the effects on man, and the world as a whole, and in some cases I have suggested ways in which both local and global dangers may be avoided.

2
Introduction to Solid Waste Pollution

Introduction

Environmental problems include air pollution, water pollution, and solid waste disposal. It is unfair to classify solid waste as being deleterious to a single part of the environment; rather solid waste problems transcend traditional environmental boundaries and contribute to air and water pollution as well as land pollution.

At all levels of development human beings produce domestic wastes: at the very minimum these comprise kitchen wastes, ashes from fires, broken utensils and worn-out clothing. In a small agriculture community nature readily accepts these wastes into her natural cycle: animals consume food residues, and other materials are rapidly incorporated into the soil.

The industrial revolution in the temperate countries concentrated people in urban areas of very high population density and added new sources of wastes : shops, institutions and factories. Apart from any question of public health it was impossible to accommodate these wastes, arising at a rate of up to one ton/family/year, within the urban areas. Thus, in those countries where industrialization came early, services for the regular removal of domestic and trade wastes have been in operation for a hundred years or more.

Many changes have taken place during this period : the character of the wastes has altered in line with rising living standards, changes in retail distribution methods and fuel technology. The volume has greatly increased; storage methods have evolved from open heaps through portable containers to expendable containers; transport has changed from horses to motor vehicles and from open trucks to compactor vechicles. New problems have arisen—high—rise buildings, supermarkets, industrial wastes of many kinds—and usually efficient solutions have been devised.

At this stage in their development, most industrialized countries exhibit the following common features:

- a high level of national wealth which is still increasing;
- a taxation system which provides a reliable source of funds for public services such as refuse collection;
- a long tradition of regular and efficient removal of refuse direct from dwellings and other premises, and usually operated by a municipality ;

- a high level of sensitivity to aesthetic standards within private premises (which does not always extend to standards in public places);
- willing acceptance by householders shopkeepers, and others of statutory duties such as providing and maintaining storage containers, sometimes of a specified type and size, and assisting in making containers available for emptying by municipal employees at prescribed times.

During the past twenty years the main thrust of technical development has been towards mechanization, because of rising labour costs, even to the point of installing underground pneumatic transport systems at a few high-rise developments. Although manual emptying of containers continues to be employed almost everywhere, the ratio of vehicles to men is rising and service is becoming more capital intensive with investment up to Rs. 400000 per collector in some cities.

In the industrialized countries, therefore, basic health and environmental problems have been solved in the storage and collection of solid wastes, although major problems remain in regard to resource recovery and disposal. Despite recurring conflicts over the allocation of national resources as between public and private expenditure, sufficient money is available normally to provide an efficient public service and the citizens co-operate well with the system. The technology of wastes handling is now highly developed; substantial sectors of industry are engaged in the production of equipment; there are established institutions for technical training; there is a wealth of technical literature much of which is devoted to the efficient deployment of labour in relation to capital equipment— another indicator of western pre-occupation with rising labour costs in the service industries.

By contrast India and many other developing countries are suffering all the problems of urbanization, often with population densities much higher than in any western city, but without the financial resources needed to provide solutions of the kind used in the West.

The developing countries are not unaware of the importance of avoiding the environmental pollution which shadowed urbanization and industrialization in Europe and North America. The quality of urban environment is a matter of growing concern to them and the importance of efficient solid wastes management is increasingly recognized. The West has often been able to provide developing countries with technical guidance in environmental matters, but solid wastes may prove to be an exception; there are too many climatic, economic and social differences for systems to be successfully transplanted. Even the technical literature of the West may be of little value for either training or operational guidance.

Sources and Characteristics

''Solid wastes'' is the term now used internationally to describe non-liquid waste materials arising from domestic, trade, commercial, industrial, agriculture and mining activities, and from the public services. "Non-liquid" is a relative term because sludges of certain kinds fall within the scope of solid wastes management; these arise primarily from industrial sources and from sewage treatment plants.

Solid wastes comprise countless different materials: dust, food wastes, packaging in the form of paper, metals, plastics or glass, discarded clothing and furnishings, garden wastes, and hazardous and radioactive wastes.

Domestic wastes have three main characteristics: weight generated, density and constituents; these vary not only from country to country, but from town to town within a country, according to the level of industrial development and other factors :

- the range of weight generated/person/day usually lies between 250 and 1000 g. worldwide,
- density varies from 100 kg/cu metre to 600 kg/cu metre,
- thus volume may range between 1/2 litre and 10 litres/person/day,

The main constituents of domestic wastes vary as follows:

- vegetable/putrescible matter, 20% to 75%,
- intert matter, 5% to 40%,
- paper, 2% to 66%,
- glass, 0% to 10%,
- metals, 0% to 15%,

No rational decisions on solid wastes systems are possible until data of this kind are available. The method and capacity of storage, the correct type of collection vehicle, the optimum size of crew and the frequency of collection depend mainly on volume and density; climate also has some influence. The disposal method may be conditioned by the proportions of saleable materials that could be a source of energy.

Just as solid wastes comprise a vast number of materials, they arise from a multitude of separate sources: a city of a million inhabitants may have a quarter of a million separate sources, as well as many kilometres of streets upon which solid wastes accumulate. Thus the four main aspects of solid wastes management are :

- storage at or near the point of generation,
- collection,
- street cleansing,
- disposal.

An apt definition of solid wastes is "matter in the wrong place ", implying that a material becomes waste only when a specific owner ceases to have a use for it. Yesterday's newspaper is waste to the man who bought it, but it could be the raw material for paper mill. Current waste disposal philosophy is to endeavour to treat all wastes as resource materials : some for recycling ; some for conversion to fertilizer or as a source of energy: and the balance for land reclamation.

Solid-waste pollution has necessarily always been more advanced in the largest cities. Disposal problem become difficult with increase of population density. Simultaneously there is a greater production of waste per unit area, and a decreased proportion of land available for its disposal. Therefore, the history of solid-waste pollution is largely connected with the histories of the largest cities.

From the earliest civilization burial of wastes has always been comparatively easy in rural areas, and was specified by Moses as a required practice for Israel (Deuteronomy 23, vv 12-13).

During the flourishing of the Minoan civilization on Crete from 3000 to 1000 BC, solid wastes in the capital of Knossos were placed in large pits with layers of earth at intervals. However, the Romans had no organized system of waste removal; disposal and wastes accumulated in the streets and around the towns and villages (ibid). This practice was said to have persisted until the 19th century.

Ancient cities have yielded their secrets to archeologists partly through analysis of their refuse piles. Because so much waste materials was reused in various ways the ultimate discards tended to be long-lived. Examples are ceramic pot shards and the dust and mud from floor and highway sweepings. Despite the small quantities of these materials which were produced, the steady accumulation year after year contributed in many cases to the eventual burying of the city and to its subsequent rebuilding either on the same site at a higher level or at another site.

The old Roman city of Bath in England is between 12 and 20 feet above the level of the Roman period. The rush-covered floors of houses in Tudor times contained debris up to two- or three-feet thick and rich in nitrates. Until the first importation of nitrtates in 1610 this debris was reputed to be the sole source of the saltpetre used in the manufacture of gun-powder. Even the right to "digge for saltpetre" was sold (ibid.).

Some medieval German cities avoided the danger of being buried in their own wastes by requiring that departing wagons which had been used, perhaps, to bring farm produce into the city return with a load of wastes to be deposited in the countryside.

Few features of the environment have been more abrasive to the senses than the discarded objects of civilized life. Litter along the roadsides, trash on the beaches, heaps of refuse in the streets, the pungent piles of rotting garbage, and similar unwelcome debris are unpleasant reminders of untidy human habits. Thousands of alleys, backyards, and vacant lots contain the remains of discarded refrigerators, defunct washing machines, and parts of shattered automobiles— rusting remnants of the chrome plated glory of a technological society. Open dumps scar the landscape; and highway travelers view a countryside spotted with the incongruity of beauty and blight.

The process of civilized living has been creating mountains of solid waste, the disposal of which has created a formidable problem. In countries where the population is high and land scarce, the conventional municipal refuse tip will soon become a thing of the past, unless we are prepared to sacrifice our valuable landscape and countryside. In many advanced countries, the actual amount of refuse which is produced is equal to 300 kg. occupying 1.5 m^3, per person per year. There is every sign that these quantities would go up in the near future. Two main factors are responsible for this :

1. In the past much domestic refuse was made to burn into open fires to burn. With the spread of central and electric heating this practice has been to some extent discontinued.

2. An increasing amount of packaging has been used for the goods we buy, more newspapers are being read, etc. all increasing the quantity of refuse generated by the average family.

The composition of refuse gets varied from area to area. Where open coal fires are still being used, a high proportion of the refuse would be having partly and fully burned clinker and ash; modern housing estates supply a less dense refuse which largely consists of packaging materials. A seasonal variation also occurs in the weight of refuse produced per household ; it tends to be is at its maximum in winter and its minimum in summer.

Because of the changes of living standards and habits today, a rapid change in the nature and composition of refuse has taken place. The average composition of refuse in 1975 and the forecast for 1985 has been provided in the following table:

It is expected that by 1990, the quantity of refuse per home will rise from 15 kg. per home per week to 20 kg.

Component	*Percentage 1975*	*Percentage 1985*
Dust	24	12
Cinders	12	6
Vegetables and other rotting materials	17	13
Cardboard and paper	25	50
Metals	7	6
Textiles	3	3
Glass	8	6
Miscellaneous materials	6	4

Plastic waste has been an important and increasing component of domestic refuse. About 100 g. per week of waste plastics are given out per dwelling, but the nuisance created by waste plastics has been much greater than has been suggested by the moderate quantities involved. Plastics do not rot, like most other materials, although they can be burned. PVC (polyvinyl chloride) has been especially objectionable in that it forms highly corrosive hydrochloric acid when burned, which is an extreme nuisance in the operation of refuse incinerators. If we use an inadequately high temperature in burning waste PVC, it is even possible for highly poisonous phosgene ($POCl_3$) to get produced.

Modern man has been using an enormous quantity of basic raw materials, converting them at first into valuable products, and later, when the products get discarded, into waste. It has been estimated that the present annual rate of man-induced scatter of certain elements would be as in they table below.

Element	*Metric tonnes per annum*
Iron	345×10^6
Manganese	2×10^6
Copper	4.8×10^6
Zinc	4.3×10^6
Nickel	660×10^3
Lead	2.5×10^6
Molybdenum	70×10^3
Chromium	7×10^3
Silver	12×10^3
Mercury	10×10^3
Tin	168×10^3

According to the Law of Conservation of Mass, it means that waste materials, theoretically, remain available for use on this planet. Availability in theory and usefulness in practice have been, however, very far apart, because the cost of recovering almost all of these elements invariably becomes much greater than the cost of obtaining them from our natural resources. Until quite recently, we have realised that they can be and will be severely depleted in the near future, and that we must now make greater efforts to salvage them from our own wastes.

The 'natural history' of the motor car has been a classic example of man's recklessness in the matter of essential raw materials. In one component part or another, most of the elements shown in the table above finds use in motor manufacture. These elements are first extracted from crude ore, then refined, alloyed, treated and formed into specific parts; once all the parts get assembled together, the completed motor car has a high monetary and utilitarian value. For several years it is usefully used, until it can no longer satisfy economically the needs of any owner. 'Planned obsolescence' ensures that this point would get reached after a given number of years or a given mileage travelled, whichever gets achieved the sooner. As often as not the vehicle is then abandoned to dis-figure the landscape, since its low scrap value is hardly enough to induce its owner to do anything else with it. In consequence, large quantities of valuable materials would get virtually lost to man for ever. Since a technological endeavour, salvaging these materials has been well within the ability of man; but economically, reclamation of specific elements from composite waste would not be a viable commercial proposition. It is widely accepted that the economic balance must be adjusted so that it is always economically viable to extract useful raw materials from waste.

Not all of the components of solid waste are valuable. Some are useless and others are toxic. For example, selenium, a poisonous element that occurs naturally in the soil, is taken up by plants that are used for marking paper and is present in newspaper in the amount of about 8.6 ppm. This element is found in incinerator stack gas, raw solid wastes, incinerator residue, and in water that has been used for removing fly ash. Many other toxic chemicals of various kinds and amounts are present in municipal and industrial waste.

The adverse economics of the disposal of solid wastes as commonly practiced are far-reaching. Mineral resources such as copper, zinc, lead, and tin are limited and the ores are nonrenewable. These methods are not destroyed by disposal but are often scattered beyond recovery. Even iron ore is in finite supply in economically desirable deposits. The squandering of these resources that occurs when wastes are dumped indiscriminately in spoil areas will inevitably produce a disruptive economic impact at some time in the future unless improvements are adopted. Development of salvage techniques will, therefore, have direct economic effects. Salvage and recycling will also alleviate other problems associated with the rapidly increasing density in populations, which results not only in the generation of greater quantities of wastes but also in wastes of greater complexity.

Today they disposal of garbage and other solid wastes appears to be much simpler: urban dwellers merely place them in bags or other containers, put these in garbage cans, and have them picked up by a public or private collection agency. Only when one passes an open dump sees a landfill outside a big city as outer Ring Road in New Delhi the real magnitude of the solid waste problem becomes apparent.

The agriculture and food processing industry is considered to be the largest single contributor to the total annual production of solid wastes, but mining operations and some industries are also important sources. Postconsumer wastes (those generated by end user, not by operations that make consumer products, and excluding junk vehicles) were responsible for solid wastes. Of this, 37% consisted of paper products alone, the largest postconsumer source of solid waste. Table 2.1 gives a breakdown of postconsumer wastes by material categories, together with quantities recycled and the net quantities disposed. Urban refuse annually includes about 60 million cans, 360 million bottles, 53 million tons of paper, 4 million tons of plastics, and about 1 million abandoned vehicles.

The enviromentalists expect the quantities of total wastes to increase greatly by 1990, not only because of the increased population but also because an increased amount will be discarded per person. Even though larger quantities will be recycled, by 1990 the India will have 23% more postconsumer wastes to dispose than presently. Information about solid wastes have been comparatively scantly for a number of reasons. The need for this information could not be recognized a number of years ago and

many communities made no weight or other measurements. Measurements have been difficult to make due to the complexity of the material, and techniques for sampling and making laboratory measurements have not been standardized. The quantities and character of solid wastes also vary greatly with geographical location, food patterns, etc. Data collection has started to improve, however, with the holding of national conferences and with the solid wastes program of the Central Government.

A typical classification of solid wastes is the following :

1. ***Garbage*—purtrescible (decomposable) wastes from food, slaughter-houses, canning and freezing industries etc.**
2. ***Rubbish*—non-putrescible wastes, either combustible or non-combustible wastes would include paper, wood, cloth, rubber leather, and garden wastes. Non-combustibles would include metals, glass, ceramics, stones, dirt, masonry,and some chemicals.**
3. ***Ashes*—residues (such as cinders and fly ash) of the combustion of solid fuel for heating**

Table 2.1

Postconsumer Solid Wastes, Residential and Commercial Sectors, (millions of tons)

Material Categories	*Gross Discards*	*Material Recycled* Quantity	*Material Recycled* Percent Recycled	*Net Waste Disposal* Quantity	*Net Waste Disposal* Total Waste
Paper	53.0	8.7	16.5	44.2	32.8
Gas	13.5	0.3	2.1	13.2	9.9
Metals	12.7	0.2	1.6	12.5	9.3
Ferrous	11.2	0.2	1.4	11.0	8.2
Aluminium	1.0	0.04	4.0	1.0	0.7
Other non-ferrous metals	2.4	0.0	0.0	0.4	0.3
Plastics	5.0	0.0	0.0	5.0	3.7
Rubber	2.8	2.0	7.1	2.6	1.9
Leather	1.0	0.0	0.0	1.0	0.7
Textiles	1.9	0.0	0.0	1.9	1.4
Wood	4.9	0.0	0.0	4.9	3.6
Total non-food products	94.8	9.4	9.9	85.5	63.4
Food waste	22.4	0.0	0.0	22.4	16.6
Yard waste	25.0	0.0	0.0	25.0	18.5
Miscellaneous Inorganic waste	1.9	0.0	0.0	1.9	1.4
Total waste	144.0	9.4	6.5	134.8	100.0

and cooking or the incineration of solid wastes by municipal, industrial, and apartment house incinerators.

4. *Large wastes*—demolition and construction rubble pipes, lumber, masonry , brick, plastic, roofing and insulating materials). automobiles, furniture refrigerators and other home appliances, trees, tires, etc.

5. *Dead animals*—household pets, birds, rodents, zoo animals, etc. There are also anatomical and pathological wastes from hospitals.

6. *Sewage treatement process solids*—screenings settled solids, sludge.

7. *Industrial solid wastes*—chemicals, paints, sand, explosives etc.

8. *Mining wastes*—"tailings, slag heaps, culm piles at coal mines, etc.

9. *Agricutural wastes*—farm animal manures, crop residues, etc.

Some of these wastes have been the result of control techniques for air and water pollution, such as fly ash recovered from flue gases and sludge from sewage treatment plants. Others have been capable of producing air or water pollution quite readily, such as organic matter that can decay to produce odors or farm animal wastes that can enter water supplies. Much of the solid waste produced is collected or otherwise disposed of but much is permitted to litter the landscape— highway litter, dog wastes in cities, discarded oil drums etc.

The sources of solid wastes have been (1) municipal street sweepings, sewage treatment plant wastes, wastes from schools and other institutions; (2) domestic—garbage, rubbish, and occasional large wastes from homes (and, in years past, large amounts of ashes from the use of coal); (3) commercial from stores and offices; (4) industrial—from manufacturing plants: (5) mining— from coal mining, strip mining, etc; and (6) agricultural.

Solid wastes could be bound out in terms of their mass or their volume. Their densities generally vary over a wide range — typically 0.05 to 0.1 g cm^3 — and mass (or weight) measurements are felt to be more useful. Solid wastes are rather low in density due to the large amounts of air present in them and it is usually not difficult to compact them to several time their original density, as done in sanitary landfill operations or by home compactors.

The composition of refuse seems also to vary from area to area. Where open coal fires are still being used, a high proportion of the refuse consists of partly and fully burned clinker and ash; modern housing estates supply a less dense refuse which largely consists of packaging materials. There is also a seasonal variation in the weight of refuse produced per household; this is at its maximum in winter and its minimum in summer.

The composition of collected refuse gets changed over the years. with the quantity of ashes decreasing as coal heating of households gets replaced by gas, oil and electric heating and with the amount in consumer packaging.

In its place the initial step dealing with refuse will have to be incineration. Against the capital cost of constructing incinerators on the scale needed there have been several advantages to be considered. Most organic waste gets oxidized by combustion to CO_2 and water. After incineration the volume of refuse substantially gets reduced. Valuable elements such as chromium, molybdenum and lead can be recovered from the residue with relatives ease, and the heat of combustion can be usefully employed. The ultimate residue has, of course, very little value, consisting mainly of compounds of silicon and aluminium, of which the earth's crust has a superabundance. Approximately 25.7 per cent of all mineral matter is having silicon and 7.4 per cent of aluminium. Wastage of some iron is not a particular tragedy, for at 4.7 per cent of the earth's crust iron has been the fourth most widely distributed element. A proportion of the ultimate residue can already be employed as an aggregate in building construction and a ballast in road or motorway construction provided that the temperature of incineration has been sufficiently high. A further proportion can be employed in earth-filling works like dams, embankments and land reclamation. The rest-amounting to less than 10 per cent of its original volume—can only be dumped; we have, therefore, to consider where best it may be placed.

Geophysical research has revealed that a number of 'tectonic trenches' exist on the ocean floor, where a kind of convection occurs within the earth's crust itself. In one of these, near the mouth of the River Fraser, Canada, the ocean floor has been found to be sinking continuously at the the rate of some six metres per annum over an area of 12 km^2. If we deposit residual inorganic refuse in these tectonic trenches, it would rapidly sink and disappear. It is possible to defect tectonic positions with relative ease because they are associated with unusual fluctuations in gravity and with deep-seated seismic disturbances. Bostrom and Sherif, whose geophysical findings were published in *Nature* (10 October, 1970), have concluded that there are sufficient areas exhibiting tectonic phenomena to cater for all the waste materials produced by man for many years to come. The initial costs of compressing and transporting residual waste to the tectonic trenches would be somewhat but, it is thought, not unduly high once the practices became accepted and the techniques properly developed.

The simplest method has been crude *tipping* or *open dumping*, a common method which finds use in most Indian cities. More satisfactory has been *controlled tipping* or the *sanitary landfill*, which has been recently adopted in Delhi for solid waste disposal. In sanitary land fill, a layer of about 2 metres of refuse has been covered by at least 23 cm of earth, ash or other inert material, upto the level of the hole chosen. The surface can then be used for housing or sports fields. Before such filling, the wastes could be pulverized by machines to a uniform particle size : by this means the volume could be reduced and thus the life of the tip gets extended, and some of the refuge has been more quickly biodegraded.

Much of the discarded waste is swept under the rug, so to speak, in accordance with the anaesthetizing but naive philosophy, ''out of sight, out of mind." But less visible parts of the iceberg-like problem remain. Solid wastes are dumped in the ocean: mining wastes are produced at the rate of millions of tons a day; slag heaps and mill tailings accumulate near processing operations ; and industrial refuse contaminates streams and lakes.

3

Properties of Solid Waste Materials

Introduction

The first step in solving solid waste pollution problems is to discard the belief that ''waste'' is so heterogenous in its composition and so variable in its properties that problems cannot be defined, let alone solved. To be sure, waste streams exhibit great variability. The design of solid waste disposal systems must, therefore, provide for more operating flexibility, reserve capacity. and materials 'stamina" then conventional process equipment. Nonetheless, characterizations of the average composition and properties of waste streams provide the starting point for design. An appreciation of the range of excursions from these averages has been also a necessary input. In addition, forecasts of likely changes in waste characteristics should be regarded in the design. Unlike manufacturing processes, the raw materials fed to be solid waste disposal systems are usually specified by constraints and priorities beyond the influence of the waste manager.

Several properties of potential importance in waste characterization are shown in Table 3.1 Although it becomes seldom necessary to characterize wastes in all the areas in Table 3.1 such a checklist can be useful in alterting the waste manager to processing and opportunities or safety hazards.

Table 3.1

General, Physical and Chemical Parameters of Possible Significance in the Characterization of Solid Wastes

General Parameters

Compositional weight fractions			
		C.	Agricultural
A.	Domestic, commercial and institutional		Field
	Paper (broken into subcategories)		Processing
	Food waste		Animal raising
	Textiles	D.	Industrial
	Glass and other ceramics	E.	Mining/metallurgical
	Plastics	F.	Special
	Rubber		Radioactive
			Munitions, etc.

(contd.)

Leather
Metals
Wood (limbs, sawdust)
Bricks, stones, dirt, ashes

B. Other municipal
Dead animals
Street sweepings
Catch-basin cleanings

Pathogenic
Moisture

Process weight fractions

Combustible
Compostable
Processable by landfill
Salvageable
Having intrinsic value

Physical Parameters

Total wastes

Size
Shape
Volume
Weight
Density
Density stratification
Surface area
Compaction
Compactability
Temperature
Color
Odor
Age
Radioactivity
Physical state
Total solid
Liquid
Gas

Solid wastes

Soluble (%)
Suspendable (%)
Combustible (%)
Volatile (%)
Ash (%)
Soluble (%)
Suspendable (%)

Hardness

Particle characteristics

Size distribution
"shape"
"surface"
"porosity"
"sorption"
"density"
"aggregation"

Liquid wastes

Turbidity
Color
Taste
Odor
Temperature
Viscosity data
Specific gravity
Stratification
Total solids (%)
Soluble (%)
Suspended (%)
Settlable (%)
Dissolved oxygen
Vapour pressure
Effect of shear rate
Effect of temperature
Gel formation

Gaseous wastes

Temperature
Pressure

Volume
Density
Particulate (%)
Liquid (%)

Chemical parameters

General

- pH
- Alkalinity
- Hardness ($CaCO_3$)
- MBAS (methylene-blue-active substances)
- BOD (biochemical oxygen demand)
- COD (chemical oxygen demand)
- Rate of availability of nitrogen
- Rate of availability of phoshorous
- Crude fibre
- Organic(%)
- Combustion parameters
  - Heat content
  - Oxygen requirement
  - Flame temperature
  - Combustion products (including ash)
  - Flash point
  - Ash-fusion characterization
  - Pyrolysis characterization
- Toxicity
- Corrosivity
- Explosivity
- Other safety factors
- Biological stability
- Attractiveness to vermin

Inoroganic and elemental

- Moisture content
- Carbon
- Hydrogen
- (P_2O_5 and phosphate)
- Sulfur content

Heavy metals

- especially Mercury
- Lead
- Cadmium
- Copper
- Nickel

Toxic materials

- Chromium
- especially Arsenic
- Selenium
- Beryllium
- Asbestos

Eutrophic materials

- Nitrogen
- Postassium
- Phosphorus

Organic

- Soluble (%)
- Protein nitrogen
- Phosphorus
- Lipids
- Starches
- Sugars
- Hemicelluloses
- Lignin
- Phenols
- Benzene oil
- ASB (alkyl benzene sulfonate)
- CCE (carbon chloroform extract)
- PCB (Polychlorinated biphenyls)
- PNH (Polynuclear hydrocarbons)
- Vitamins (e.g B-12)
- Insecticides (e.g. Heptochlor),

Alkali metals	DDT, Dieldrin, etc.)
Alkaline-earth metals	
Precious metals	

When designing waste-processing systems it will often be found that desired data have been lacking. Many industrial and municipal-waste studies have shown the large errors possible from armchair estimates of the generation rate, composition or properties of waste. It has been strongly recommended, therefore, that the results of especially commissioned waste surveys and analyses be incorporated in the problem-definition phase of the design effort for waste-management systems. For systems disposing of municipal waste, for example, the local mix of waste-generator types, climate, level of construction and demolition activity, and existing waste-management practices and expectations can greatly affect the characteristics of the wastes which have been collected. For industrial wastes, unique processing methods or integrated plant designs can result in net waste-production rates which have been greatly from "norms" predicated on other plant configurations within the same industry.

Beyond the deviation in average properties resulting from local influences, the waste manager should also prepare for significant short-term or more modest long-term excursion from these averages. For instance, municipal waste may average 8% mental, but recently in Delaware County, Pennsylvania, a full packed-truck of magnesium alloy shavings was delivered and charged (almost intact) into a county incinerator furnace. The resulting damage to the equipment had been reported to be extensive. Although unexpected, this problem would have been moderated by an established regimen of refuse mixing. Long-term excursions could be typified by the large increase in the yardwaste content of municipal waste associated with leaf collection in the fall. Such changes in characteristics must be provided for in the design or the operating protocols of waste-processing systems. Even with such safeguards, however, the worst should get anticipated (live ammunition) cans of flammable solvent, containers of toxic chemicals, etc.)

Solid Waste Properties

Establishing the basic composition of mixed-waste streams has been regarded as the first step in analysis of solid-wastes-management systems. In order to design equipment and processes to cope with the problem, however, detailed understanding of the physical, chemical and biological properties of the mixed refuse and its components have been needed. It has been the objective of this section to provide this understanding.

Recognizing that the properties of waste materials are not necessarily influenced by their source, it is evident that many of these data have direct applicability to non-municipal waste—management analysis.

Chemical Composition

Mixed Municipal Refuse. The chemical analysis of mixed municipal refuse gets varied throughout the year and from location to location. In an extensive study of solid-waste problems, average compositions for the various categories of waste were developed (Table 3.2).

Analysis for Specific Elements in Refuse. Due to toxicological hazard, enhancement of corrosion in refuse-processing equipment, air-or water-pollution potential, and similar problems, special attention has been drawn to the concentration of selected elements in refuse. The sections below concern several

Table 3.2

Estimated Ultimate Analysis of Refuse Categories (%). (Dry Basis)

Category	*C*	*H*	*O*	*N*	*Ash*	*S*	*Ee*	*Al*	*Cu*	*Zn*	*Pb*	*Sn*	P^b	Cl^-	*Se*	*% Fixed carbons (dry basis)*
Metal	4.5	0.6	4.3	0.05	90.5	0.01	77.3	20.1	2.0		0.02	0.6	0.03			0.5
Paper	45.4	6.1	42.1	0.3	6.0	0.12									Trace	11.3
Plastics													0.01		51.	
Leather and Rubber	59.8	8.3	19.0	1.0	11.6	0.3				2.0						6.4
Textiles	46.2	6.4	41.8	2.2	3.2	0.2				2.0			0.03			3.9
Wood	48.3	6.0	42.4	0.3	2.9	0.11							0.05			14.1
Food wastes	41.7	5.8	27.6	2.8	21.9	0.25							0.24			5.3
Yard wastes	49.2	6.5	36.1	2.9	5.0	0.35							0.04			19.3
Glass	0.52	0.07	0.36	0.03	99.02											0.4
Miscell-aneous	1.30	2.0	12.0	3.0	70.0											7.5

such elements.

Chlorine. Chlorine, usually as metal chlorides, such as sodium and calcium chloride, occurs in many refuse components. As such, it appears in the fly ash of municipal incinerators. These solid chlorides may take part in corrosion reactions with hot metal surfaces in incinerators of boilers. Also, the chlorides tend to accelerate metal corrosion in incinerator wet-scrubber systems which recirculate the scrubber water. In landfills, metallic chloride (most of which are soluble) leach out into percolating rain water, thereby adding to the dissolved-salts content of the receiving waters.

In addition to the importance of the metallic chlorides, however, considerable attention has been assigned to the chlorine appearing in polyvinyl chloride (PVC) and vinylidene chloride (Saran etc.). These material appear in refuses as housewares, packaging and film, toys, etc. Pure PVC resin (i.e. without plasticizers, fillers, pigments etc.) burns in air to quantitatively convert its chlorine content to hydrogen chloride (hydrochloric acid). Each kg. of PVC thus yields about 0.59 kg of hydrogen chloride. Although some portion of the hydrogen chloride undoubtedly reacts with basic oxides in the fly ash, the majority remains in the gaseous phase. This gaseous hydrogen chloride attacks high-temperature metal surfaces in incinerator boilers like superheater tubes and is absorbed in scrubber waters to yield highly acidic, corrosive scrubber liquor.

Estimates of the fraction of PVC in plastics are shown in Table 3.3

Table 3.3

Anticipated PVC Content of Municipal Refuse.

Year	*Estimated plastics content of refuse (weigh %)*	*Estimated % PVC resin in refuse plastics*
1968	1.1	7.8
1970	1.3	10.0
1975	1.8	10.6
1980	2.7	8.7
1990	3.5	10.9
2000	4.2	11.2

Selenium. As a large fraction of municipal refuse is paper and as selenium in known to be present in almost all types of paper, the possibility of exposure to selenium in waste-management activities has been significant. Selenium, a toxic and perhaps carcinogenic element, occurs in trace quantities in refuse and in the by-products of refuse combustion.

Agricultural Wastes. Due to their entirely organic origins and water-pollution potential, analyses of agricultural wastes have been reported in dimensional units quite different from the combustion process oriented analyses of municipal and industrial refuse.

Density Characteristics

Probably no other property of solid waste has the engineering and economic importance of bulk density. The low initial density and poor compaction characteristics (at moderate pressures) of solid wastes has been found to limit carrying capacity and thus have been contributing to the high cost of collection and hauling. Indeed, few collection vehicles approach their maximum allowable payload weight when they must depart from their routes to be emptied. Incineration systems, built and operated at high expense, have been justified primarily as a volume-reduction process to extend landfill life and/or reduce long-haul transportation costs for refuse disposal.

Bulk Density. The bulk density of most solid waste has been appreciably lower than the ultimate density of its components.

Many of these wastes have been readily susceptible to two- or three-fold increases in density in conventional hydraulic compactor refuse-collection vehicles or in stationary compactors. The effect of high-pressure compaction, as in baling equipment, has been described as follows :

Compaction Effects. With increasing solid-waste-generation rates and decreasing urban landfill capacity many municipal and regional agencies have been implementing transfer-station and railhaul operations to distant disposal areas. Also, attention has to be paid to the use of baling operations as a precursor to long-haul landfill or marine disposal. Each of these operations has been found to depend importantly on substantial refuse densification, usually by using hydraulic compaction equipment. Because to the importance of compaction, the various studies have been made about the refuse densification. The following is largely drawn from the results of that effort.

The data presented below got obtained on a modified metalscrap baler shown in Fig. 3.1 In operation, ram 1 got advanced and locked to gather the material and close the charging box. Ram 2 then advanced and locked, followed by ram 3. Hydraulic pressure and ram displacement had been measured. Experiments were run on raw, shredded and synthetic municipal refuse and oversized bulky waste. A small number of tests got incorporated water-soluble adhesives which had been intended to improve the binding properties of wastes.

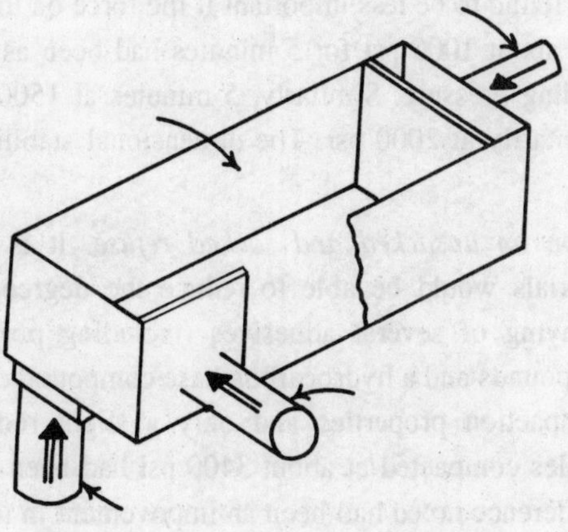

Fig. 3.1. Compaction press.

Although compaction data have been generally reported as a volume-reduction ratio, such "data" have been significantly less useful than documentation of ultimate density. It occurs due to the extreme variability in starting density occasioned by previous compaction and storage history and mechanical/operational limitations of the compactor loading hopper and loading crew. For instance, reported compaction ratios at 3500 psi (maximum pressure) ranged from 7 : 1 to 2 : 1, a 300% variation. The same data reveal a range in ultimate density from 93 to 108 pounds per cubic foot (lb/cu ft), a 15% variation. These considerations have been especially important in evaluating compaction-equipment vendor- preformance claims.

Ultimate Density. As the density of different papers is somewhat between 43 and 71 lb/cu ft, it is clear that high-density waste components (glass, metals, dirt) contribute importantly to bale density. Prepackaging of refuse in paper bags provide almost identical results to those for unpackaged refuse.

Liquids and sludges expressed during compaction. Expected increases in bale density with wet refuse could not be realized due to the expulsion of moisture during compaction and the increased bulkiness of wetted papers. Weight losses of up to 37% had been recorded for wet refuse exposed to rain, Moderately wet refuse lost only 1 to 3% of its initial weight and no loss had been recorded for dry refuse.

The expressed liquids had been greenish in appearance. The settlable sludges were gray. Both had an offensive odor. Chemical analysis of the leachings yielded the data in Table 3.4. Review of these data reveal that no unusual pollutants will occur in the refuse pressing liquors.

Volume expansion after compaction. Compacted refuse undergoes expansion upon release of the compaction pressure.The degree of expansion had been following the degree of pressure application: the greatest expansion occurs in the direction of ram 3 and the least in the direction of ram 1. Although expansion continues for several hours after the release of pressure, volume increases taking place in the initial 2 minutes were used as a benchmark measurement.

Expansion effects had been found to be less important if the force on the bales got held for several minutes. The bales compacted at about 1000 psi for 5 minutes had been as stable as those compacted at 1500 to 1750 psi without holding pressure. Similarly, 5 minutes at 1500 psi gave to product bales similar to those held only momentarily at 2000 psi. The dimensional stability of high-moisture refuse bales had been consistently poor.

Effects of adhesive addition on unsacked and sacked refuse. It is speculated that adhesives effective in bonding paper materials would be able to reduce the degree of post-compression bale expansion. The addition by spraying of several adhesives (including polysaccharide-, protein- and cellulose-base water--soluble compounds and a hydrocarbon base compound dissolved in mineral spirits) offered no improvement in compaction properties and only a slight reduction in short-term bale spring-back. The expansion of bales compacted at about 3400 psi had been about 8% less than that of the unsprayed bales. The main difference noted had been an impovement in the appearance and stability of the bales after standing. The sprayed bales got less distorted and more rugged than those not sprayed.. The degree of improvement, however appeared too little to justify the use of adhesives in commercial operation,

Table 3.4

Chemical Analysis of Leachings

Sludge	*Sample 1*	*Sample 2*
Organic matter	38.9%	76.7%
Silica as SiO_2	30.8	6.9
Aluminum as Al_2O_3	15.2	5.8
Phosphates as P_2O_5	8.9	4.4
Calcium as CaO	2.0	1.3
Magnesium as MgO	1.4	0.0
Iron as Fe_2O_3	0.5	0.6
Sulfates as SO_3	Trace	Trace
Carbonates as CO_2	0.0	0.0
Liquid		
pH	7.8	4.5
Total dissolved solids	6090 ppm	6040 ppm
Organic matter	1720	4480
Bicarbonates	90	610
Sulphates	250	200
Chlorides	21	218
Heavy metals (suspension)	Large % amount	Small % amount

Some improvement in the stability of bales of plastic sacked refuse had been noted with addition of adhesives to the outside of the sacks. It appeared importantly associated with the inadequate strength of the plastic-sacked refuse bales due to the lack of adhesion across the plastic film.

Effects of refuse shredding prior to compaction. Experimental compaction tests were undertaken with a single load of shredded refuse. The shredded refuse was obtained from the shredding facility in New Delhi. This limited test series do not show any improvement in ultimate bale density in the machine or after expansion had taken place. Data showing the volume increase of compacted, shredded refuse with time have been given in Table 3.5.

Gas formation in bales. In order to get information regarding the hazards of gas formation in compacted refuse, gas samples had been taken from six bales. The bales were stored at 85 ± 5°F. Two bales had been wrapped in plastic immediately after leaving the press. After one day, they were warm to the touch and beads of moisture had been observed on the plastic cover. The unwrapped bales remained cool and dry to the touch. After three days of storage, the remaining four bales had been wrapped. Gas composition data had been recorded. Although the gas compositions suggest aerobic bacterial activity, the short duration of the tests precludes definitive characterization of the long-term decay process and effluent gas composition.

Table 3.5

Volume Increase of Compacted Shredded Refuse with Time.

Sample	*Time (min)*	*Volume increase (%)*	*Compaction pressure (psi)*
	1	44	
1	5	49	3400
	30	58	
	1440	71	
	1	60	
	5	69	1850
2	30	82	
	1440	95	

Moisture effects on compaction and post-compaction expansion. Several sets of experiments had been made to run with mixed-paper and with newspapers of controlled moisture content. Beyond the addition of readily deformable mass with water addition, there appears to be further increase in density, perhaps associated with enhanced pliability in the paper.

The degree of bale expansion following discharge from the compaction device was also shown to get affected by refuse-moisture content. Bales of more than 40% moisture by weight had been tended to disintegrate immediately upon release of moisture.

Compaction properties of selected refuse components. A series of compaction tests were made to run on a variety of homogeneous or mixed refuse components. The test results, from experiments at compaction pressures of about 3500 psi have been reported in Table 3.6.

Combustion Properties

Combustion processes (which include on-site and municipal conventional incineration, incineration with heat recovery), combine firing of refuse and fossil fuels in utility boilers, and municipal and industrial waste-management activities.

Heat content. The most fundamental combustion property has been probably the heat content or heating value. In the tables which follow, the quantity reported is the higher heating value (HHV) expressed in Btu/lb. The HHV may be defined as the heat released from the complete combustion of the material in question starting with the reactants at 20°C and ending with all products at 20°C. Thus, with this definition, the latent heat of condensation of the water vapour formed by combustion has to be included in the heating value.

Recently a correlation based on thermochemical principles got developed to predict the heat of combustion (dry basis) from the ultimate analysis of the waste. The correlation neglects metal oxidation but the resulting error must be small in conventional incinerators. The correlation needs the knowledge of the proportion (lb per lb) of organic and inorganic (carbonate) carbon, oxygen, sulfur, nitrogen and hydrogen, denoted C_{org}, C_{inorg}, O, S, N and H respectively), all on a dry basis, in the refuse sample. The final equation has been as follows:

$$\text{Btu/lb (dry basis)} = 14{,}096\, C_{org} + 55{,}750\left(H - \frac{O}{8}\right) + 3982S - 6382\, C_{inorg} + 4274\left(\frac{O}{2}\right) + 1040N$$

Table 3.6

Baling Characteristics of Refuse Components (3500 psi Compaction Pressure)

1. *Rabber tires.* Several samples of rubber tires were subjected to compaction. However, they could not be compacted. After release of pressure the tires always separated.
2. *Rubber tires enclosed in wire mesh.* The compacted bundles of tires enclosed in the wire mesh always expanded rapidly. The wire mesh, depending on its strength, usually burst open at a few or many welded spots.
3. *Shredded rubber and residential wastes.* Residential waste, containing about 25% of small rubber pieces, produced good bales. Shredded rubber on its own could not be compacted.
4. *Metal cans and oversized metallic wastes.* All metallic waste products, including refrigerators, washing machines, stoves and appliances, produced excellent bales.
5. *Oversized wastes.* Mixtures of refrigerators or other oversized metallic wastes, together with furniture and textiles produce very stable bales.
6. *Oversized wastes and household refuse.* These mixtures produced very good bales.
7. *Plastic scrap.* Compacted mixtures of plastics did not hold together after release of pressure.
8. *Very dry papers.* Such papers did not produce stable bales.
9. *Vegetables, fruits, meat scraps.* These wastes could not be compacted on their own or in combination.
10. *Vegetables, fruits and residential refuse.* An appreciable amount of food pulp was formed by compacting residential wastes which contained a large amount of fresh vegetables or fruits. Addition of about 25 weight % of vegetables to residential wastes produced a large

amount of fine pulp which was partially released from press openings at about 900 psi. The pulp which adhered to the outside of the compacted bale fell off after removal of pressure.

11. *Cooked meats and residential refuse.* Similar results results as those obtained with vegetables and fruits were obtained with meats containing a large amount of moisture.

12. *Vegetables and wood shavings.* The addition of vegetables to wood shavings produced unstable bales.

13. *Wood shavings.* Wood shavings, alone, produced relatively good bales.

14. *Wood chips and residential refuse.* Mixtures of residential wastes and wood chips procuced good bales.

Table 3.7

Sample Computation of Stoichiometric Air Requirement from Ultimate Analysis.

1. Using the refuse ultimate analysis the refuse composition and oxygen requirement on an as-fired basis are :

Component	*Pounds per 100lb of refuse*	*Molecular weight*	*Moles*	*Required Moles O_2*
Moisture	28.16	18.0	1.564	0
Carbon	25.62	12.0	2.135	2.135
Hydrogen (bound)	2.65	2.0	1.326	1.326
Oxygen (bound)	12.21	32.0	1.326	1.326
Hydrogen	0.80	2.0	0.399	0.199
Sulfur	0.10	32.0	0.003	0.003
Nitrogen	0.64	28.0	0.023	0.0
Metal (assumed iron)	8.70	55.8	0.156	0.010
Ash	12.12	—	—	0.0
	—	—	—	—
	100.00			2.347

2. Stoichiometric air requirement
Moles oxygen = 2.35
Moles nitrogen 8.83

11.18 moles air = 322 lb per 100 lb of refuse

Air requirement. The stoichiometric (theoretical) air requirement has been an important variable in many combustion calculations. The air requirement could be readily calculated from the ultimate analysis, taking proper account of the oxygen contained in the refuse. Oxidation of metals could be assumed to occur to the extent of about 10% . For computational purposes, it can be assumed that the refuse metal is iron. A sample calculation of the stoichiometric air requirement is presented in Table 3.7.

As an alternative, it can be noted that, for many fuels the air requirement per million Btu has been approximately constant. For municipal refuse, an air requirement of 724 pounds per million Btu (HHV) has been an acceptable estimate. For industrial wastes, especially when a component other than paper has been present in high concentrations, it may be prudent to compute the air requirement from the ultimate analysis.

Residues from the combustion of municipal refuse. In the incineration process, four solid residues have been produced generated: siftings, dust, grate residue, and fly ash. Each of these residues has been a solid waste in its own right and needs disposal, usually via landfill. In the following, the chemical and physical properties of incinerator residues will be presented.

Siftings. In the operation of municipal incinerators which incorporate a grate, a fraction of the refuse and grate residue falls through the air openings in the grate. This material, called siftings, is characterized in tests on a four-section, Flynn and Emrich rocking grate.

The siftings got collected from beneath the first (feed) grate section and a composite sample got obtained from stoker sections 2, 3 and 4.

Dust. In an incinerator operation large amounts of dust have been to be found. These dusts are characterized as to heat content as a consequence of concern regarding their explosive hazard. Using techniques developed by the Bureau of Mines, the Explosivity Index (relative to coal) of incinerator dusts was determined. The results are shown in Table 3.8. In general, the dusts were found not readily susceptible to initiation and propagation of an explosion.

Table 3.8

Calorific Value Explosivity Index of Incinerator Dusts.

Dust source	*Heating value (cal/g-dry)*	*Explosion Concentration (mg/litre)*	*Explosivity index*
Airborne	3390	—	—
Charging floor	2770	—	0.33
Walls	2030	—	<0.22
Platform by pit	2280	—	<0.25
Railing near pit	2614	—	<0.28
Ground refuse	3430	70	0.8
Pittsburgh coal	6667	55	1.0

Grate residue. The material discharged from the grate of a municipal incinerator has been comprised about 25-33% of the weight but only 5-15% of the volume of the incoming refuse. These residues have been generally landfilled although interest is increasingly being shown to processing these wastes for the recovery of ferrous and non-ferrous metals and glass.

In connection with studies of by-product recovery from incineration plants, analyses had been made of the ferrous-metal residues. The high copper content, (far higher than would be expected in the eron alloys anaylized) is presumed to result from the following sequence of reactions :

(*i*) Polyvinyl chloride in the refuse burns to hydrochloric acid.

(*ii*) Hydrochloric acid attacks hot copper to form cupric chloride.

(*iii*) The cupric chloride dissolves in the quench water at the end of the grate.

(*iv*) The copper plates out on the iron by replacement.

The formation of clinkers in the refuse fire and the sintered and fused deposits of furnace enclosures has produced great interest in the fusion properties of the ash of refuse and refuse components.

Fly Ash. Fly ash, the fine particulate matter entrained in the gases leaving the burning refuse bed in an incinerator, has gained importance from two standpoints. Fly ash has been the material which gets accumulated as slag on the walls and in the flues of incinerator furnaces and has been the material which presents the most significant air-pollution-control problem of incinerator systems.

The importance of the slag-forming nature of fly ash makes its chemical analysis important. The fly-ash composition data reveal the strong similarity between the compositions of refuse ash, incinerator slag and fly ash. Thus the ash-fusion data must be useful guide in assessing the wall-temperature maximums which avoid fly-ash adherence and build-up.

The air-pollution problem due to fly ash needs an understanding of the particle-size distribution of this material.

RFFUSE CHARACTERISTICS

The generation of domestic and trade wastes is a remorseless and continuing process. Throughout the day a housewife is performing such tasks as sweeping floors and trimming vegetables; the shopkeeper is discarding packing materials and spoiled goods; the shoemaker is paring leather; almost every member of the population is continuously creating wastes of one kind or another. In the case of human body wastes, a sewage system provides facilities for transfer from the production point to the disposal place by a continuous flow process in pipes containing water. Despite the successful use of pneumatic transfer in underground pipes at a few residential developments in Europe, no such method is generally practicable for solid wastes. Thus refuse collection must be batch process, or a series of batch processes, whereby wastes are stored at the point of origin for a given period, before being transferred to a vehicle, the contents of which represent one batch of many arriving each day at the diposal point.

The main sources of those solid wastes for which a municipality normally assumes responsibility are domestic premises, shops, offices, hotels, institutions, and small factories, together with refuse swept from the streets. Domestic wastes often account for about 75% of the total.

The main constituents of solid wastes are similar throughout the world, but the proportions vary widely from country to country and even within a city, because the variations are closely related to income levels.

As personal income rises, paper increases, kitchen wastes decline, metals and glass increase, total weight generated rises and the density of wastes declines. This will be more clear from the table given below.

There are often local variations in wastes generation and constituent proportions over weekly and seasonal cycles; the former is related to the pattern of work and leisure and the latter to seasonal food products and sometimes to fuel residues arising from space heating in winter.

Clearly, the amount of work involved in refuse collection depends upon the weight and volume of wastes generated and the number of collection points from which the wastes have to be removed. The extent to which these factors may vary is shown in the following table:

The following important issues emerge from these figures:

- the importance of density determining the volume to be stored and collected: about 4 litres from an Indian dwelling with a daily collection compared with about 130 litres from a European dwelling (many houses have two 70-litre bins) for a weekly collection;
- the higher rate of occupancy in India helps to reduce the number of sources, but this is far outweighed by the need for more frequent collection;
- the use of communal containers can reduce the total number of collection points to only 0.5% of the total sources; this enormously simplifies the organisation of the work, but may cause serious environmental problems.

Item	*Indian city*	*N .European city*
Paper	4%	27%
Vegetable/purrescible metter	75%	30
Dust etc., under 10 mm	12%	16%
Metals	0.4%	7%
Glass	0.4%	11%
Textiles	3%	3%
Plastics	0.7%	3%
Others, Stones, ceramics etc.	7%	3%
Weight/person/day	414 g	845 g
Weight/dwelling/day	2.5 kg (6 persons)	2.5 kg (3)
Density kg/cu.m.	570	132

Work components of refuse collection

A refuse collection service requires vehicles and labour. For their efficient deployment, it is necessary to have a clear understanding of the three main components of refuse collection:

- travel to and from the work area;
- the collection process, which includes: transfer of the wastes from storage to the vehicles and travel between successive collection points;
- the delivery process whereby the contents of the vehicles are transported to the disposal site.

Item	*Indian City*	*N. European city*
Generation:		
Volume/person/day	0.72 litres	6.4 litres
Volume/person/day	(6) 4.32 litres	(3) 19.2 litres
Sources/1,000 population	166	333
Collection points:		
(per million population)		
House-to-House collection daily	166,000	
House-to-house twice weekly, per day	56,000	111,000
House-to-house weekly, per day		56,000
Communal storage points/million population		
(for average contents 1. cu.m)		
points to be visited daily	720	

During non-working hours vehicles will be kept at a depot with enclosed parking space. The distance between the depot and the collection area should be kept to a minimum because time spent in travelling is unproductive. In the case of motor vehicles this requirement may have to be balanced against the need to centralize facilities for maintenance and stores, particularly fuel supplies, and to centralize the allocation and control of drivers and vehicles.

The slower speed of animal carts and handcarts requires the provision of closely spaced district depots, and these are also much more efficient for the control of the collectors, who can book on and off within a short distance of the area where they will be working.

The many methods of transferring wastes from storage to a collection vehicle fall into three main categories:

- the direct emptying of a portable storage container into the vehicle, normally used when the vehicle can be positioned close to the containers;
- the emptying of a portable storage container into a transfer container (usually a large basket) which is then emptied into the vehicle, normally used when the storage container is some distance from the vehicle, in order to avoid the return walk with the empty container.
- the transfer of wastes stored at ground level, which usually requires that the wastes be raked or shovelled into the vehicle.

These three categories have been set out in ascending order of work content. Thus the first is the most efficient in terms of labour and vehicle productivity; it is also the method minimizes human contact with the wastes.

Travel between successive collection points depends first upon the spacing. When these are some distance apart, as with large communal storage sites, travel by motor vehicle, although it represents lost time, will be at normal road speed and the collectors will ride on the vehicle. It is thus an eficient method of transporting the workers between sites.

When collection is from house to house, however, the collectors will walk the short distances between containers and the vehicle will move slowly and at intervals. For this element of travel the motor vehicle is very ineffectively employed; it incurs heavy transmission wear and high fuel consumption. Handcarts and animal carts are much more efficient : they operate at their optimum speed and no energy is used while they are stationary.

For collectors walking from house to house, the distance to be walked is proportional to the number of men in a team. One man alone walks from one house to the next, in a three men team each man collects from every third house, thus labour productivity declines as team size increases. Vehicle productivity, however, increases with team size as it is loaded more quickly.

The delivery process, when a full vehicle travels at normal road speed from the collection area to the disposal site represent maximum productivity for the vehicles, but lost time for the collectors if they accompany it. Handcarts and animal carts are very inefficient for this operation because of their slow speeds and limited capacity.

The following conclusion can now be drawn :

- minimum physical infrastructure requirements are a central workshop with parking space for motor vehicles, and district depots for mustering and controlling collectors, handcarts and animal carts; the locations of depots should minimise travel time between depot and working area;
- systems which provide for the direct emptying of portable storage containers into a vehicle offer the highest productivity and the lowest health risks to workers;
- for the collection of wastes direct from houses, large teams give low labour productivity and higher vehicles productivity;
- handcarts and animal carts may be more efficient than motor vehicles for the house-to-house collection phase;
- motor vehicles are the most efficient for transport of complete loads from the collection area to the disposal site.

Four basic collection systems have been evolved in relation to the amount of work imposed upon the householder :

- communal storage which may require delivery of the wastes by the householder over a considerable distance;
- block collection, where the householder delivers the wastes to the vehicle at the time of collection;
- kerbside collection, where the householder puts out and later retrieves the bin;
- door-to-door collection where the collector enters the premises and the householder is not involved in the collection process.

4

Health and Injury Hazards of Solid-Waste Treatment

Introduction

There are potential risks to health and to the environment from inproper handling of solid wastes. Direct health risks concern mainly the workers in this field, who need to be protected, as far as possible, from skin contact with wastes. (The National Environmental Engineering Research Institute, India, has carried out studies that show a higher incidence of intestinal parasites in collectors of domestic wastes than in the population as a whole.) There are also specific risks in handling wastes from hospitals and clinics.

For the general public, the main risks to health are indirect and arise from the breeding of disease vectors, primarily flies and rats. Improper storage and disposal provide the conditions under which these risks arise.

The most obvious environmental damage caused by solid wastes is the countryside by uncontrolled dumping of city wastes. More serious, however, and often unrecognized, is the transfer of pollution to water, which occurs when the leachate from a refuse dump enters surface water or wastes, either in the open air, or in plants that lack effective treatment facilities for the gaseous effluents.

Industrialization introduces dangers of a different kind: hazardous wastes from industry present risk during transport and disposal. Traffic accidents can result from toxic spilled wastes : they have caused death and injury to people in the vicinity. Improper disposal of such wastes has resulted in the death of humans and animals through contamination of crops or water supplies.

Finally, there is the specific danger of the concentration of heavy metals in the food chain, a problem that illustrates the relationship between solid and other wastes. It has sometimes happened that liquid industrial effluents containing heavy metals have been discharged to a drainage system and have contaminated the sludge leaving the treatment plant. These metals can be taken up by the plants growing on land on which sludge has been deposited, creating risks to the animals which graze and the humans who consume the animals.

As solid waste may potentially have any of the solid materials found in nature and in addition many of the man-made materials, they have been regarded as the most heterogeneous collection of substances possible. Laws and ordinances should prohibit certain materials from being put into the solid wastes stream. Domestic refuse has, with alarming trequency, been found to have, for instance, live ammunition or explosives; highly toxic substances; or diseased cultures.

Industrial wastes from any one location is generally having a lesser degree of heterogeneity than that of domestic refuse, but the variations of compositions and properties from one operation to another and one type of industry to another has been very large.

The scale of the potential dangers in solid wastes, has only recently been realized, and consequently data are extremely scanty.

Solid wastes could be hazardous as generated; they may subsequently combine to produce secondary hazards; or they could be hazardous by reason of their location or the method of processing used. These categories will find use in the following discussion.

Initially Hazardous Wastes

Solid wastes which have been dangerous at the point at which they have been discarded, such as explosive or incendiary devices, or toxic substances require little discussion here because this constitutes the most obvious category of hazard.

A related but distinet category has been including the broad range of wastes which, while no toxic, have the potential of causing injury. A plate-glass window becomes solid waste at the moment it has been broken, and at that point it has obvious injury potential for men and animals. Machine-shop turnings belong in the same category, with additional dangers because they have been generally hot enough to cause burns or alternatively may be contaminated with cutting oils which may bring about adverse physiological reactions if the skin has been punctured by such turnings.

A wide variey of substances can bring about harm or disfigurement either in general or to sensitive individuals. Acids and alkalis have been universal problems; though not strictly solid wastes they appear along with many other liquid substances in discarded containers. Pesticide containers have caused intoxication and death after being played with by children. Minute quantities of foaming or hardening agents used for plastic resins bring about acute reactions in some individuals. In some cases death may take place in otherwise fit individuals within 30 minutes of exposure to, for instance, pencilin.

Although organisms which are able transmit disease are organic and will live on organic refuse, the quantity or organic material necessary to maintain a culture could be an invisibly small coating on plate or container. Or the culture might be maintained at the tip of a hypodermic needle. The careless discarding of hypodermic needles after use into wastes-baskets has unfortunately frequently resulted in janitors becoming infected when they employ their hands to clean out the residue from a waste container.

A German study on illnesses caused by refuse revealed that the practice of 70% of medical laboratories and institutes of allowing their wastes to be collected along with regular domestic wastes has been particularly dangerous. All medical wastes should be kept in special closed, plastic bags and collected and disposed of separately.

Half the outbreaks of salmonellosis were found to occur in hospitals. Regular focal examination of food handlers in hospitals and institutions and action to eliminate salmonellae from animal feeds has been recommended.

Pathogenic bacteria can be and have been present in very high concentrations in hospital solid wastes. The highest coliform counts were found by Trigg to occur in wastes from the intensive care, pediatric, and psychiatric wards.

Hospital refuse chutes were found by Armstrong to be potentially transmitters of airborne bacteria. The use of closed plastic bags in refuse has been especially important, and an air removal and purifying system to remove the smallest particles which can penetrate the respiratory tract is necessary.

The use of disposable needles, knives, swabs, diapers, examination gowns, thermometers, transfusion fittings, bed pads, sheets, plates and utensils, and so forth, in hospitals and institutions requires special handling if injuries and cross-infections have been to be avoided. A particular danger arises if a contaminated non-woven item gets laundered along with washable linen. The non-woven item will probably distintegrate and spread over all other clothing theory potentially spreading infection over the whole institution.

Airborne Pathogens Associated with Solid Wastes Handling

Air samples taken from municipal incinerators and municipal compost plant showed that the potential health hazard from the pathogens associated with the dust has been somewhat significant. The dust was found to carry a large number of microorganisms including pathogens of intestinal and respiratory-tract origin. The degree of hazard is dependent on the concentration of the dust and on the type of organisms carried.

Table 4.1 shows microbial characteristics of representative environments and of the ambient air in various locations in municipal incinerators. The dumping-floor values can be regarded to representative of dumping floor of other types of solid-wastes-handling plants; the microbial levels might be somewhat lower for incinerators because, in good designs, the combustion air gets extracted from the dumping area to reduce dust, smell and pathogen levels.

Table 4.1.

Representative Environmenal Microbial Levels

Environment	*Total microbial levels (colonies/cu ft air)*
Country air	56
General offices and schools	95
City streets	72
Factories	113

Fungus spores also responsible for discase (mycosis) through inhalation and passage to the throat and lungs.

Asbestos Dust

The high incidence of lung and other cancers among asbestos workers, including those in the demolition industry, has been recognized. Stringent regulations have been introduced in many countries to control one form of asbestos—the blue crocidolite—which has been found to be principally responsible.

Those engaged in the manufacture of asbestos products and those who both use and work with such products have been mainly aware of the health hazards in question and take every precaution to safeguard the health of employees. The scarp industry is nevertheless vitally involved, because its members have so often to handle and process scrap materials which are contaminated with asbestos thermal lagging, e.g.., pipes, boilers, tanks, etc., which are received into their yards, and in addition often have to deal with such materials on demolition/dismantling sites, and the regulations quite specifically cover all operations which causes the liberation of asbestos-containing dust to such an extent as may result in the emission of dangerous quantities of dust.

The health hazards involved have been primarily the risk of contracting asbestos is through the inhalation of heavy concentrations of asbestos dust over a long period. This sometimes gives rise to the development of lung cancer, particularly if the person concerned has been a heavy smoker. Another rare form of cancer, mesothelioma, affecting the chest or stomach wall, sometimes takes place chiefly from exposure to certain types of 'blue' asbestos.

Types of asbestos. The main types of asbestos used in the past have been :

Amosite	—Colour	—grey/brown
Anthophyllite	— "	—white
Chrysotile	— "	—white
Crocidolite	— "	—lavender blue

Crocidolite has been the most dangerous types of asbestos and is likely to be found in acid fume areas such as chemical plants and gas works, and it is also found in power stations. Total emphasis upon the ability to recognise crocidolite by colour alone cannot be relied upon because it has been sometimes found mixed with the other types of asbestos, and although the significant proportion of admixture of crocidolite may be as low as 12%, even in these proportions it could be highly hazardous if to the dust gets inhale and unless the original owners of asbestos lagged materials can certify that such lagging has been free from crocidolite, the only sure identification of its presence has been by x-ray diffraction of samples.

Asbestos Dust — Threshold limit value (TLV). The Asbestos Regulations have been based upon the concentration of respirables asbestos dust in the atmosphere of the working area together with the duration of exposure to such dust, and threshold limit values have been set as follows :

Chrysotile (white) and Amosite (grey/brown)	—2 fibers per cc average over 4-hr working period and not above 12 fibers per cc average in any 10 min. period.
Crocidolite (blue) asbestos	—0.2 fibers per cc average over 10 min

The Stripping of Asbestos Lagging. The removal of old lagging has been the cause for serious concern as this is generally an extremely dusty process and the following paragraphs attempt to deal with the problems associated with the striping of such lagging in the varying circumstances likely to be encountered by the scrap processor.

Preparation Prior to Stripping Old Lagging. The first requirement is to establish the types of asbestos in the lagging that is to be stripped and where no record of this exists it will be necessary to determine if it contains 'blue' asbestos. If it is not apparent by inspection, samples should be taken from a number of different points for examination. Normally, positive-pressure respiratory equipment will be needed where 'blue' asbestos is present. In other cases a simpler form of approved dust respirator may suffice.

Protective Screens. Where stripping has been to be done occupied premises, screens or tents made from plastic sheets erected around the plant may be sufficient to reduce the asbestos dust level in adjacent working areas to an acceptable level.

The wet method of stripping. Thorough saturation of asbestos lagging wherever possible prior to stripping can greatly reduce the dust generated. Each job has been likely to require its own technique, e.g., non-absorbent surfaces will need to be punctured to allow to be introduced into the insulation. This operation could be by a hallow probe drilled along its length to permit water to penerate into the

Table 4.2

Indentification of Possible Sources of Origin of Asbestos Materials

Industry	*Asbestos Usage*
Electricity generating	Thermal lagging and delagging (Special risk from 'blue' asbestos).
Steel	Thermal lagging and delagging.
Heavy engineering	Furnace insulation.
Locomotive building	Heat and sound insulation. (Special risk from 'blue' asbestos used in the past.)
Railroad-car building	Heat and sound insulation. (Special risk from 'blue' asbestos used in the past.)
Boiler making	Heat insulation.
Paper making	Manufacture of filter papers and grinding of rollers.
Motor-vehicle repair	Repairs to brake and clutch parts.
Building trades	Asbestos cement sheets and insulation boards, asbestos spraying.
Electrical insulation	Insulation systems.
Chemical plants	Thermal insulation. (Special risk from acid-resistant 'blue' asbsestos.)
Gas work	Thermal insulation. (Special risk from acid-resistant 'blue' asbestos.)
Other Sources of Asbestos Lagged Materials	
Dockyards	
Shipbuilding	
Shiprepairing	
Installation of plant in heavy industries, *e.g.*, steel.	
Large building projects—industrial and domestic—or insulation and repair of heating apparatus, *e.g.*, schools.	

insulation, such probes parallel being coupled to a suitable water supply. Alternatively, a fine low-pressure water supply may be used in such a way that dust does not arise from the spray hitting the surface of the lagging. Spraying in this manner has been not as effective as total saturation in controlling the dust and the arrangement may require to be augmented by portable exhaust ventilation equipment and the use of respirators.

On removal, the lagging should not be permitted to fall but placed immediately in suitable plastic bags (200-gauge polyethylene double-sealed at bottom has been found satisfactory) for removal from site and subsequent disposal by burying. The slurry created should not be allowed to dry out should be removed from all resting places while still wet. Special additional precautions are needed when the waste contains 'blue' asbestos.

Landfil disposal of asbestos wastes. The Asbestos Research Council has laid down a code of practice for landfill disposal given below in part.

Asbestos waste should be disposed of in such a way that no dust is emitted into the air during transport, in the act of dumping, or after dumping. Where the waste has been derived from blue asbestos, the receptacles containing it should be clearly marked to show this. Waste which is adequately kept in impermeable bags or other suitably sealed receptacles does not require special transport.

Special containers can be obtained which, when filled, have been removed by specially fitted road vehicles direct to a landfill. Whatever type of container has been used, steps should be taken to prevent emission of dust from containers when in use.

Table 4.3

Non-hazardous Asbsetos Wastes

Linoleum	
Floor tiles	
Rubber	
Paints	Used as a filler and locked in
Plastics	with the orher components
Adhesives	
Roofing compounds	
Motor assembly	Grinding in assembly of brake and clutch parts.
Light engineering	Asbestos washers and gaskets.

Wet waste should be transported to the disposal area in sludge tankers or similar vehicles. After use all vehicles and containers should be cleared of loose fiber or dust by vacuum cleaning or other dustless method.

All waste deposited should be covered with at least 9 in. of consolidated earth. Covering should be done promptly (immediately in the case of waste derived from blue asbestos), No waste should be left uncovered at the end of a working day.

The disposal of wet waste on a dry waste landfill has been acceptable provided that the quantities are not excessive. It should be remembered that, even if waste has been wet when dumped, it must still be covered so that dust cannot escape from the dried-out material.

5

Radioactive Wastes

Introduction

These are materials with half lives of more than 100 days. U-236 has the longest half-life, over 23 million years.

The problem of dealing with such wastes could not be satisfactorily solved at the time of writing (mid-1970s). Radioactive wastes cannot be "disposed of" but must be stored in perpetuity. Liquid wastes are converted to a solid or a greaselike form. Groundwater is being monitored.

While all agree upon the absolute requirement of extremely stringent standards, they believe that there is no safe way to handle radioactive wastes.

However, the minimum requirement for locations where radioactive wastes have been housed temporarily includes primary and secondary containment, with continuous monitoring of the air in the secondary containment (*e.g.*, a building or a cave) to detect leakage, and the digging of observation wells to monitor groundwater quality continuously. Radioactive wastes should never be incinerated, nor accepted for any other form of conventional solid-waste disposal.

Atomic Waste Disposal Problems

In the operation of nuclear power station, radioactive waste presents ever more pressing problems. *Mildly* active atomic waste consists of residues from filter and purification plants, from contaminated laboratory equipment and gloves and of sewage sludge from waste water separation. Its disposal has been effected hitherto by dumping in the sea, by burial in the ground, by depositing in salt mines, or in certain cases by discarding it irresponsibly in refuse dumps.

Moderately active atomic waste consists of component parts of nuclear power stations rendered radioactive by neutron radiation, radioactive residues from purification processes (such as ion-exchange resins) and waste from nuclear research. Moderately active atomic waste was formerly dumped in the sea and deposited in salt mines. As recently as 1972, 3800 tons of mildly to moderately radioactive waste packed in 7600 containers was sunk in the Atlantic by the European Atomic Energy Authority. As no containers will stand up to exposure to sea water for centuries, it must be expected that this and other radioactive waste deposited in the sea in large quantities will one day be released, be enriched in the food chain in the ocean, and sooner or later crop up in the foodstuffs of people.

The real problem, however, concerns *highly radioactive* atomic waste. This is left behind as a liquid upon the separation of uranium and plutonium from burned fuel in reprocessing plants. The quantity of long-life highly active atomic waste is estimated for the year 2000 at about one trillion curies. No satisfactory solution for the disposal of highly active atomic waste has yet been found. Atomic waste cannot be destroyed. In contrast to chemical waste, which can be transmuted and deprived of poison by chemical reaction, radioactive substance disintegrate according to a time scale (which cannot be varied) related to their half-life value. The principal problem thus concerns the disposal of the fission products which have a long halflife value, such as strontium-90 (with a half-life of 26 years), cesium-137 (30 years), samarium-151 (100 years), plutonium-239 (24, 400 years), and iodine-129 (17,200,000 years). The atomic waste being produced today must therefore be stored away so securely, for periods

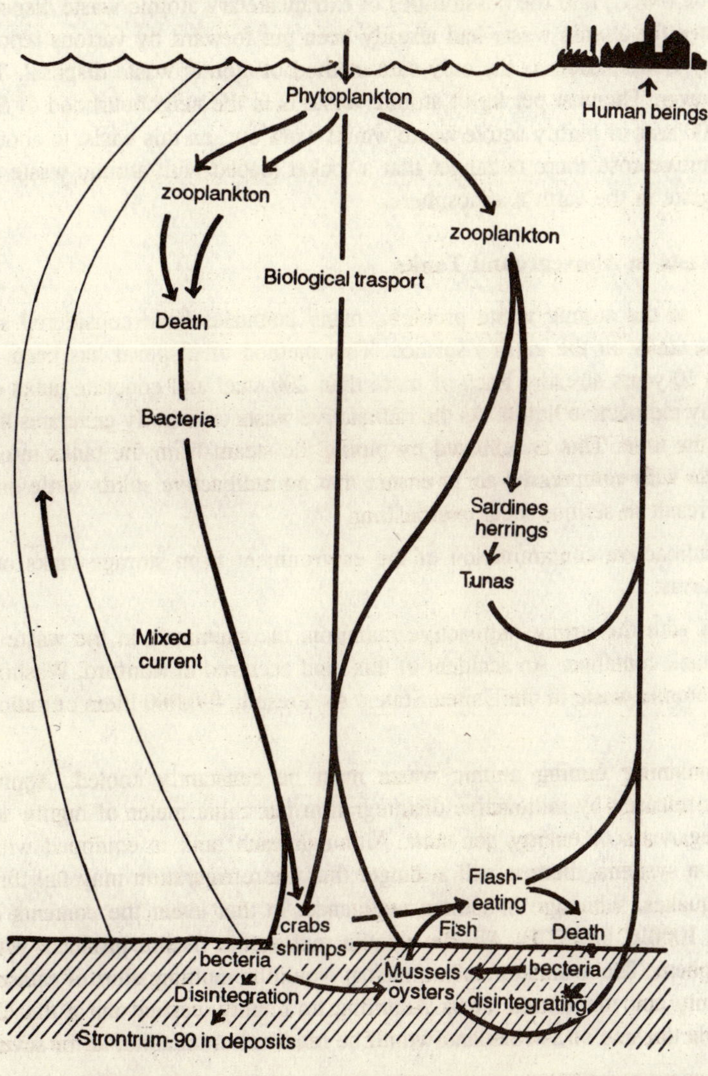

Figure 5.1. Biological consequences of sinking radioactive substances.

hardly conceivable to man, that none of it can escape into the biosphere and so seep into the food chain of any living creature.

Some countries are exploring the possibility of sinking radioactive waste through the ice in the Antarctic. Existing international agreements concerning the Antarctic provide for special amendments if all the signatories agree. The heat released by radiation would make the atomic waste containers sink gradually into the ice and melt their way through. In America it is proposed to ship the containers and transport them overland into the Eastern Antarctic, where the ice is particularly thick.

In view of the long-term nature of the problem of disposing of atomic waste on the earth, the National Aeronautics and Space Administration (NASA) has carried out a study on behalf of the Nuclear Regulatory Commission (NRC) into the possibilities of extraplanetary atomic waste disposal. Using the universe as a depository for atomic waste had already been put forward by various representatives of the Nuclear Regulatory Commission as the only safe method of atomic waste disposal. Two obstacles stand in the way, however. The cost per kg of atomic waste is in the neighbourhood of $200,000. The cost of disposing of 100 tons of highly active waste would work out, on this basis, to about $20 billion Apart from this prohibitive cost, there is danger that a rocket loaded with atomic waste could make a false start and disintegrate in the earth's atmosphere.

Storage of Atomic Waste in Aboveground Tanks

As a "solution" to the atomic waste problem, many countries have considered storing highly active atomic waste in tanks on the earth's surface. This method of disposal has been in use in the United States for over 20 years already. Each of more than 200 steel and concrete tanks contains over 3 million liters of highly radioactive liquid. As the radioactive waste constantly generates heat, the tanks have to be cooled all the time. This is achieved by piping the steam from the tanks into a condenser and mixing the contents with compressed air to ensure that no radioactive solids settle on the floor of the tank, which could result in serious local overheating.

The danger of radioactive contamination of the environment from storage tanks on the earth's surface takes several forms:

1. In conjunction with the strong radioactive radiation, the chemicals in the waste solution can cause corrosion of the tank container. An accident of this kind occurred in Hanford, Washington, which is the main deposit for atomic waste in the United States. As a result, 490,000 liters of radioactive waste seeped into the ground.

2. The tanks containing fuming atomic waste must be constantly cooled. Approximately 9 kilowatts of energy are released by radioactive disintegration per cubic meter of highly active atomic waste, or about 31 megawatts of energy per tank. Although each tank is equipped with two quite independent refrigeration systems, there is still a danger that the refrigeration may fail through power shortage, floods, earthquakes, sabotage or human negligence. In that event the contents of the tanks would heat up to over 1000°C (1832°F). All the volatile radioactive fission products would then be released into the atmosphere. The results of such a mishap would be nothing short of catastrophic, both for the immediate vicinity and for remote areas. Acording to a study carried out at the University of California an area double the size of Switzerland would be rendered uninhabitable for several decades.

3. Owing to radioactive radiation, water molecules are split radiolytically into hydrogen and oxygen. The rate of this hydrogen production is so high that in the event of failure of the ventilation

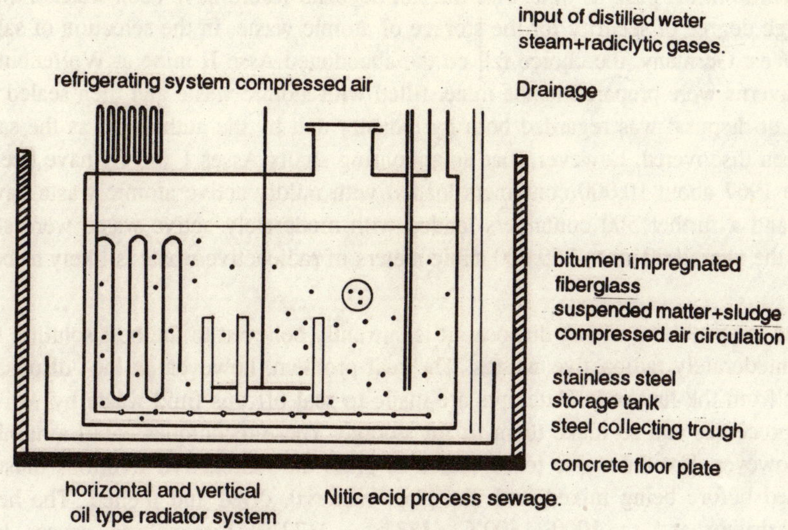

Figure 5.2. Tank for aboveground storage of atomic waste.

system the lower explosion limit of 4% for hydrogen in the air would be reached in a few hours. Through spontaneous heating in this way, an oxyhydrogen explosion could result, by which the tank would be destroyed together with the refrigeration system. Attempts have recently been made to provide more secure storage conditions for the radioactive residues in liquid form. But the problem of the radiolytic disintegration of water and the contingent risk of an oxyhydrogen explosion remains. The effects of a failure of the refrigeration system would be roughly the same.

The storage of atomic waste in aboveground tanks is thus attended by substantial risks, the effects of which cannot be finally assessed.

Table 5.1
Accidents recordel in connection with the storage of atomic waste

Year of accident	*Amount of radioactive liquid released (in liters)*	*Release of cesium-137 (kCi)*
1960	130000	4
1965	190000	40
1969	260000	51
1973	440000	40

Storage of Atomic Waste in Salt Mines

The idea was developed in West Germany of storing away radioactive waste material in salt mines.

This was based on the fact that salt mines have no connection with groundwater (a fact that must have been equally true in the past, as otherwise the salt deposits would have been washed away), so that the provide a high degree of security for the storage of atomic waste. In the selection of salt mines for this purpose in West Germany, the choice fell on the abandoned Asse II mine at Wolfenbuttel. Many large and small caverns were prepared in the mine, filled with atomic waste and then sealed off. For a time, this method of disposal was regarded both by industry and by the authorities as the safest solution. It has since been discovered, however, that neighbouring shafts Asses I and III have been permeated by water. Since 1967 about 10,000 containers loaded with mildly active atomic waste have been stacked in Asse II, and a further 500 containers loaded with moderately active waste were stored in special caverns. By the year 2000 up to 250,000 cubic meters of radioactive waste is likely to be stored in Asse II.

Despite the problems of salt disposal, it is currently believed to the best solution for disposing of mildly and moderately radioactive wastes. The real problem, however, is the disposal of the highly active waste from the fuel rods. Attempts are made to seal off this fluid waste by various glazing and calcination processes and so make them fit for storage. These techniques entail a number of unsolved problems, however. For the waste to be sealed in glass the radioactive solutions must be neutralized and denitrated before being mixed with the glass material, dried and melted. The heating to 200°C (392°F) for drying and to 1000°-1300°C (1832° - 2372°F) for melting causes loss of volatile radionuclides. About 20% of cesium, for instance, is released, and nearly 100% of ruthenium in the form of ruthenium oxide. The value of the process is reduced by the releases of these nuclides; they play an important part in the biologically relevant fission product activity, but their separation and long term storage pose additional problems.

The frequently discussed and varied methods of moving bed calcination—by which fission products interstratified for a long time are converted into oxide form at burning temperatures of 600°C to 1000°C (1112° to 1832°F)—also present a number of unsolved problems. The volatile fission products must further be given special treatment. A not inconsiderable portion of the fission product can be washed out of the oxide substance itself, in the course of time, according to the level of the burning temperature.

As regards the storage of such calcined waste in salt mines, the problem of heat development arises in connection with humidity of the air penetrating the shafts or percolation of water. In the case of storage of atomic waste in steel tanks, the separation of condensation water in consequence of thermoconvection in the deposits can result in the formation of corrosive salt solutions which, in conjunction with radiation corrosion, can disintegrate the container materials. A further possibility of damage can come from earthquakes, whereby a large part of the radioactive waste stored away in salt mines over the years could be released. This problem is increased by the fact that the highly active waste previously stored could work its way into the salt as a result of the heat developed and could not be retrieved.

In Canada, where storage in salt mines was also projected, the Atomic Energy Control Authority vetoed this form of atomic waste disposal and recommended storage aboveground.

Hazardous Industrial Waste

The following notes on some particularly troublesome metals and organic wastes are meant to be illustrative rather than all-inclusive of the health hazards that can result from incorrect handling of as generated wastes.

Inorganic Materials

Lead. Lead poisoning causes brain damage, especially in children, and interferes with the ability of people of all ages to resist infection. Lead is dissolved out of lead water-supply pipes to enter the diet through the drinking water. Children take additional lead from eating paint chips in old lead - painted buildings. The disposal of metallic lead and demolition debris in landfills therefore has created a potential hazard if the leachate has been allowed to pass into ground-water feeding a water-supply system.

Mercury. Metallic mercury has been so valuable that it has been unlikely to appear as such in solid wastes to any extent. Organic and inorganic compounds of mercury have been widely used, however, and can appear in industrial and agricultural wastes. All mercury compounds have been found to be highly toxic to man and animals, giving severe mental and motor symptoms, neurological symptoms, kidney damage, gastrointestinal and pulmonary symptoms and genetic damage.

Beryllium. It is a low-density metal which finds use in many industries, principally in alloy form, and as a rocket fuel in the pure form. When ingested as atmospheric particles it causes severe disease involving respiratory-tract irritation, dermatitis and conjunctivitis and possibly involves other organs. Mortality has been 35% in its chronic form. Whether long-term, low-level exposure can lead to cancers is still uncertain.

Manganese. It finds use principally as an alloying element in steels. Inhalation of manganese oxides may cause chronic manganese poisoning or manganic pneumonia.

Organic Materials

Benzopyrene and benzofluoranthene. 3,4. benzopyrene and 3, 4.benzofluoranthene have been formed when organic substances have been heated above 500°C in incineration and in other processes. These hydrocarbons have found to be carcinogenic to animals. The substances occur in soils, incorporated from sediments from flue emissions, and in root vegetables (potatoes, carrots, radishes etc.) grown in such soils.

*DEHP. Di-2-*ethylhexyl phthalate finds use as a plasticizer for polyvinylchloride packaging. It has been found to be toxic to animals and man. It is suspected that deaths in patients receiving blood transfusions from blood stored in PVC bags had been due to this plasticizer being transferred from the bag to the blood. Handling and disposal of quantities of waste DEHP must be done with stringent precautions.

PCB. Polychlorinated biphenyls had been previously used for no-carbon copy paper. A small proportion recycled into combination paper board used for packaging breakfast cereals allowed transfer of PCB to the foodstuffs and caused measureable levels is human fat. PCBs or an associated contaminant, chlorinated dibenzofuran, is responsible for liver malfunction and damage, severe skin eruptions and hydropericardium.

Fiber polymers and polyurethane foams. The open burning of some fiber polymers, such as Orlon, and foamed plastics, such as polyurethane, can give rise to the formation of potentially toxic concentrations of hydrogen cyanide.

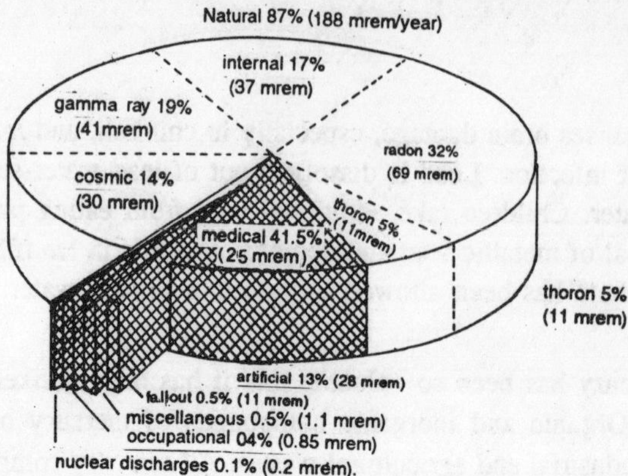

Figure 5.2. Average proportions of background radiation from different sources in Britain (From Fremlin 1987).

Fiber Remark

There is no doubt that high doses ionising radiation kill, and that even moderately high doses cause cancer and other forms of sickness. It is equally certain that nuclear power stations produce considerable quantities of radioactive waste, and that under some circumstances the radiation from this waste could cause people or animals exposed to it to die or to develop cancers. For the purposes of this book, we need to know whether radiation, particularly what we may call 'waste radioactive substances from the nuclear industry, from accidents and from military uses of radioactive substances. Moreover, there are the effects of natural radiation, the so-called 'background' radiation to which we all exposed and with the effects of natural radiation which is present in different locations, but for which human activities bear no responsibility.

While I do not wish to get bogged down with considerations of theoretical physics, and discussions of the apparently complicated units used to describe different types and levels of radiation, the topic cannot be entirely ignored, if we are to understand the ways in which radiation may affect living organisms, and the possible effects of different intensities of radiation.

The subject is somewhat complicated. Radiation consists of electromagnetic waves or bits of atoms moving at high speeds, and these can damage living cells. Different radioactive substances produce different types of radiation, which have their own biological effects. Some radioactive substances give off rays, which affect the tissues outside the body, others produce energetic electrons the so-called x rays, which only penetrate a few cell layers, and so are only effective when actually on, or in, the body. Finally, some produce particles, which are actually nuclei of helium atoms. These deposit a large amount of energy over a very short path in the tissues, and are generally the most damaging. Radioactive substances taken into the body may be deposited in sites (e.g the thyroid gland) where they are in a position to do maximum harm.

Radioactive substances do not go on emitting radiation for ever, they all 'run down' eventually. Some of the radioisotopes produced in a nuclear explosion give out all their radiation in a fraction of a second. Some continue for thousands, even millions, of years. The concept of 'half life' is useful in this connection. Thus strontium-90 which gives off a rays, has a life half life of 28 years. This means that at the end of 28 years the level of radiation given out falls to half, by 56 years to a quarter, and after 112 years, to an eighth of the original levels. Plutonium-239 has a half life of 24,000 years, and so is often said to be very long-term hazard to the environment. However, the amount of radiation given out by a sample in a given time may be trivial. Uranium-238 has a half life of 4,470,000,000 years. This is roughly the age of the earth, so only half the uranium-238 that existed when the earth was formed has decayed away.

The many units used to describe radiation may seem confusing, but as we are concerned here with the biological effects of radiation on living organisms, the 'absorbed dose' is perhaps the most relevant conception. Until recently the unit most commonly used was the rad, which is still employed by many writers. In the S.I. system the Gray (Gy) is now used, and the conversion is simple: 1 Gy=100 rads.

The former unit of radioactivity was the Curie (Ci). This was very large unit, and activity often had to be measured, for instance from the fall out of a distant bomb, in picocuries ($pCi = Ci \times 10^{-12}$). The S.I unit is the Becquerel (Bq) and $1 Bq = 2.7 \times 10^{-11} Ci$.

The Bq is thus a very small unit, and this may be confusing. After the Chernobyl accident, when lambs in Britain picked up some radiation the British government forbade consumption of meat containing more than 1000 Bq of caesium-137 in a kilogram which seems quite a large amount. According to Professor J.H. Fremlin, however, anyone consuming a whole kilogram themselves would receive a dose which, in terms of carcinogenicity, would only be the equivalent of smoking a single cigarette.

The biological effect of radiation obviously depends on its intensity. Thus a really high dose of, say, 100 Gy will be immediately fatal. A substantial dose, in the order of 2-6 Gy, will cause burns and other acute effects, and some of those exposed will die within six weeks. We are mainly concerned with much lower doses, such as may have been experienced in this country, either as a result of leaks from nuclear power plants, fall out from testing of nuclear weapons, or as the result of such incidents as the Chernobyl Accident, when radioactivity from Russia was transported to Britain.

Before considering other sources, let us look at the background radiation to which we are all exposed. Fig. 5.3 shows that 87 per cent of the radiation comes from 'natural' sources, and only 13 per cent is manmade. The natural background radiation comes from rocks and from potassium within the body, as well as smaller amounts from various radioactive materials, The greater part of the man-made radiation is from medical sources, particularly X-rays, so this is very unevenly distributed. Some people, who never visit hospitals, will receive very little from the discharges from power stations is tiny. The figure, incidentally, makes use of a unit the millirem. This is a measure of the so-called dose equivalent, which allows for variations in the way different types of radiation (*i.e.* X-rays, gamma rays, electrons and alpha particles) affect living organisms. The background radiation fluctuates in different parts of the world, and even in different parts of Britain. The cause of the variation is mainly the type of rock. Granites are much the most radioactive, and thus the background radiation in Cornwall, or near Aberdeen, is much higher than in the rest of Britain. These contrasts may provide a useful clue as to the effect of low levels of radiation.

All types of radiation so far considered are called 'ionising radiations'. They have property of knocking out electrons from the atoms in the substances through which they pass, and thus producing 'ionised' atoms which have great chemical activity. When this happens in a living cell, the usual effect is for the cell to be damaged. Thousands of cells in our bodies are damaged during the time it takes to read this chapter as we are continually bombarded with background radiation. Damaged cells may be killed, and then will be absorbed into the tissues; or they may be partially damaged, as with the DNA in the chromosomes. Such a damaged cell can be the focus of a cancerous growth. If the damage is to the germs cells, which produce sperm or eggs, these cells may, when they develop into recognisable organisms, differ to some extent from their parents; in other words, *mutation* has taken place.

Fortunately, the vast majority of the cells which are damaged by radiation are automatically repaired, and go on to lead a normal life. But a very small proportion are not repaired, and may cause cancer or mutations. The greater the extent of the radiation (assuming it is not sufficient to prove lethal) the greater the risk that a cell or cells not be repaired. Older people develop cancer more commonly than younger members of the community, because over the years they have more cells damaged by radiation. As already stated, high levels of radiation may prove fatal, or at lower levels, very damaging to the tissues. What we need to know is the probable effect of exposure to very low levels.

At one time it was believed that low levels of radiation which produced no instant results were either harmless or actually beneficial. Consequently many workers were exposed to doses which would be considered very dangerous today, and some developed leukemia or other forms of cancer. The permitted dose has therefore been greatly decreased to levels which some people now consider unnecessarily low. We no longer permit the sale of such commodities as radioactive bath salts or mineral waters that claims to be the most radioactive in the world. It is probable that such products actually carried little or no radioactivity, but not so long ago such claims were widely believed. Nor is it impossible that in the future we may once more argue that low levels can be beneficial.

At present legislation and regulations are based on the linear hyhothesis. We know approximately how dangerous a large dose is. The hypothesis assumes that damage will be proportional to the dose, and that there will be no threshold below which damage does not occur. This theory can be defended. The number of cells damaged must be proportional to the dose, and there is no reason to assume that

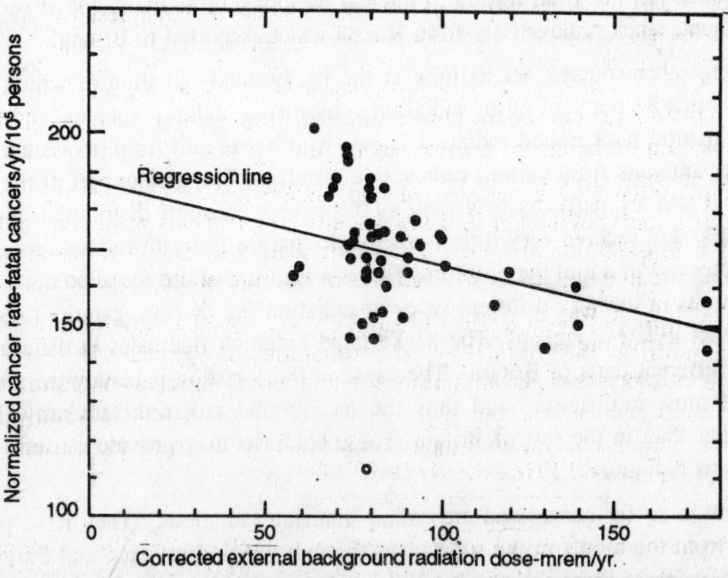

Figure 5.3. Total cancer incidence against background radiation, USA

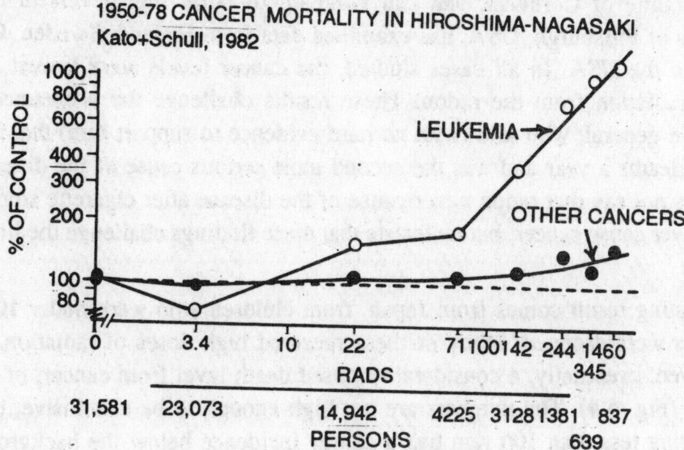

Figure 5.4 Mortality rates from leukemia and other cancers among Nagasaki and Hiroshima survivors as a function of radiation received (From Luckey 1988).

where damaged cells are few they will be less likely to show signs of damage. Radiation levels that have been permitted have been where the number of cases of cancer are so small as to be unrecognisable, taking the normal rate of cancer in the population involved.

However, there is no proof that the linear hypothesis is correct. In Japan, the long-term radiological of the atomic bombs dropped Hiroshima and Nagasaki in 1945 was the production of fewer cancers after a period of 45 than had been expected. Studies of the effects of the Chernobyl accident by the International Atomic Energy Agency in Vienna in 1991 found that, 5 years after the accident, there was no increase in any radiation-related illness, such as leukemia, birth defects or thyroid cancer in the villages near the reactor which had been contaminated by radiation.

In fact, there is growing evidence that low levels of radiation may produce fewer cancers than the linear hypothesis would predict, and even that radiation up to several times the background level may in some way be beneficial. It is well known that many poisons, lethal at high levels, may be beneficial in low doses, and may actually be used in medicines and drugs. There is now a considerable amount of evidence to suggest that the phenomenon of hormesis may operate in the case of ionising radiations. There is no general agreement on this but the observations of many who work in this field suggest that there is very little evidence that low levels of radiation are as harmful as is generally assumed.

The first general observation, based on investigation of various places with different background levels of radiation, is that nowhere has any area with high background radiation proved to have a higher than average level of cancer. In the United States several surveys have shown the very opposite results, depicted in Fig. 5.3. It has been suggested that these results could be explained differently. The lowest level of cancer in the state if Utah, where is a high proportion of Mormons who do not drink or smoke. The other states with high natural radiation and below average cancer are mainly rural with comparatively little air pollution. So if this were an isolated observation, it would need to be treated with caution.

A second set of data concerns the effects of the radioactive gas radon which bubbles up out of rocks such as the granite of Cornwall, and can build up to fairly high levels in houses with poor ventilation, B. Cohen of Pittsburgh, USA, has examined data from Finland, Sweden, China and no less than 415 countries in the USA. In all cases studied, the cancer levels were lowest in the areas with greatest amount of radiation from the radon. These results challenge the pronouncement of the US government's surgeon-general, who said (with no hard evidence to support him) that in the USA radon was causing 20,000 deaths a year and was the second most serious cause of the disease after cigarette smoking. Cohen does not say that radon never cause of the disease after cigarette smoking Cohen does not say that radon never cause cancer, but he insists that these findings challenge the linear, no-threshold theory of cancer risk.

Another interesting result comes from Japan, from children who were under 10 years old at the time the atom bombs were dropped. Many of these received high doses of radiation. Those receiving above 100 rem showed, eventually, a considerably raised death level from cancer, of which about half were from leukemia (Fig. 5.4). The numbers are not high enough to be conclusive, but they do show that the children getting less than 100 rem had a cancer incidence below the background for Japan as a whole.

Some animal experiments support the same conclusion. Mice receiving low doses of a few rem in experiments made by E. Lorenz (low doses, but well above the background) generally lived longer than controls. Of course, more heavily exposed mice died in a shorted time. R.J Pentreath found that salmon derived from irradiated eggs survived better than those whose eggs had not been so treated D.J.Jefferies found that if the Saw-toothed Grain Beetle (*Oryzaephilus surinamensis*) is subjected to gamma radiation, pupae receiving up to 4 rads survived better than controls, though 10 to 20 rads proved fatal.

It is too early to be dogmatic about these results, but there is a growing volume of similar data, much of which was summarised in a special issue of the journal Health Physics in May 1978. There have also been experiments which have given negative results. It has been suggested by J.H Fremlin that any radiation, even low doses, must initiate some cancers. However. this same radiation may also produce an improvement in the immune system, the way in which damaged cells are repaired. This would support the observations, and also explain why there may not always be a 100 per cent agreement between exposure to radiation and its beneficial, or harmful, effects.

For many people, the greatest objection to nuclear power is that it produces dangerous radioactive waste. It is said by some, and widely believed, that we have no satisfactory means of disposing of this waste, which will remain dangerous to life for long periods. If this waste were to be treated as waste from the LeBlanc process was treated 100 years ago, i.e, simply dumped in any convenient waste site with no regard to anyone living in the vicinity, the results would be serious. But no one proposes to do this. There are very detailed plans for waste disposal, and the fact that they are not all operative today is due to uniformed apposition, often politically inspired.

The final waste from a nuclear power station, from the reactor core and everything which has come into contact with it is radioactive and emits α and β particles and gamma rays at levels which could be very dangerous. Wastes are classified as low medium or high level, depending on the level of radioactivity. Low-level wastes are the least dangerous, and the easiest to dispose of. They have also been the cause of the greatest political uproar. Several sites in Britain were proposed for the burial of such low-level wastes. These substances would have been buried in trenches 9 m deep, and covered with 2 m of clay. The radiation reaching the surface from such a site filled with this waste has been

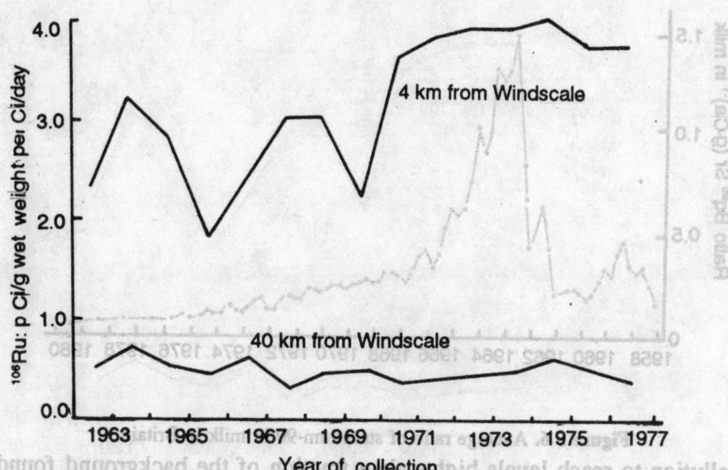

Figure 5.5. ^{106}Ru in seaweed Porphyra umbilicalis from area off Cumbrian coast, where low level discharge takes place.

calculated as less than a millionth of the natural potassium radiation reaching one person from another sitting in the same room. No α or β particles would pass through the clay cover. Even if someone fell into a filled trench, assuming all the most dangerous estimates or risk, the increased chances of developing cancer are so small as to almost incalculable. On this subject, Professor Fremlin, one who has done much to combat the dangers from radiation said: 'The callous cruelty of the irresponsible propagandists who have successfully terrified uninformed people in areas that might be used for storage has not been exceeded since the witch hunts of the 17th century.''

Medium and high-level wastes would need to be stored in special containers and kept for a long period, until decay had rendered them harmless. The high level waste is very hot and intensely radioactive, and needs to be stored carefully, perhaps for as long as 50 years untill it can be classified as medium-level. Medium-level wastes must be treated with caution. Probably the best solution is to convert them to rock-like substance called 'synrock', which could be stored in deep mines or even in the deep ocean where very little mixing with shallower waters takes place. Propagandists have claimed that these wastes will remain radioactive for perhaps as long as 100,000 years, and so will need to be contained for that period. There is no guarantee that global changes during such a period would not bring such material to the surface, even if deeply buried today. This argument is based on a misunderstanding. It is true wastes would contain, among other substances, some plutonium-249 with a half life of 24,000 years. But the quantities are no small that even the most radioactive waste would, in 500, or at worst 1000 years, be quite safe to handle, and would be emitting less radiation than naturally occurring rocks which are mined without endangering the miners. So if nuclear waste is a problem, it is one which the opponents of nuclear energy have exacerbated. Until now, in Britain nuclear waste has not affected the environment sufficiently to have had any effect on our species or on the flora and fauna.

There have been a number of much publicised 'leaks' of radiation from nuclear power stations, and these have been reported in such a way as to cause the maximum public concern. In no case has any leak, except in a few cases within the station itself when operating staff have been affected, caused

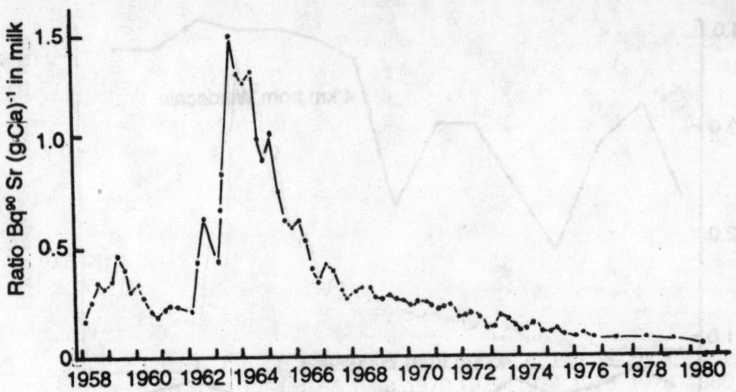

Figure 5.6. Average ratio of strontium-90 in milk in Britain.

any radioactive pollution to reach levels higher than fraction of the background found in the vicinity. Some low-level radiation has been discharged to the sea from several power stations. This has never been shown to damage any aquatic organism. Radioactive substances have been found to be accumulated in some living organisms, and concern has been voiced that if people ate there they might ingest a harmful quantity of radiation.

Two examples, which have had the effect of modifying the amount of material discharged from two power stations, illustrate what precautions have been taken. It was decided that no one who ate any of the organisms which picked up radioactivity should receive more than the equivalent of one-tenth of the background radiation from this source. From the Sellafield area the seaweed Porphyra umbilicalis is harvested, and made into 'layer bread' which is consumed in South Wales. Fig. 5.5 shows how ^{106}Ru was taken up by this seaweed; the levels of radiation near to the discharge were raised, but there was little effect further out to sea.. Calculations were made as to the highest level of radiation if the most radioactive seaweed were harvested. The maximum amount eaten by anyone in Wales was also estimated, and these figures were used to calculate the level of radiation permitted in the power station's output.

In the case of discharge from the Winfrith generator, the lobster *Homarus gammarus* was the species most involved. Few people consume more than a few grams of this expensive crustancean in a year but one man near Winfrith claimed to eat lobsters daily. This individuals' unusual diet caused the permitted discharge to be substantially reduced. There was no evidentally that the radiation had the slightest effect on the seaweed or on the lobsters, and in my opinion it would not have mattered much if their body levels of radiation had been much higher. Such radiation levels could hardly have had any harmful effects on the consumer even if layer bread and lobsters had been eaten in even greater amounts.

The Institute of Biology has been concerned with the possible dangers of radiation. As early as 1950 it organised a symposium on 'The Biological Hazards of Atomic Energy, which drew attention to possible dangers and how these might be avoided. Then in 1979 it held a second symposium, this time on 'The Biological Implications of Expanded Nuclear Programme.'' This was.to some extent a stocktaking exercise. It showed that there was no evidence of any harmful, biological effect from the small amounts of radiation arising from nuclear power, and reached the general conclusion that an expanded programme need not pose any dangers provided existing standards of safety were maintained. Mention was also made of radioactivity from other sources, which confirmed that these were small compared with background levels. As Fig. 5.6. shows, in the 1960s there was a significant rise in radioactive strontium in milk, as a result of testing bombs many thousands of miles away. The graph looks alarming, but the levels are still very low. When the Test Ban Treaty took effect, the levels fell and still and remain virtually insignificant.

In recent years, the main concern has been caused by fall from the Chernobyl accident, which was blown over to Britain, some of it being washed out in rain over the Welsh mountains and in the Lake Districts where elevated levels were found in sheep and which remained in some animals even after two years. The British Govermment acted quickly and banned the sale or consumption of sheep with more than 1000 Bq of caesium-137 in a kilogram of their meat. Although some compensation has been paid, this action has caused considerable hardship to many sheep farmers, as animals thought to be affected had almost no cash value. In my opinion, the Government's action was totally unnecessary. As mentioned above (p. 149) meat with 1000 Bq per kilo would, if anyone ate this amount increase his or her risk of developing cancer as much as smoking one cigarette, but only if the linear hypothesis applied. Put another way, an individual flying the Atlantic would be exposed to a higher level of radiation by flying at 10,000 metres above sea level for seven hours.

The fall out from Chernobyl also affected Sweden and other parts of Scandinavia. When reindeer were found to be slightly radioactive, the Swedish Government had the animals slaughtered even though this upset the whole culture of the Swedish Lapps. The danger from this source was tiny (if indeed it was a risk at all), far less than permitting the Lapps to smoke tobacco. The Norwegians had the same problem. Very sensibly, they took no such action. There is no evidence that this has had any harmful effects.

One other cause for worry about radiation has not been discussed. This is that the germ cells may be affected, which may cause harmful mutations. Professor Fremlin has suggested that had the background radiation been lower, mutations would have been less frequent, and the human race as we know it would not have evolved by this time. The worry is that most mutations are harmful, and produce imperfect organisms which seldom survive in nature, although nowadays were they human we might succeed in keeping them alive. In theory, all radiation which reaches the developing germ cells could produce mutations the number being, presumably, proportional to the intensity of the radiation. As the background radiation does not seem to be causing many mutations, any tiny increase will probably produce even fever. There is little or no evidence of a great increase in mutations in Japanese survivors of the nuclear bombs at Hiroshima and Nagasaki, although some received considerable dose of radiation. In an experiment in the Brookhaven Forest in the U.S.A, a source of radiation sufficient to kill all trees and vegetation within 50 metres has been running for many years. There is a fairly sharp cut-off where trees seem to grow normally, though they must be subject to levels of radiation many times those which would be tolerated near a nuclear power station or a waste dump. I know of no reports of great number of mutations in any of the animals which roam freely in this site. The concern about mutations from man-made nuclear sources appears now to be much less than it was 20 years ago.

My general conclusion must be that, in Britain today nuclear power the wastes from nuclear power stations, any accidental leaks from such stations, and any radiation from overseas deposited in Britain have so far had no recognisable effect on the human population, or even on wildlife. I believe that even with a vastly increased nuclear industry, if reasonable precautions are observed, there should be no significant effects in the future.

6

Nuclear Waste Disposal

I. Introduction

It is a paradox that although nuclear wastes have been generated since the 1940s, there are relatively few disposal sites in existence; and it is only in the last two decades that serious attempts to deal with these wastes have been made. In this brief chapter the different kinds of nuclear wastes will be discussed, and the current philosophy toward their disposal will be covered.

The disposal of nuclear wastes is one of the most pressing problems confronting mankind. In 1995 approximately 550 nuclear power plants will operate in 35 countries, each generating nuclear wastes. Of more staggering proportions are the nuclear wastes generated by the defense sectors of several world powers, roughly 10 times as abundant as the wastes from commercial development. Radioactive wastes from medical facilities and research institutions are also increasing dramatically.

During the last few years the responsibility for the disposal of radioactive wastes in the United States has been divided between the Department of Energy (DOE), the Environmental Protection Agency (EPA), and the U.S. Nuclear Regulatory Commission (NRC). The DOE is responsible for the selection and building of the repository, the EPA establishes the standards, and the NRC provides licensing and adherence to the standards. The regulations are extremely rigid, and great pains are being taken to ensure that a high-level repository will be environmentally acceptable.

II. Types of Nuclear Wastes

The four principal types of nuclear wastes are (not including uranium mill tailings) nuclear fuel rods and high-level, transuranic, and low-level wastes. Nuclear fuel rods, after use in a reactor and not subjected to reprocessing, are referred to as spent fuel and contain uranium, minute amounts of transuranics, actinide daughters, and abundant amounts of fission products. Many fission products such as strontium and cesium (^{90}Sr and ^{133}Cs) have a short half-life and thus are intensely radioactive.

High-level wastes result when the fuel rods are reprocessed to recover fissionable uranium (^{235}U) and small amounts of plutonium (239Pu). More than 90% of the ^{235}U from a spent fuel rod can, in theory, be recovered by reprocessing. The fission products that are separated from the actinides make up high-level waste. Reprocessing spent fuel rods is practiced almost routinely in Europe and Japan; France and Great Britain operate commerical reprocessing plants that serve six other nations. In the United States, reprocessing is not carried out for commercial reactors, but it is performed for materials generated by the military.

Transuranic waste is generated exclusively from the military during reprocessing and consists of machine and tool parts, lab coats, gloves, sludge, glass, and other routine garbage that has been contaminated with small amounts of transuranic element. Small amounts of fission product contamination may be present. If the waste contains more than 100 nanocuries of alpha-emiting radionuclides, it is classified as transuranic waste. Most of this, 98%, is weakly radioactive and is designated CH for contact handled; the smaller portion with some contaminant fission products as well as transuranice is called RH for remote handling.

Table 6.1

Amounts of Nuclear Wastes in the United States 1985-2020[a]

	1985	*2000*	*2010*	*2020*
Waste from the defense sector				
High-level	355	346	370	379
Transuranic	286	376	437	497
Low-level	2181	4043	5159	6256
Waste from the commercial sector				
Spent fuel	5	16	25	39
High-level	2	8	8	8
Low-level	1160	2441	3545	4972
Uranium mill tailing	100,000	146,500	197,300	265,900

Source U.S. Department of Energy.
[a] Values given in thousands of cubic meters.

Low-level waste is the largest component and includes all wastes that are not spent fuel, high-level waste, transuranic waste, or by-product material. These wastes are largely derived from the defense industry in the United States but significant contributions come from the public sector—mainly from nuclear reactors and institutional and medical facilities.

Uranium mill tailings are, by volume, the largest source of nuclear waste materials. Interestingly, they contain only naturally occurring actinides and actinide daughters—this will be discussed separately from other categories of wastes. Typically, uranium ore worldwide may contain only 0.5% uranium by weight, and the milling processes are usually about 95% efficient; therefore, the tailings become a mix of materials from the milling operation that contain up to several hundred parts per million uranium. While this amount is small, it is nevertheless well above background for uranium values, which can be 1 to 6 ppm for many crustal materials. In a number of sites the actual risk from the tailing comes not from their radioactive budget but from the other heavy metals present.

III. Quantities of Nuclear Wastes in the United States

Information for the world budget of nuclear wastes is uncertain. Table 6.1 shows the figures available from the United States for the diffierent kinds of wastes on hand in 1985 along with the projected wastes expected for the years 2000, 2010. and 2020. For convenience, no reprocessing of commerical waste is assumed, and wastes from decommissioning nuclear reactors will be assumed low-level wastes. Uranium mill tailings are shown in the commercial sector, although much of the uranium recovered from the milling operations in the United States was for defense purpose.

It is clear that the greatest amount of high–level and lowlevel wastes and all the transuranic wastes generated in the United States are from the defense industries. When projected to the year 2020, although commercial spent fuel will increase nearly eightfold from 1985, commercial spent fuel will still be about one-tenth as abundant as high-level defense wastes. Similarly, although commercially generated low-level wastes will increase more than fourfold by 2020 and defense wastes by threefold, the latter will still be significantly more abundant.

IV. Interim Storage

There are no permanent repositories for high-level waste or spent fuel at the time of this writing, and decisions or anticipations of a repository for the nine major countries generating nuclear wastes will not be made or realized until well into the next century. There is controversy that a mined repository could be expedited; certainly, such a resolution cannot be made in haste. Presently, most countries store their nuclear wastes at or near the reactor sites, some ship the wastes to a reprocessing plant, and others use independent facilities away from the reactor (see Table 6.2).

In the United States, most spent fuel units are stored in water pools. The water cools the fuel rods and also shields workers in site against radiation. These pools are intended for short-term storage, and many may well have reached their optimum capacity. Because of this, plans are in place to increase the density by racking spent fuel rods stored at interim sites. Further, by carefully designing the racks and adding neutron-absorbing material between the racks, it is possible to increase the number of fuel rods stored at any one site. When a storage are reaches capacity, some fuel rods may be transshipped to another facility.

The interim storage space available in the United States is quite limited, as it is for other countries. Figure 6.1 shows the anticipated need for additional at-reactor storage until a mined repository is available for the spent fuel. It should be emphasized that this projection takes us only until the year 2000. Since it is likely that a mined repository will not be a reality until well after that date, the amounts shown in Fig. 6.1 (vertical bar) are minimum estimates. Government and industry are also examining other methods for interim storage and then considering (in addition to increasing the number of spent-fuel assemblies in pooled areas) dry storage technologies that include casks, silos, drywells, and vaults.

Casks, large metal containers built to provide good radiation shielding, are stored above ground. Silos are cylinders constructed above ground of concrete and will, below grade, provide sealed secondary containment for the spent-fuel assemblies. Drywells, located below grade, are designed to hold spent fuel; these wells are lined with concrete and steel, and the surrounding earth acts as an additional barrier. Vaults, which use air convection for cooling, average structures built below grade out of concrete.

Table 6.2
Storage of Commerical Spent Fuel and High-Level Waste in Nine Countries

Country	Electric energy generated (%)	Number of reactors (12/31/89)	Reprocessing (where)	Interim storage: Location[a]	Interim storage: Duration (year)	Repository schedule
Canada	15.6	18	No[b]	Reactor sites	>50	>2010
France	74.6	55	Yes	Reactor sites;reprocessing plants	~1 ~2 >20	>2010
West Germany	34.6	24	Yes (WG, Fr, UK)	Reactor sites; independent facility(Gorlben,Ahaus)	3-10 Undecided >10	~2008
Japan	27.8	39	Yes(Fr,Ja)	Reactor sites; reprocessing plants	2-3 ~4 30-50	> 2030
Spain	38.4	10	No	Reactor sites; independent facility-(not cited)	40 40	~2020
Sweden	45.1	12	No	Reactor sites; independent facility (CLAB, at Oscharschamn reactor site)	~1 15-40	>2020
Switzerland-	41.6	5	Yes(Fr,UK)	Reactor sites; independent facility (Wurenlingen HLW; SF not under reprocessing contract)	<12 ~40	>2025
United Kingdom	21.7	39	Yes	Reactor sites; independent facility for AGR fuel (at Heysham reactor Undecided site)	~1 Undecided	>2040
United States	19.1	110	No	Reactor sites	Undecided	>2003

Source: U.S. Department of Energy

[a] *AGR, advanced gas-cooled reactor; ClAB, central storage for spent fuel; Hi,W high-level wastes, SF, spent fuel.*

[b] *No decision about future reprocessing has been made.*

[e] *A storage facility that does not affect the safety of a nuclear power or fuel reprocessing plant*

The first successful dry storage system, shown in Fig. 2.2 was demonstrated by the Carolina Power & Light Company. Spent fuel is kept inside a sealed steel canister flooded with nitrogen or helium, and the steel container is then schielded by a concrete structure.

Rod consolidation techniques involve dismantling the fuel assemblies and placing them in canisters. This is proposed to facilitate more dense racking in pooled storage areas, but the technique can also be applied to dry storage systems. Even with wet and dry methods for increasing available interim storage, these approaches are nothing more than stop-gap attempts to handle the growing amount of spent fuel until a permanent repository is available.

Table 6.2 shows a summary of the nuclear generating capacity statistics for nine countries whose yields were 15% or more of total electrical energy. Table 6.2 also gives the number of reactors, information on reprocessing commitment, interim storage (locations, methods, duration), institutional responsibility, and a tentative schedule for repositories. This table, compiled from international and United States government sources, is subject to revision. Further, for the United States, the table only contains commercial nuclear waste information—not the much large segment of wastes generated by the defense industry.

V. Underground Storage

Underground storage of spent fuel and high-level waste is preferred worldwide, although no repository is open at present. Belgium, France, Italy, Swtizerland, and the United Kingdom are considering clays; Germany, France, The Netherlands, Switzerland, the United Kingdom, and the United States have investigated bedded salt and/or anhydrite. Canada. Finland, France, Japan, Sweden, Switzerland, the United Kingdom, and the United States are looking into crystalline rocks; and volcanic ruffaceous rocks are being considered by the United States and Japan.

In the United States, the Nuclear Waste Policy Act (1982) and the Nuclear Waste Policy Amendments Act (1987) address high-level repository effort. Prior to 1987, sites in basalt (near Handford, Washington), bedded salt (Palo Duro Basin, Texas), and tuffaceous rocks (Nevada Test Site at Yucca Mountain, Nevada) were being characterized for consideration for storage of spent fuel and high-level waste. In 1987 the decision was made to fully evaluate only the Yucca Mountain, Nevada site, and work on the basalt and bedded salt sites for spent fuel and high-level waste was halted.

VI. Multibarrier Concept

A series of natural or engineered barriers is required to isolate nuclear wastes from the environment and to prevent the buried waste from being breached; however, if breaching occurs, barriers will act to retard the movement of the radionuclides to the surface. The U.S. Environmental Protection Agency (EPA) requires that a disposal system must prevent cumulative releases of radionuclides to the accessible environment for 10,000 years in amounts not to exceed one-tenth of the calculated quantities. At the same time, the U.S Nuclear Regulatory Commission (NRC) requires that stored waste and the immediate engineered surroundings provide essentially complete containment for 300 to 1000 years. Accordingly, both EPA and NRC require the installation of the multibarrier disposal system that consists of the waste package, the repository, and the surrounding host geologic medium.

The waste package, the first component of the multibarrier system, consists of the waste and any material that separates it from the surrounding environment; all types of containers, seals, packaging

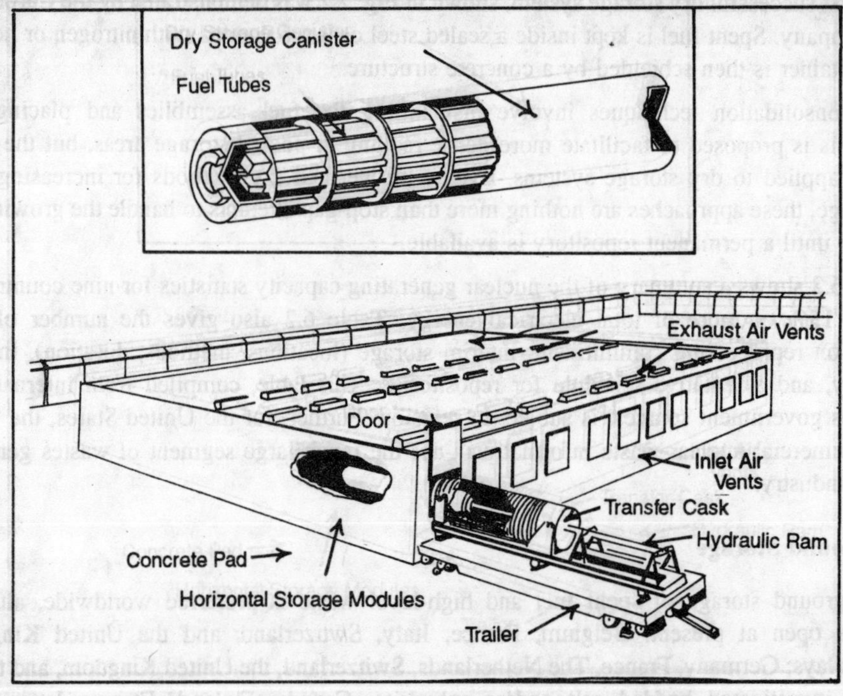

Figure 6.1. Generic horizontal modular storage system (NUHOMS).

barriers, absorbents, shielding, etc., are included. The waste form for both spent fuel (small pellets of uranium oxide that have been irradiated) and high-level waste will only be solids. The pellets are sealed in metal tubes, called cladding, that consist of a zirconium-rich alloy that is heat and corrosion resistant. Hundreds of these tubes are bound together into a fuel assembly. The result is a waste fission product with actinides daughters that become the first isolation barrier.

At present, high-level nuclear waste concentrated in the United States will be vitrified into a boron-rich (borosilicate) glass — the preferred waste form because it offers a favourable medium to immobilize the fission products and other radionuclides. The borosilicate glass is strong, withstands normal leaching conditions, and can be manufactured in large-scale operations by remote means. In its molten state, the nuclear waste-bearing borosilicate glass will be poured into stainless steel canisters and cooled until solid. The containers, known as pour canisters, will be plugged and welded shut and then tested for leaks. They will be decontaminated and moved to some short-term storage vault. Another canister, known as the disposal canister, will surround each pour canister and separate the pour canister from the surrounding rocks. Materials such as copper alloyed steel, carbon steel, and stainless steel, chosen for their ability to resist corrosion on geologic media, are being considered for the disposal canister.

At Yucca Mountain (see below), the site under consideration for a repository is above the water table and rests in unsaturated rocks; therefore, no additional barner between the disposal canister and the surrounding rock is contemplated. In other countries where repository media are below the water table, an engineered barrier of some highly absorbing material, such as bentonitic clay, will be used to separate the disposal container from the repository rock. This material minimizes chemical attack on

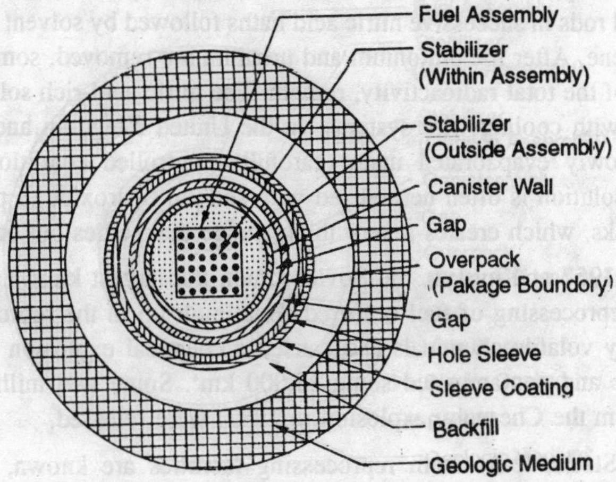

Figure 6.2. Conceptual design of the multibarrier isolation system.

the container, decreases radionuclide solubility, sorbs many radionuclides, and because of its plastic-like nature will adjust to rock stress.

The repository portion of the multibarrier system is not associated with the waste package. Backfill material will be placed around the disposal containers to fill existing space as well as the space that remains after the maximum number of containers have been placed in the repository. Backfill material will primarily prevent waters from reaching the canisters and also provide structural support, take up stresses in the repository area, and conduct heat from canister to surrounding rock. At Yucca Mountain, backfill material will consist of crushed host rock, but it may be modified to include additive of clays or chemicals.

Shaft seals and borehole seals are not officially part of the multibarrier system, yet when successfully deployed they will prevent water from entering and will further isolate the site.

The ultimate barrier is the repository medium. The repository rock must be suitable for construction of the facility as well as containment of the waste. Its hydrologic and geo-chemical characteristics must be such that breaching of the respository is minimized. Important hydrologic features are low groundwater flow rates and very long pathways from repository rocks to the accessible environment. In the unlikely event of a breach, transport of radionuclides must be sluggish and predictable and more in less than the release limits allowable. Figure 3 is an artist's conception of the multibarrier system.

The sorbing capacity of the host rocks for radionuclides must be high, and the rocks must be able to rapidly dissipate heat as well as withstand stresses induced both naturally and from the repository. The tectonic activity and volcanism of the area should be slight. Ideally, a site can be selected by the absence of potential economic resources, where future human intrusions are unlikely.

VII. Reprocessing

Fuel rods from defense and commercil sectors are reprocessed to remove the isotopes of interest.

In military operations, plutonium and duranium are carefully removed from rods by a complex dissolution of the fuel rods in successive nitric acid baths followed by solvent extraction using tri-n-butyl phosphate and kerosene. After the plutonium and uranium are removed, some dozens of radionuclides, which possess 98% of the total radioactivity, remain. The nitric acid-rich solution is then transferred to stainless steel tanks with cooling wire systems. In the United Kingdom and in France, the high-level waste liquids are slowly evapaorated under carefully controlled conditions. In the United States, however, the acidic solution is often neutralized with sodium hydroxide to prevent undue corrosion of the stainless steel tanks, which creates a very thick sludge that settles out on the bottom of the tank.

Historically in 1957 at Khystyn, the Soviet Union the largest known noncritical nuclear-related explosion from the reprocessing of fuel occurred. Heat built up in the reprocessed liquid waste in the presence of extremely volatile chemicals and caused a chemical explosion that blew the roof off the containment structure and contaminated some 15,000 km^2. Some two million curies of radioactivity (about 4% of that from the Chernoby explosion in 1987) were released.

In the United States, leaks from reprocessing facilities are known, including minor leaks at Hanford, Washington and Savannah River, South Carolina, and at Sellafield in the United Kingdom.

As shsown in Table 6.2, reprocessing plays an important role in France and the United Kingdom, and a large reprocessing facility is being built in Japan.

VIII. Yucca Mountain Site, Nevada

In the United States, Yucca Mountain, Nevada is being investigated for the first high-level water repository site. This repository will be above the water table and lies on the edge of the Nevada Test Site (NTS) and Nellis Air Force Range. The obvious advantage of this selection is that these rocks are unsaturated hence the likelihood of water-promoted chemical attack on the stored waste canisters is minimal. Yet at the same time other problems are posed. For example, since these rocks are unsaturated by water, the estimates on rates and directions of any water flow that may take place in such media are unreliable. Further, the rocks are highly fractured and thus water may be channeled in specific directions at rates greater than predicted for homogeneous, nonfractured rocks. For these reasons close investigation of the hydrology, paleohydrology, paleoclimatology, and meteorology, of this site along with the more traditional studies of geology, geochemistry, and geophysics are being considered.

The tuffaceous volcanic rocks of Yucca Mountain consist of devitrified glass, potassium feldspar, soda feldspar, silica (quartz, tridymite, cristobailite), and zeolites. The repository rocks are the tuffs of the Topopah Springs Member. A cross section of the area is shown in Fig. 6.4. The presence of zeolites in the tuffaceous rocks is important, for these minerals have excellent sorbing capacities. Thus if radionuclides were to be lost from breached nuclear waste canisters, the zeolites would retain or strongly retard their movement toward the accessible environment.

A. Tuffaceous Rocks of the Nevada Test Site

The general geology of the NTS, which includes granite, argilite, and tuffaceous rocks, is shown in Fig. 6.5. The focus of the National Waste Terminal Storage programme has shifted somewhat to place a great emphasis on tuffaceous rocks.

The tuffaceous rocks at NTS are extremely thick, with variable composition, mineralogy, and other physical properties. Examination of these rocks has focused on Yucca Mountain in the southwestern part of the NTS; several drill holes have penetrated these rocks.

The general lithologic and mineralogic description is directly applicable to the NTS rocks and for silicic tuffs in general. The stratigraphic section for the rocks is shown in Fig. 6.6, and variation of authigenic phases in this section is given in Fig. 6.7.

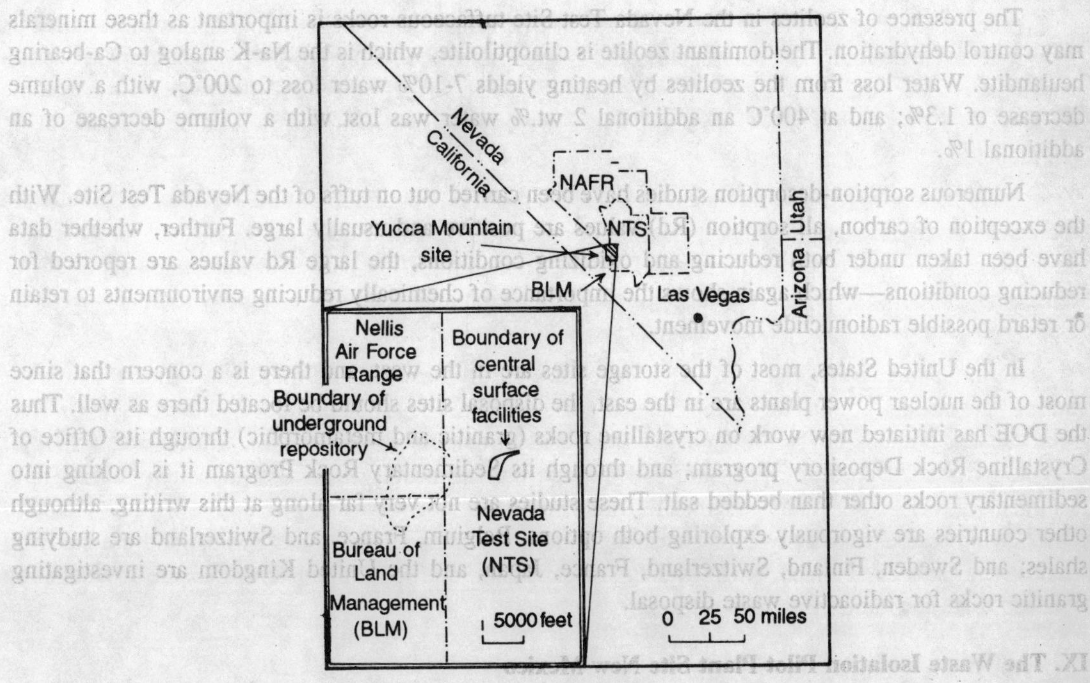

Figure 6.3 .Location of the Yucca Mountain site on southern Nevada.

There is considerable lithologic variation among the rocks—but this is expected, considering the dynamic origin of tuffaceous rocks. The oldest unit, of probable thickness in excess of 200 m, is a nonwelded ash-flow tuff below the Crater Flat Tuff, which contains the Bullfrog member and the Prow Pass Member. The former member is a welded ash-flow tuff containing abundant biotite.

The Prow Pass Member is a vitric crystal tuff in which the lower part consists of nonwelded air-fall and ash-flow units with an upper welded ash-flow unit. (See Fig. 6.8).

The Bedded Tuff of Calico Hills is a sequence of air-fall and ash-flow tuffaceous rocks and volcaniclastic sedimentary rocks whose thickness varies from 95 to 140 m.

The Painbrush Tuff includes the basal Topapah Springs Member (370 m thick) in which the high-level waste repository is intended, followed by the Yucca Mountain Member (8 m thick) and the uppermost Tiva Canyon Member. It is a zone dash-flow tuff with nonwelded basal units grading upward into welded vitrophyre, which is in turn overlain by devitrified welded tuff. Phenocrysts increase to near 17% at the top and consist of sanidine, plagioclase, quartz, biotite, and opaques. Lithic fragments are common near the base but less frequent upward in the unit.

Alteration is variable in the unit, and both the basal vitrophyre and the upper welded zone are essentially unaltered. Much of the upper welded zone, however, has devitrified to intergrowths of cristobalite and potassium feldspar. Lower nonwelded tuffs are highly zeolitized to clinoptilolite, and heulandite occurs as fracture fillings. Minor amounts of erionite and phillipsite also occur.

The youngest tuffaceous unit in the area is the Tiva Canyon Member, which contains a basal bedded air-fall tuff and an upper ash-flow tuff.

The presence of zeolites in the Nevada Test Site tuffaceous rocks is important as these minerals may control dehydration. The dominant zeolite is clinoptilolite, which is the Na-K analog to Ca-bearing heulandite. Water loss from the zeolites by heating yields 7-10% water loss to 200°C, with a volume decrease of 1.3%; and at 400°C an additional 2 wt.% water was lost with a volume decrease of an additional 1%.

Numerous sorption-desorption studies have been carried out on tuffs of the Nevada Test Site. With the exception of carbon, all sorption (Rd) values are positive and usually large. Further, whether data have been taken under both reducing and oxidizing conditions, the large Rd values are reported for reducing conditions—which again shows the importance of chemically reducing environments to retain or retard possible radionuclide movement.

In the United States, most of the storage sites are in the west, and there is a concern that since most of the nuclear power plants are in the east, the disposal sites should be located there as well. Thus the DOE has initiated new work on crystalline rocks (granitic and metamorphic) through its Office of Crystalline Rock Depository program; and through its Sedimentary Rock Program it is looking into sedimentary rocks other than bedded salt. These studies are not very far along at this writing, although other countries are vigorously exploring both options. Belgium, France, and Switzerland are studying shales; and Sweden, Finland, Switzerland, France, Japan, and the United Kingdom are investigating granitic rocks for radioactive waste disposal.

IX. The Waste Isolation Pilot Plant Site New Mexico

In the 1950s, the Atomic Energy Commission (AEC) requested the prestigious National Academy of Science-National Research Council to investigate the matter of disposal of radioactive wastes in rocks. After a long and detailed study, the Council recommended salt beds for the following reasons:

- Favourable thermal conductivity (salt can readily absorb heat).
- Low porosity (water can's flow).
- Virtual absence of included water.
- Plasticity (if a fracture such as a fault were to cut the salts, salt would flow into and seal the fracture).

Studies of bedded salt were begun in Kansas and continued in New Mexico. In the mid-1970s, the Waste Isolation Pilot Plant (WIPP) project was officially endorsed. This site continues to have many favorable features both geologic and socioeconomic. Hydrologically, the rocks contain water only in the form of microscopic fluid inclusions, water of crystallization of some minerals, and isolated brine pockets. The fluid inclusions and brine pockets are saturated with salt. When a solution is saturated with respect to a solute, no more of that solute can dissolve in that solution; thus the brines and fluid inclusions are incapable of dissolving their host bedded halite.

For more than 50 years there has been active potash mining in the area, and technology to mine

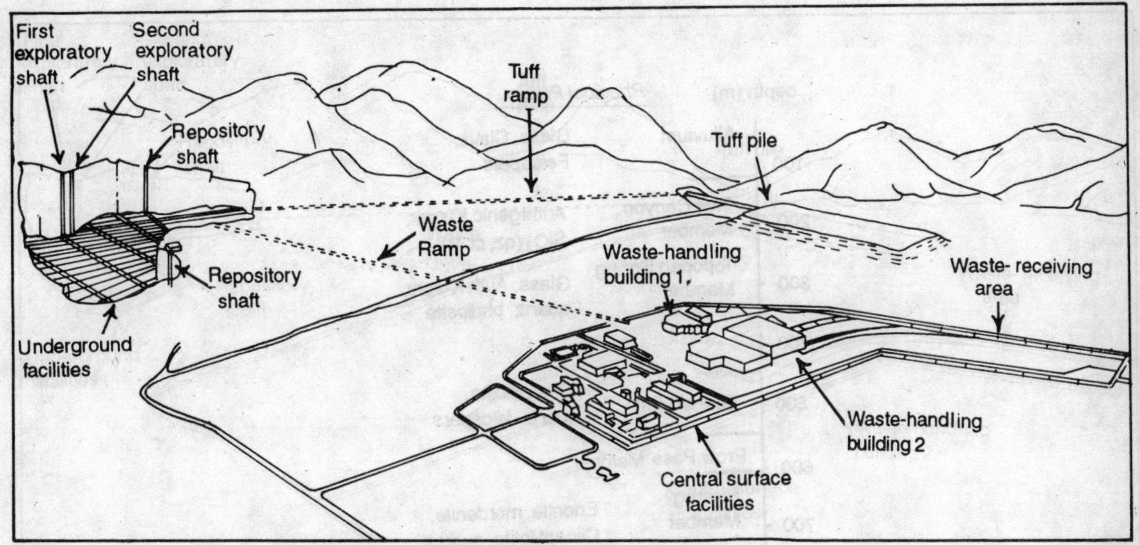

Figure 6.4. Conceptual drawing of a repository at Yucca Mountain.

and maintain openings in bedded salt at depth is very well developed. Geochronologic studies show that the bedded salts have been closed chemical and isotopic system for 200 million years. The surface value of the land is minimal. The land is not suitable for growing crops, and the WIPP site occurs in an area of low population density. Aso, it only requires about one-eight of one square mile to be permanently withdrawn from other use. Access to the site by both rail and road is excellent. Although potash is present in the WIPP site, it occurs in small amounts that will not impact on the reserves in the area. Even so, the potash could be mined up to within one kilometer of the WIPP shaft.

In the late 1970s the WIPP site, a Department of Energy (DOE) facility, was chosen to be used only for the disposal for government (mainly defense) generated low-level wastes that contain one or more transuranic elements (plutonium, neptunium, americium, curium) as well as uranium. The material to be stored is essentially garbage: old gloves, bottles, packing and other containers that contain small

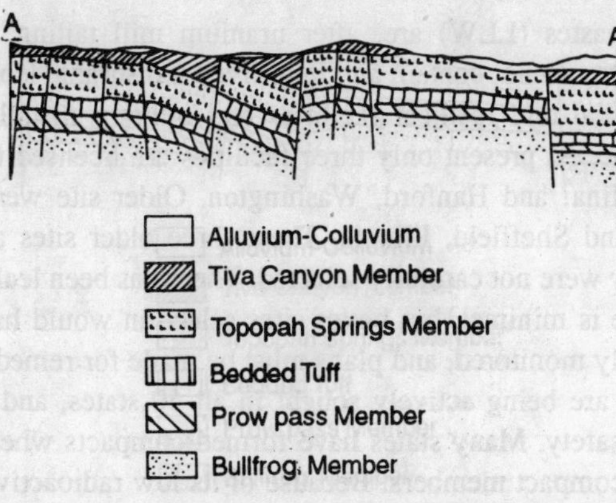

Figure 6.5. Generalized cross section A–A′.

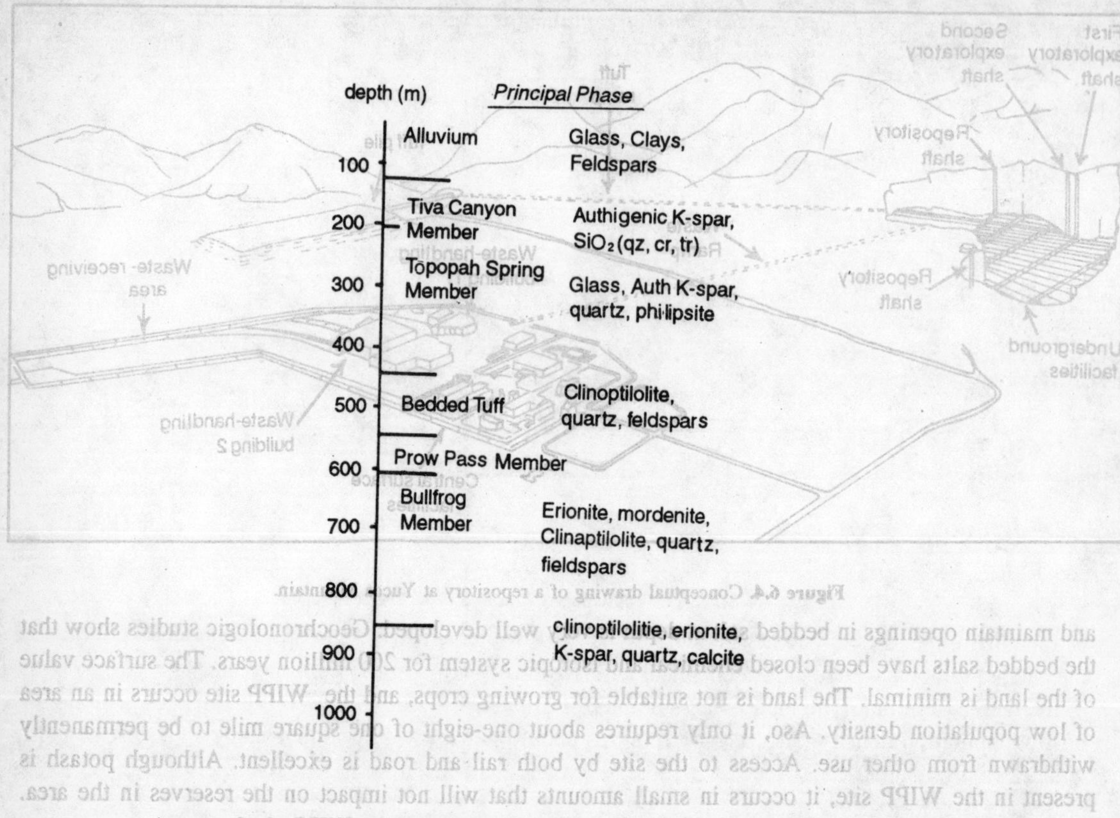

Figure 6.6. Generalized stratigraphic section, test well J-13, showing major authigenic phases.

amounts of TRU. Some high-level waste (HLW) and spent fuel rods might be temporarily emplaced at the WIPP site to study their effects on salt.

X. Other Storage Sites

Low-level radioactive wastes (LLW) are, after uranium mill tailing, the largest category of radioactive materials; these wastes do not contain any long-lived transuranic isotopes. LLW is generated everywhere, including educational and medical facilities and military installations. Disposal of this mountain garbage is problematic. At present only three facilities are licensed to receive LLW: Beatty, Nevada; Barnwell, South Carolina; and Hanford, Washington. Older site were at West Valley, New York; Maxey flat, Kentucky; and Sheffield, Illinois. These three older sites are in areas that receive considerable precipitation—they were not carefully selected. There has been leakage of tritium at Maxey Flats, for example. The leakage is minimal, but better sites selection would have prevented a mishap. Now these sites must be carefully monitored, and plans must be made for remedial action should further leackage occur. Sites for LLW are being actively sought in all 50 states, and the utmost care is now taken to ensure environmental safety. Many states have formed compacts whereby each state takes its turn to dispose of LLW for all compact members. Because of its low radioactivity levels, LLW may be disposed of in steel drums emplaced in 30–foot-deep trenches and then covered by earth fil.

An interesting idea for LLW disposal involves the poisoned land/inland sea concept which argues

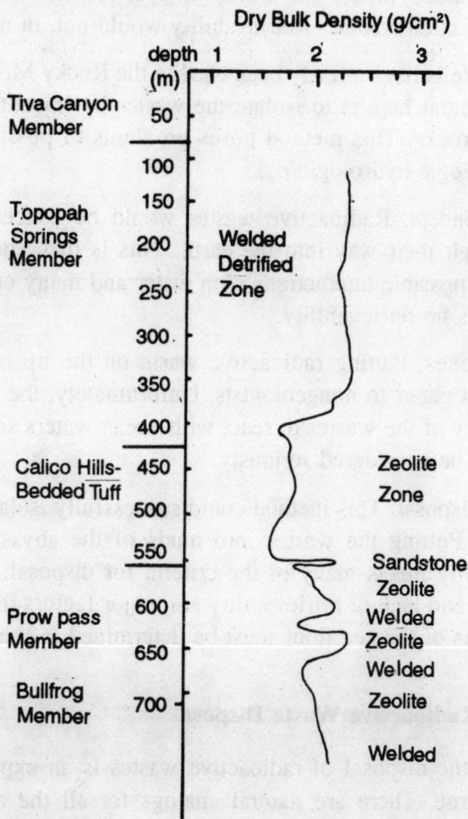

Figure 6.7. Generalized stratigraphic section, showing major zeolite zones.

that there exist many areas in the United States that have been naturally or anthropogenicaly; poisoned; many of these areas are in closed hydrologic basins, where there are no economic resources, and are isolated from population centers. For example, there are playas in Nevada and New Mexico where arsenic has accumulated and the toxicity has damaged plant and animal life. If these playas are already closed hydrologic basins, the addition of LLW to these sites makes use of otherwise unproductive land and at the same time eliminates part of the LLW problem.

XI. Alternate Methods of disposal

Numerous methods for disposal of radioactive wastes have been proposed during the last 30 years.

1. Rocketry. The idea is to encapsulate the waste and fire it into extraterrestrial orbit. The force for this option is to remove the waste from the earth; the negatives are the expense for transportation, the risk of rocket failure, and the lack of retrievability.

2. Antarctic. Use of the Antarctic for radioactive waste disposal would require a United Nations reversal of an old resolution that barred such disposal. Further, there are severe logistical drawbacks, including safety on the sea lanes and a need for dual port facilities

to load and unload waste; and transport over ice and land in the Antarctic would be expensive and cumbersome. Retrievability would not, in most scenarios be possible.

3. Deep wells. Wells have already been used in the Rocky Mountain area for some chemical wastes. The intent here is to isolate the wastes by injecting them into deep, essentially impermeable rocks. This method poses problems of possible contamination of aquifers and other geologic-hydrologic risks.

4. Rock melt concept. Radioactive wastes would be concentrated and encapsulated and allowed to melt their way into the earth. This is theoretically possible but not a good option due to possible interactions with water and many other negative geologic factors. Again, there is no retrievability.

5. Subduction zones. Putting radioactive waste on the lip of land about to be subducted looks good on paper to nongeologists. Unfortunately, the rate of subduction is so slow, the opportunity of the wastes to react with ocean waters so great, that this method is not likely ever to be considered seriously.

6. Ocean floor disposal. This method could successfully isolate the wastes from the surface environment. Putting the wastes into muds of the abyssal plains in areas of extreme tectonic stability meets many of the criteria for disposal. Technically, expense, risk of sea transport, and lack of retrievability are major factors that must be considered. Again, the legal status of the sea floor must be determined in the United Nations.

XII. Natural Analogs for Radioactive Waste Disposal

It is often stated that the disposal of radioactive wastes is an experiment for which there is no precedent. But this is not true. There are natural analogs for all the elements (and isotopes of the elements). For example, the branch of geology that deals with the dating of rocks and minerals according to their mix of isotopes is called geochronology, and geochronologists have dated rocks and minerals as old as the earth itself (4.55 Ga) by any number of methods. Specifically, 238-uranium decays to 206-lead, 235-uranium decays to 207-lead, 232-thorium decays to 208 lead, 87-rubidium decays to 87-strontium, 40-potassium decays to 40-argon and 40-calcium, and 147-samarium decays to 143-neodymium. Therefore, if a rock has been dated by all these methods and the results are concordant, we know that this particular rock has remained a closed system — it has not lost or gained parent or daughter isotopes since it formed.

We know that rocks do behave as closed systems in many cases to uranium and thorium, which, due to their similar characteristics as actinides, are analogs for the transuranics as well. They have also remained closed systems to actinide daughters (protoactinium, radium, radon, polonium, bismuth, lead) and the eventual stable lead isotopes. For if this were not true, the ages would be too young and wildly discordant. The rocks have also remained closed systems to rubidium and strontium and thus to cesium as well, since rubidium and cesium behave nearly identically in rocks and in solution. Further, barium is chemically very similar to strontium, so if a rock is a closed system to strontium, it is also probably closed to barium. Potassium—argon dates tell as that the noble gas argon is also retained in many rocks. Since krypton and xenon would also be retained. The date from the samarium—neodymium method involves two rare earth elements; hence all the rare earths, with their nearly identical chemistries, must have behaved the same.

Collectively, radiometric age determinations rell us that rocks and minerals can and have retained their chemical and isotopic integrity for very long periods of geologic time, much longer than the isolation time for buried radioactive waste now required. These systems are only disturbed when exposed to chaotic conditions of thermal metamorphism or plutonism at very high temperatures and pressures over tens of thousands of years. Buried waste will have moderate initial temperatures near or less than 250°C, and this will drop rapidly as the highly radioactive species decay away.

Another approach is to examine the zones of contact metamorphism surrounding igneous dikes and plutons to see what effect the emplacement of molten rock at extremely high temperature (800-1200°C) and high pressure (always without benefit of barriers) has on elemental distribution in these affected zones. For the most part, the redistribution of elements in these cases is by no more than millimeters, even when the thermal regime is maintained in excess of 10,000 years and in the presence of fluids. This argues that at the much lower temperatures of our own radioactive waste, with the barriers described earlier, with the absence of available fluids for reaction with the waste, and with the decay of the radioactive species, there would be essentially no migration of elements from the emplaced canister to pose a threat to public health.

The best natural analog is the Oklo Natural Reactor, a very ancient uranium deposit found in the Republic of Gabon near the small town of Oklo, where a body of uranium ore sustained nuclear fission reactions two billion years ago, and criticality lasted up to 500,000 years. During that time hundreds of tons of fission products formed, and today all of these "wastes" can be accounted for in the local rocks. The uranium of the rocks was deposited in concentrated form about two billion years ago, when it contained 3.25% of the 235 uranium isotope. (Because 235-uranium has a shorter half-life than 238-uranium, the percentage of 235-uranium is progressively greater the older the rock.) Thus at Oklo two billion years ago the ore was actually rich enough to serve as a fuel.

Other criteria had to be met as well. Evidently the original low- grade ore was deposited in sandstone, but the sandstone fractured at depth and the surrounding shale wedged into the fractures between the slabs. During this time some of the low-grade uranium was dissolved and reprecipitated in these shale-filled fractures— in very rich amounts, sometimes to 70% uranium. Thus another important criterion, high uranium content, was realized. Yet the story doesn't end there. For a reactor to work there must be a proper moderator, and apparently just enough water was present, about 15%, to serve as moderator. Further, there must have been a lack of neutron poisons. All of the conditions were favourable at Oklo, and consequently a long-time nuclear reaction was started soon after the uranium was deposited and remobilized.

The evidence that Oklo went critical first surfaced at the French gaseous diffusion plant at Pierrelattee, where a sample of African ore was found to be significantly depleted in 235- uranium. The ore was traced to Oklo, and a high grade sample was tested at 0.4-0.5% 235-uranium rather than the usual 0.71–0.72%; in other words, the particular samples were strongly depleted in 235-uranium. The only natural process by which this can happen is nuclear fission. The French Atomic Energy Commission immediately argued that the Oklo ores had fissioned, and that, if so, the evidence for fission products in the rocks would have to be found. This can be determined quite easily in many instances because the isotopes of many fission elements are produced in different amounts than are found in the natural element. Rubidium, for example, contains isotopes of mass 85 and 87, and the natural ratio of 85/87 is 2.59. But nuclear fission produces rubidium with an 85/87 ratio of about 0.35. Today's mass spectrometers can measure such isotopic ratios to better than ± 0.00001, so even very small anomalies can be detected. The French and their co-workers found that they could account for the presence and retention of most of the fission products as well as for the actinides. These studies were expanded to include investigators from outside France, including the United States, and major conferences on Oklo were held in Libreville, Gabon in 1975 and in Paris in 1977.

The results of these studies, which are still ongoing, show that the Oklo body acted not only as a nuclear reactor but also as a nuclear waste repository, and, most important, all the radionuclides of concern were kept in the rocks without any appreciable migration. The Oklo data convincingly show that rocks can and do retain radioactive wastes; and it should also be noted that the Oklo rocks are not especially good candidates for a repository as they are fractured, impure, somewhat porous, and permeable. Yet they have retained their waste for two billion years. All in all, Oklo is a fascinating scientific gold mine. Yet Oklo did not act as a high-powered reactor. It sustained fission reaction for at least 500,000 years, produced some 16,000 megawatt years of energy, and produced hundreds of kilgrams of fission products, all of which stayed in the rocks at a depth of 500 to 1000 meters and migrated no more than a few meters. Studies have shown no migration for zirconium, molybdenum, niobium, cadmium, strontium, rubidium, uranium, thorium, lead, antimony, tin, barium, rare earths, and others. Oklo is the strongest testimony for the retention of radioactive wastes in rocks.

XIII. Uranium Mill Tailings

Uranium mill tailings are the waste rock generated by processing uranium ores. In the abundant sandstone-type uranium deposits of western United States, only 0.5 to 0.2% of the rock is uranium oxide; the rest consists of quartz, feldspars, and rock fragments. Many metals are also fixed with the uranium in and with organic carbonaceous matter and pyrite.

The tailings generated in the United States through 1984 total some 200 million metric tons, and cover over 1300 hectares at 51 sites—26 abandoned sites and 25 active sites. Many of these sites are on standby at the present time, but they have not been officially closed. An acid leach method, shown in Fig. 6.9, is most commonly used for extraction of uranium from ore.

The abandoned sites are covered by the Uranium Mill Tailings Remedial Action (UMTRA) program and are referred to as Title I Sites. Except the site at Cannonsburg, Pennsylvania and two mills in Florida associated with uranium by-product recovery from phosphate mining, all of these sites are in the western United States (Colorado, New Mexico, Wyoming, Idaho, Arizona, North Dakota, Oregon, Utah and Washington). The active or standby sites are called Title II Sites.

Sands constitute some 65% of the tailings by volume and slimes 35%; but the slimes, because they contain residual uranium and most of the radium, contain 85% of the total radioactivity. The slimes also contain most of the molybdenum, vanadium, selenium, arsenic, and other trace metals. Furthermore, the solutions accompanying the slimes and tailings are high in chloride, sulfate, sodium and some organics and possess pH of 1-2.

Some negative aspects from uranium mill tailings are

1. Generation of 222-radon from the piles: Both 230-thorium and 226-radium, the grandparent and parent to 222-radon, have long half-lives and are present in the piles; hence there is a potentially large radon flux. The concern is that 222-radon flux from the piles constitutes a health hazard.

2. Removal of particulate radioactive particles from the piles by wind: The concern is that

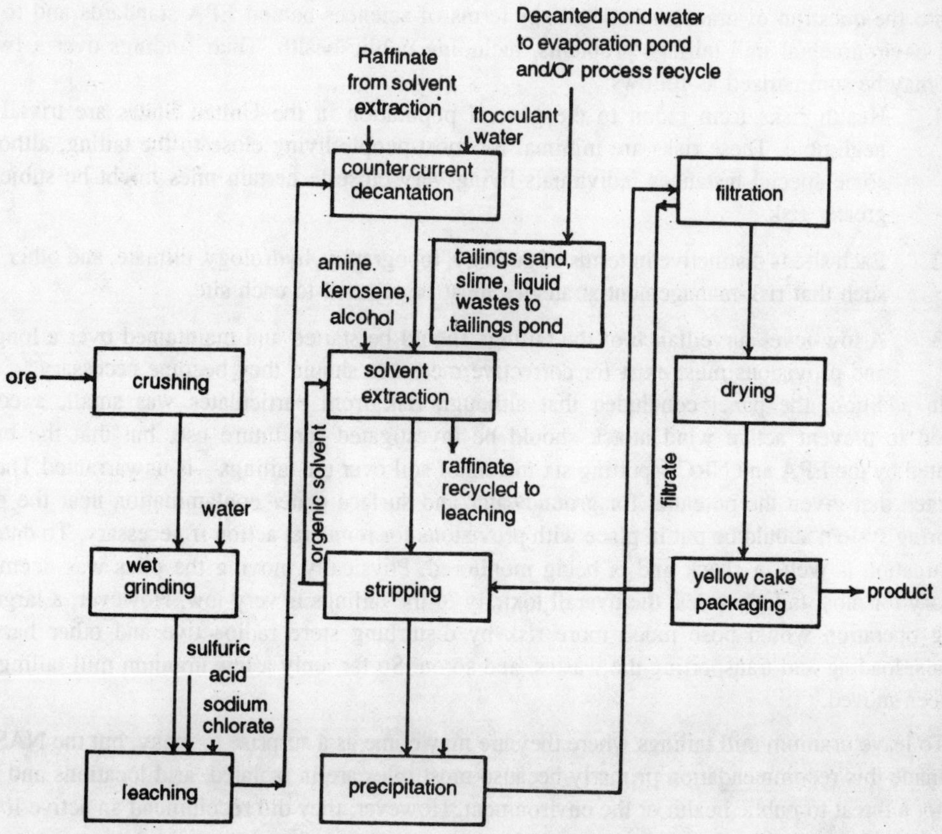

Figure 6.8. Flow diagram for acid leach process.

these particles constitute a health hazard as they would be inhaled by people living near the piles.

3. Contamination of water by dissolved heavy metals, radionuclides, or other toxic materials (such as organics) seeping into the ground beneath the piles: The very acidic tailings waters contain high amount of dissolved toxic materials and it needs to be established just how far some of these materials may be transported.

4 Tailings used for construction or other industrial indoor purposes: There is a potential for creating a new indoor radon source that may be a potential threat to public health.

As with any part of the nuclear fuel cycle, there is disagreement on the magnitude of the threat to public health. At one extreme, tailings are considered to be the most serious type of radioactive waste, because of their large volume and the presence of long-lived radionuclides. At the other extreme, a negligible radiological hazard to public health is advocated because of the very low radioactive content of the piles (at or just barely above background) and, in most cases, their location far from populated areas.

In 1984 the National Academy of Science, National Research Council (NAS-NRC) was asked to look into the question of uranium mill tailings terms of sciences behind EPA standards and to assess overall environmental mill tailings problems, including public health. Their findings over a two-year period may be summarized as follows:

1. Health risks from radon to the general population in the United States are trivial if not negligible. These risks are minimal for most people living close to the tailing, although in some special instances individuals living very close to certain piles might be subject to a greater risk.

2. Each site is distinctive in terms of geology, topography, hydrology, climate, and other factors such that risk-management strategies must be tailored to each site.

3. A low-level surveillance of the tailings should be started and maintained over a long term, and provisions must exist for corrective measures should they become necessary.

In addition, the panel concluded that although risk from particulates was small, a covering designed to prevent active wind attack should be investigated for future use, but that the measure advocated by the EPA and NRC—putting six meters of soil over the tailings—is unwarranted.The panel also noted that given the potential for groundwater and surface water contamination near the piles, a monitoring system should be put in place with provisions for remedial action if necessary. To date, such contamination is well in check and is being monitored. Physically moving the piles was deemed not necessary for most tailings piles; the overall toxicity of the tailings is very low. However, a large-scale moving operation would pose much more risk by disturbing store radioactive and other hazardous materials, loading and transporting the wastes, and so on. So far, only a few uranium mill tailings piles have been moved.

To leave uranium mill tailings where they are may come as a surprise to many, but the NAS-NRC panel made this recommendation primarily because most piles are in isolated, arid locations and are, as such, not a threat to public health or the environment. However, they did recommend an active lowlevel monitoring program with accompanying record keeping to ensure that there is no human intervention. They also recommended that tailings not be used for building materials and, if used for other types of construction, that the sites be monitored.

One concern is that of flooding, which is of greater significance at some tailings piles than others. Short of heroic measures, it is probable that future 100-year floods will have a pronounced effect on at least some of the piles. In the late 1950s part of the Green River, Utah pile v.as washed away by a flood. Studies in the area find no trace of the tailings removed showing that nature's action in this case effectively distribute the tailings and their constituent chemicals over such a wide area that all were infinitely diluted.

In summary, uranium mill tailings are large by volume and should be monitored, but the threat they pose to public health and to the environment is very small.

Glossary

Actinide. Very heavy element with atomic number from 89 and higher. All isotopes of the actinides are radioactive.

Actinide daughter. A daughter isotope, or decay product, from a parent actinide. These actinide daughters may be radioactive (e.g., Ra, Rn, Fr, Ac, etc.) or stable (e.g., Pb and Bi).

Alpha particle. Positively charged helium nucleus possessing high ionization capability but low penetrating power. A sheet of paper is sufficient to stop an alpha particle.

Beta particle. Negatively charged particle, essentially an electron, with little ionizing capability but greater penetrating power than an alpha particle. A thin sheet of metal will stop a beta particle.

Fissile isotope. Isotope that undergoes fission.

Fission. Process by which a radioactive element spontaneously disintegrates into daughter isotopes accompanied by the release of neutrons, other particles, and energy.

Gamma rays. Short wavelength electromagnetic radiation, very low ionization capability, very large penetrating powers. Several feet of cement and metal are required to block gamma rays.

Transuranic. Pertaining to elements with atomic number greater than 92.

7

Health and Injury to Sanitation Worker

Introduction

Many organizations believed that there was on particular problem with hazards to sanitation workers. One reason for this generally false belief has been that people who work in solid-waste handling have been often undereducated, having little bargaining power, in largely transient employment, and they often leave the job when they are injured or become oil. Tables 5.1 and 5.2 show data from two cities on reported injuries to their sanitation employees.

It was also reported that sanitation workers suffered average or below-average illness rates for industry in general. Other scattered reports state that ''refuse removal workers (in Germany) have been found to be astonishingly immune against infectious diseases, while street cleaners often fall ill with toxoplasmosis''. Another disease to which street cleaners and others have been exposed has been cutaneous larva migrans, a form of dermatitis characterized by linear lesion marking the migratory paths of larval parasites. As an occupational disease cutaneous larva migrans is most frequently seen in plumbers, electrician, and construction workers who are required to be in contact with damp or wet soil

Table 7.1

Injury Analysis for 1964

Item	*City I*	*City II*
Number of employees	23	60
Number of employees injured	11 (48%)	44 (73%)
Once	6 (26%)	29 (48%)
Twice	5 (22%)	10 (17%)
Three times		5 (8%)
Total number of injuries	16	64
Frequency (f)	348	532
Total number of lost time injured	12	36
Frequency (F)	252	300
Total lost time	216	334
Severity (S)	4,700	2,780

contaminated with feces of cats or dogs. The only practical preventive measure has been avoiding contact with infective soil or water.

Thrift G. Hanks, "Solid waste/disease relationships a literature survey," U.S. Public-Health Service, U.S. Department of Health, Education and Welfare, report SW-lc, 1967.

Table 7.2

Fungus Diseases (Mycoses) Associated with Solid Wastes

Fungus Diseases
Coccidioidomycosis
Sporotrichosis
Histoplasmosis

Occupation mycoses—infections from fungi—also affect workers who handle certain types of wastes. Histo-plasmosis can be contracted, most readily by contact with chicken feces. Cryptococcosis has been a fungus disease carried in pigeon feces, and can afflict demolition workers particularly. Some deaths are attributed to this mycosis.

8

Secondary Hazards from Solid Waste

Introduction

After disposal, solid wastes normally are allowed to pass first to a container, either immediately or via a refuse chute; then the wastes are generally transported by vehicle either in the original container or after transfer to a larger container; then there may occur a processing stage such as incineration, compaction shredding, or reclamation; and so-called ''final'' disposal (if not by recycling) must by definition be back to the environment in solid, liquid, or gaseous form.

During this general sequence of events, non-toxic wastes could become poisonous; sterile wastes may become capable of transmitting disease; nonexplosive wastes may cause explosions; and wastes which apparently have no injury-causing potential may in fact produce injury and even death. The mechanisms involved, and steps which can be taken to combat the problems, have been discussed below.

Public Health Hazards

Within a matter of hours in a warm temperature, sterile organic matter like cooked meat, can become a potentially lethal source of toxic or disease-producing organisms. The organisms do not have to be originally present in the host material as the environment is normally well provided with spores, bacteria, viruses, insects, vermin and other vectors awaiting a favorable site on which to multiply.

The mere presence of these cultures of potentially disease-producing organisms in solid wastes has been not enough to cause a major health hazard. The blame for disease transmittal must be placed especially on the flies (Table 8.1), mosquitos (Table 8.2) and rodents (Table 8.3). It is estimated that 90% if the urban house-fly population breeds in the contents of open trash barrels. Rats also rely on readily available scraps of food such as those which are kept on open trash barrels. Refuse overflowing from or carelessly disposed around containers outside restaurants or supermarkets or the use of containers with illfitting or bent closures have been equally responsible.

Alternative forms of treatment have been available which might be practicable in certain cases. One has been compaction. Reasonably high-density compaction (perhaps defined as that producing a density of about 40 lb/cu ft or 1000 lb cu yd) eliminates most of the crevices and void on which insects prefer to lay their eggs; greatly reduces the surface area available; and squeezes out many of the juices

Table 8.1

Fly-Borne Diseases

Typhoid
Bacillary dysentery
Amoebic dysentery
Diarrheas
Asiatic cholera
Helminth/infections (worm)
Myiasis
Loiasis
Onchocerciasis
Ozzard's filariasis
Leishmaniasis
African sleeping sickness (trypanosomiasis)
Yaws
Tularemia
Bartonellosis
Cararrhal conjunctivitis
Sandfly fever

Table 8.2

Mosquito-Borne Diseases

Dengue
Encephalitis
Filariasis
Malaria
Yellow fever
Tularemia
Lymphocytic choriomengitis
Melioidosis
Rift-Valley fever

so leaving the surface dry, or perhaps at most uniformly moist without the local areas of high organic concentration which have been necessary for optimum hatching of fly larvae.

A second approach which has been less desirable, has been to process refuse within two days of it having been discarded so that fly larvae do not have time to emerge (2.3 days being required). Rats can still feed if containment is not needed and there is no presently available method of treatment other than denying rodents access to food.

A third approach is somewhat contradictory to the first in that it involves shredding solid wastes. Shredding increases the surface area and decreases the density. In these circumstances an interesting

Table 8.3

Rodent-Borne Diseases

Echinostomiasis
Hemorrhagic septicemia
Histoplasmosis
Llymphocytic choriomeningitis
Plague
Rar-bite fever
Rat-mite dermatitis
Rat-tapeworm infection
Rocky Mountain spotted fever
Salivary-gland virus infection
Salmanellosis
Schistosomasis
Bilharziasis
Sporotrichosis
Swine erusipelas
Trichinosis
Leptospirosis
Leishmaniasis
Relapsing fever
Tularemoa
Rickettsial pox
Murine ryphus

transformation occurs in organic wastes. If shredded refuse is not placed in closed containers, and if the volume in any one place is not such as to give a depth of refuse greater than perhaps three feet, suffcient air is able to reach all parts of the mass is that aerobic rather than anaerobic bacteria grow vigorously. These bacteria have been found to work many times faster than the anaerobic and give off combustion heat as well as being thermophilic. Carefully controlled tests have revealed that insects are not able to hatch on aerobically decomposing refuse, and the material is also unattractive to rats and other vermin. The rate of air movement possible within such a mass has been small, and fires, if set, tend to smolder rather than to burn. Since most fire deaths have been due to smoke inhalation rather than to flame-produced injuries, the shredding of refuse could be considered as an acceptable way to reduce many of the secondary hazards of solid wastes so long as the shredded refuse has been not stored within a building or other enclosure in which people might be trapped.

Fire and Explosions

There has been an extreme danger of fire in most categories of untreated solid wastes simply since a large proportion of the material has been paper products, wood or plastics. The potential for these materials to be ignited by carelessly disposed of smoking materials has been high. However, whenever organic materials have been amassed together there has been the danger that these and less flammable materials may undergo spontaneous combustion.

Spontaneous Combustion

Spontaneous combustion takes place when organic materials decompose, ususlly initially by action of aerobic (oxygen-loving) bacteria and the heat released has been not all able to escape so that the temperature of the mass rises. Thermal insulation has been usually provided by the neighboring mass of nonreacting material, so that in an analogy to a nuclear reaction, a critical mass of organic material must be present before heat transfer outwards from the reaction zone has been sufficiently reduced to commence.

Spontaneous combustion is a problem in agricultural materials like hay; in coal; and in other finely divided fuels like woodchips and saw-dust. A recent example emphasizing the need to avoid large masses of any materials which have combustion potential has been the fusing together of a pile of steel turnings which were awaiting reclamation in an open-air yard. The turnings had been wet with cutting oil. The plant supervisor maintained that the inside of the pile got red hot and subsequently cooled down to form a solid mass. The resulting pyramidical block became too thick to be cut with torches and all other attempts to reduce it to moveable pieces failed.

There are not scientifically useful data to give guidance for avoiding spontaneous combustion in all types of solid wastes. Common-sense solutions have been to simply avoid storing solid wastes in warm moist conditions for several days when the smallest dimension of the waste mass becomes greater than 6 to 10 ft (2 to 3 m). Smaller quantities, or very dry, or very wet conditions, have been the inhibitors of spontaneous combustion.

Explosions

While it has been obvious that explosive materials, like live ammunition, have very liable to be detonated during handling and processing of mixed solid wastes, it has been less well known that combinations of, or large quantities of apparently innocuous materials can also cause explosions in certain circumstances. Incinerator explosions, for example, have resulted from the charging of large masses of plastic wastes. It is assumed the combination of gaseous pyrolysis products with the available oxygen reached explosive proportions. The gas produced by anaerobic decomposition of organic waste can also give rise to certain circumstances to explosive mixtures, and this is described as follows:

Gas Production

The main gaseous product of anaerobic decomposition of organic wastes has been methane, CH_4, although other gases like carbon dioxide, CO_2, nitrogen, N_2, and hydrogen sulfide, H_2S, are also present. Wastes buried in landfills undergo predominantly anaerobic decomposition, and the gases normally take the shortest or easiest route to the surface. Sometimes the presence of fissures in or surrounding the filled land, possibly together with the presence of a surface barrier, like a roadway or parking area, will make the evolved gases to travel large distances horizontally. In certain cases these gases enter the basements of houses and buildings, and deaths and injuries have resulted from asphyxiation, poisoning, or from the detenation of explosions of air-methane mixtures.

Sanitary-landfill regulations need that ducts be incorporated into landfills to lead the evolved gases to be discharged at a safe location.

Leachate Production

During decomposition, organic wastes normally produce a liquid called a ''leachate'', In landfills, some of this liquid gets resulted from the pressure applied by the fill above squeezing out juices from, for instance, garbage. Another principal component of leachate has been rain water which, during its passage through the landfill, is able to dissolve a large range of materials. Leachate may emerge from a landfill with a biological oxygen demand of over 20,000 mg/litre, which has been about 100 times stronger than raw sewage. Studies have revealed that the leachate concentration from a landfill might still be significant after 20 years.

9

Danger Points in Solid-Waste Handling

In this section the progress of typical solid waste will be described through typical sequences of location and operations and particular hazards would be identified.

Wastebaskets

For typical residential, and in some cases commercial refuse, the wastebasket has been the start of the journey to final disposal. Wastebaskets are generally dry and reasonably clean and are emptied every one or two days, so that problems of fly breeding and multiplication of undesirable bacteria becomes small. The principal hazards have been fires originating from wastebaskets and from personal injury to the personnel emptying wastebaskets.

Paper products predominate in wastebaskets. The contents have been therefore usually highly flammable. As wastebaskets are usually used as receptacles for ashes from cigarettes, cigars and pipes, and frequently for the butts and pipe residues, the contents of ash trays and the matches and spent cigarette ligher cartridges there exists only a very low incidence of any of these materials being in a smoldering condition for wastebasket fires to be comparatively frequent. Wastebaskets have been often placed half hidden behind curtains and drapes and by unholstered furniture so that the fire danger becomes typically extremely high.

Some obvious methods of control have been to limit smoking within buildings; to provide a large-capacity, floor-standing ash tray near each wastebasket; or to use one of the many types of self-closing waste containers.

Danger to personnel emptying wastebaskets comes predomimantly from thoughtlessly discarded dangerous materials.

Two avenues of correction suggest themselves in these cases. One has been public education to the dangers to presonnel of dangerous substances and objects in the waste stream, along with education of the service personnel to empty wastebaskets directly into receiving containers without the use of hands. The second approach has been to promulgate regulations which make the person who discards prohibited materials into the regular waste stream legally liable for damages resulting from these actions. The second approach has been in fact a variant of the first in that education has been usually more rapidly assimilated when there have been strong incentives to learn.

Refuse Chutes

A frequently heard complaint against architects has been that the handling of solid wastes has been seldom taken into consideration in the design stage of buildings, large or small. Occasionally, however refuse chutes have been incorporated, and these undoubtedly greatly reduce handling costs and thereby reduce some of the dangers involved in solid-waste handling.

Some other danger have been introduced, such as the extreme fire danger if a refuse chute has been designed without appropriate traps to disallow the chute acting as a chimney and if the materials and construction have been insufficiently fire-resistant. Building codes in most cities need that fire traps be used. A secondary danger, which refuse chutes introduce has been that the anonymity in the disposal of refuse which they allow encourages people to dispose of all manner of prohibited, dangerous, or incriminating wastes.

Compactors

Compactors has been designed to withstand the high loads and pressures imposed on wastes. Accordingly, they have been able to withstand and have minor explosions from, for instance, vapor-propellant bottles. Fires are usually extinguished by compaction. However, fires have been comparatively frequent in collector-compaction trucks after, for example, some hot ashes might be loaded toward the beginning of a collection run, before the vehicle contents have become compressed, to any extent. If fire inside a truck body has been comfirmed, standing instructions to the crew usually have been to dump the burning load in the middle of the road insuring first, of course, that other traffic gets alerted and rerouted. A fire department can more easily put out burning refuse than exploding gasoline.

Other dangers from compactors come from the loading process when loading is done by hand. In most stationary compactors there have been interlocks to insure that the compaction device, usually a ram and sometimes a screw, get closed off before operation. Accidents to loaders of refuse-collection trucks have been fairly frequent due to imploding T.V. tubes, or the sudden movement of pieces of steel when these get pinched half in and half out of the mechanism. When workers have been on an incentive system, protective devices have been often removed in the interest of speed, and good management is required to educate what is usually a transient work force on the dangers involved.

Containers

The term ''container'' has been generally reserved for enclosures with volumes larger than those of wastebaskets or ash barrels: containers normally hold from 1 to 20 cu yd. However, containers have been suffering from many of the dangers of wastebaskets and refuse barrels: fire, dangerous materials, and the encouragement of insects and vermin are examples. In addition, the storage of large volumes of refuse in a compact mass tends to encourage biological degradation which can give rise to spontaneous combustion given the right combinations of moisture, warmth, insulation, and fuel availability.

Incinerators

Small incinerators as manufactured for apartment houses and commercial establishments have been banned in many areas due to the difficulty of reducing pollutant emissions from the stack. Larger incinerators (50 tons/day to 3000 tons/day) will normally be having a discharge or tipping floor; a refuse pit or horizontal refuse conveyors; generally an over-head crane; one or more furnaces; ash-handling equipment; gas cleaning equipment; and a stack. Such a plant is having all the dangerous characteristics of an industrial processing plant involving the mechanical handling of extremely varied materials, together with some danger of explosions resulting from intrinsically explosive materials or from a high concentration of plastics. Poisonous gases have been produced given off" during the combustion of some plastics, especially from the urethanes and the vinyls, but because furnaces have been normally maintained under negative pressure the dangers to personnel should be small.

Shredders

Solid wastes have been increasingly comminuted, the most popular device being a hammermill. Single machines possess capacities from one-half ton per hour to fifty tons per hour. No incidence of danger from shredders has been occasionally experienced. The robust construction of shredders make them to contain all but the most violent explosion.

Transfer Station

In its irreducible minimum, a transfer station has been a location where solid wastes get transferred from, usually, the collection vehicles to long-haul vehicles without any additional processing. The long-haul vehicles have been usually large capacity highway trucks, sometimes with trailers; sometimes railroad cars; and occasionally barges. Frequently, processing by compaction or comminution has been included as part of a transfer station operation. Danger points have been those with the unit operations as discussed above.

Reclamation Plants

Reclamation is done in a crude way at many town dumps by authorized and unauthorized personnel picking over refuse for possible valuable items. It is an extremely dangerous practice due to the intrinsically unsafe nature of the surface of a dump; because of the difficulty of extracting desired items from their resting position in the dump; and from the danger of being run down by vehicles while they discharge their loads. Dumps and landfills should always be adequately fenced and policed to disallow scavenging and the reasons for these restrictions be clearly stated.

Some degree of mechanization got introduced into reclamation in the early decades of the present century by loading refuse onto traveling belts which then pass by so called "picking stations" where individual personnel are having responsibility for removing items in particular categories (newspapers, corrugated cardboard, bottles and so forth). There have been obvious dangers involved, and in present day conditions it is found to be difficult to attract and but an extremely transient and poorly educated work force.

Modern automated reclamation plants have been developed which are generally marked by a lack of human contact with refuse and by the use of sophisticated control systems and protective devices. The dangers involved should therefore be small.

Landfills

Dangers from improper landfilling procedures have been described above, and fall into the categories of fires; gas production resulting in possible fires, asphyxiation and explosions in buildings;

leachate production causing possible pollution of aquifers; rat and insect, infestation, leading to spread of disease; and scavenging leading to a probability of injuries. Of these, the production of gases and leachates, and the spread of underground fires, which can persist for 20 years and more, have been the most troublesome. While easy to prevent during landfilling by following one of several approved techniques, subsequent prevention of these conditions after poorer landfilling methods have been used is extremely difficult, if not impossible. Even though, for instance, modern infrared measurements may locate with considerable accuracy the area where in underground fire has been smoldering, such as fire has been virtually impossible to extinguish with certainty. The cavern left by a smoldering fire may subsequently make a sudden collapse of the surface, with obvious dangers to people and to any buildings which are erected without sufficient design precautions.

10

Potential Health Hazards of Sewage-Sludge Composting and Land Disposal

Introduction

Recent findings at the University of Toronto indicate "alarming amounts of cadmium and lead occur in sewage sludge which is used as fertilizer for food crops. These metals could be assimilated by crops in amounts harmful to human and animal health. Medical evidence reveals a correlation between cadmium and hypertensive, cardiovascular, and respiratory problems. Lead can bring about brain damage or death".

Recent German research has also reported that there has been "no decrease in the number of pathogens in raw sewage over the last few years, and even incidence of hepatitis salmonelloses and bovine cysticercoses has been noticed."

There has been much disagreement as to what extent high quality aerobic composting kills pathogenic bacteria present in the feed material (particularly in sewage sludge, if used, and household refuse). One view, given in 1974 German report, has been that "after the temperature gets stayed for one day between 30° and 45°C the temperature may further rise to 70°C after one week all pathogenic germs are killed......." Another 1974 German report states: "The temperature of 65° to 70°C at which composting processes usually occur is not sufficiently high to kill all pathogenic germs, viruses, bacteria and bacilli found in garbage and sewage sludge. However, studies of the decomposition process in composting plants showed that microbial substances with antibiotic effects can form which kill even heat resistance forms of pathogenic germs." These antibiotic substances have been found to work effectively after the initial composting process, usually taking 3 to 7 days, during the so called "curing process." During this process, compost has to be stored in piles or windows, often in the open, for perhaps 3 months. Formerly it was considered as a subsidiary process, but with this finding it takes on primary significance.

When sewage sludge has been used for irrigation there has been no curing process Enteroviruses (poliomyelitis, ECHO, Coxsackie) have been found to be taken up by the soil and to contaminate vegetables grown in them (see Table 10.1 and 10.2)

Parasites could also be spread from sewage. Hookworm infection, prevalent over much of the world, has been reduced to a negligible point in most of Europe and the United Sates. The larvae of hookworm penetrate the skin, pass through the liver and lodge in the lungs. The pulmonary migration may bring about symptoms which have been ranging from mild, dry cough to severe dyspnoea, cyanosis, etc. Intestinal ascariasis tends to aggravate malnutrition and may cause dysentery. Hookworm infection brings about an anemia syndrome. Strongyloides infection is having a mild form characterized by catarrhal enteritis, with mucous secretions, intermediate forms characterized by edematous enteritis, and serious forms characterized by ulcerous enteritis. The VIM (visceral larva migrans) syndrome, lime massive trichuriasis and ascariasis, could be seen in dirt-eating children aged one to four years. The principal symptoms have been eosinophilia, hepatomegaly, hyperglobu linemia, fever and pneumonitis. Larval toxocariasis may bring about damage to the eye.

Table 10.1

Diseases Associated with Human Fecal Waste

Diseases
Bacterial infections
Typhoid fever
Paratyphoid fevers
Cholera
Shigellosis (bacillary dysentery)
Viral infections
Poliomyelitis
Coxsackie infection
Infectious hepatitis
(very many other enteric viruses exist)
Protozoal infections
Entamoeba histolytica
Helminthiasis
Fish tapeworm
Beef tapeworm
Pork tapeworm
Pinworm
Roundworm
Whipworm

Table 10.2

Diseases Associated with Animal Fecal Waste

Diseases
Salmonellae
Infection from pig intestinal protozoan (Balantidium coli)
Helminthiasis
Pig ascaris
Animal tapeworms
Hadatid worm

11

Hazards from Sea Disposal of Solid Wastes

Introduction

The data about the harmful effects of dumping a vast range of solid wastes and sludges at sea has been not available, except for a few specific cases. These cases, involving DDT, oil, mercury, and bacteria from sewage, alarmed and alerted the public and regulatory agencies, so that the dumping of some hazardous or possibly hazardous materials has been outlawed.

Metals

The principal metals which may offer potential hazards have been arsenic, cadmium, chromium, cobalt, copper, iron, lead, manganese, mercury, nickel, silver and zinc.

Mercury discharging from a chemical plant in Japan had been taken up by fish caught locally and extensively used for food. Deaths and serious disablement had been followed.

Sewage sludge contains metals in a quantity which can be toxic to marine life when the deposits have been allowed to build up rather than to disperse. Some of the metals, like lead and copper, dissolve out of water-supply systems; some, such as iron and lead, come from high way runoff into combined sewer systems; mercury may run into a sewer system after use as a garden fungicide; other metals like cadmium, silver and zinc get discharged, sometime illegally, into sewers by industry.

Inorganics

Inorganic materials strictly include the metals. However, the term has been more usually taken to include such materials as dredging spoils, sulfuric-acid sludges, glass from domestic refuse and demolition debris.

Organics

Wastes from the synthetic-chemical industry have been particularly complex, and new compounds are discovered and join the waste steam faster than they can be tested for long-term hazard-potential. It was the discovery in Berkeley. California that the decline of the brown pelican had been due to the effects of DDT on the shells of the eggs, and later that measurable concentrations of DDT could be detected in Antarctic penguins, which catalyzed public concern over pesticide use and the discharge of organic chemicals.

Biological Wastes

Three components of the sludge from human sewage have been of great concern. Disease organisms, particularly infectious hepatitis, may contaminate shellfish which may reinfect people. Metals dissolved from plumbing systems and from other sources could be toxic to marine growth and may concentrate in shellfish and other diet fish (e.g., mercury in swordfish) The nutrients may bring about the growth of undesirable organisms.

This third concern, the possibility of excess nutrients being present, has been quite different from the problem which arises in inland waters. The ocean is starved of nutrients, and any excess is purely a local phenomenon. Dispersal has been therefore an acceptable solution.

12

Availability of Waste in India

A detailed report on Utilisation and Recycling of Wastes (Research, Development and Extension Requirements) was submitted in 1975 by the National Committee on Science and Technology of the Government of India (NCST, 1975). The NCST is the progenitor of the Scientific Advisory Committee to the Cabinet (SACC). A lot of effort has been done in the preparation of this Report but not much cognizance has been taken of it so far. Although over a decade old, this report along with a series of other papers published by O.P. Vimal and his associates (1975-85) constitute the only available information on the extent and nature of wastes available and the R & D effort in progress as also possible on the subject. India being a developing country can ill afford to waste any wastes from agriculture, forestry, animal husbandry, industry, minerals, urban and rural community and building materials.

By proper utilisation of these materials it becomes possible to conserve the national resources, reduce the cost of production in many cases and solve the problem of pollution and generate some employment. The basic rationale on which the Report has been used is "What is waste to one industry may be raw material for another".

During the last decade, waste recycling and reuse has been a very major activity in most of the developed countries and considerable progress has been made. It has been fast becoming a hightech area. There is, therefore, need now for a very concerted national R and D and pilot level effort in this area. This should enable to conserve the resources as also energy and at the same time add to the national wealth. If a country like India does not take up serious work in this area, time will come the country will have to go round buying technology for waste recycling.

The Japanese example has been before us where work on waste recycling and utilisation has paid rich dividend. An illustrative list of types of wastes available in India is given below. The quantities wherever given are based on 1975 data.

Depending on their origin, the wastes could be grouped under seven heads namely; Agricultural, Fruit and Vegetable, Animal, Aquatic weeds, Community and Industrial and Construction materials.

I. *Agricultural Wastes*:

- Eighty species of Non-edible oil seeds(1 Mt.). Sal seeds (40,000 t), Groundnut cake after extraction of oil (2-2.5 Mt), Rice bran (2.4 Mt)and Rice husk, Rice straw; Tea waste (8-9000 t). Bagasse (5.25 Mt), Molasses (0.17 Mt), Press mud, Tobacco waste (62,000 t); Protein rich plants, Mahaua flowers; wood waste, saw dust, bark from industrial wood for venneer logs; and other agricultural wastes.

- Coconut waste by products (other than those used for constructon materials), Coconut shell (0.88 Mt), Coconut husk (0.5 Mt), Coconut water.
- Lac waste.
- Arecanut waste/by-products (other thon thase used for construction), Arecanut husk (0.1 Mt). Leaf sheath (90,000 t), Chogaru liquid.
- Cashew waste/by-products (other than those used for construction needs), Cashew apple (30 Mt), Cashew testa (3-4000 t) with 61% tannin, Cashewnut shell liquid (10-14,000 t).
- Coffee waste.
- Rubber wastes/by products, Rubber seed, Rubber wood.
- Tobacco waste, leaf scrap and stalks, seed.
- Cotton wastes/by-products. Cotton stalks (2.5 Mt), Cotton leaves; Cotton linters.
- Jute wastes and by-products, Jute sticks (2.5 Mt), Jute mill waste (50,000 t).

II. *Fruit and Vegetable Processing Industries*

- Mango peels and kernels.
- Citrus peels, pomace, etc. (0.48 Mt).
- Apple pomace.
- Pineapple peels, ears and eyes.
- Banana pseudostems.
- Waste from vegetable canning industry.

III. *Animal Wastes*

- Cattle shed waste.
- Fisheries wastes.
- Prawn wastes (10,000 t),
- Squilla (0.1 Mt).
- Frog wastes (5,000 t).
- Carcass utilisation (12.5 M carcasses).
- Slaughter house waste (27,000 t intestines, 55,000 t blood, etc.).
- Wool wax.

IV. *Aquatic Weeds*

They are very prolific in growth and a menace by blocking water ways and interfering with hydroelectric production, wasting water in evapotranspiration, hindering traffic and fishing as also help in increasing water horne diseases. The negative impacts of the aquatic weeds can be counter balanced by converting these into food (For fish, duck, donkeys, pigs, sheep and even processed animal feed for cattle), fertiliser, pulp and paper, fibre energy,.(biogas) and waste treatment (NAS,1976).

Aquatic plants including algae from marine sources need to be utilised in a planned manner.

V. *Community Waste*

- Night soil, sewage, sludge, cattle dung and urine.
- 7 Mt of dry matter containing 2 Mt of nitrogen, 0.5 Mt of each of P_2O_5 and K_2O.
- City refuse

VI. *Industrial Wastes*

- Paper waste, recycled paper products.
- Mica waste (20,000 t).
- Flyash (5 Mt in 1968-69, today around 22.016 Mt).
- By product of salt industry (50,000 t KCI).
- Wastes from steel plants (slags).

VII. *Construction Material from Wastes*

- Waste from sugar factories.
- Agricultural wastes.
- Coconut, Arecanut and Cashewnut after extracting oil.
- Redmund from aluminium industry.
- Water works silt.

The foregoing list provides an idea about the extent and nature of wastes available in the country. This is only an indicative list. These materials can be rearranged keeping in view their end use, namely animal and poultry feed ; industrial chemicals; paper, pulp and board ; fertiliser, fibre, etc. The urgency of intensifying work in this direction is no doubt obvious but in order to give a thrust, the Department of Science and Technology need to take a view on the recommendations of the NCST (1976) Report. The important recommendations have been as follows:

- Declare DST as the nodal agency to coordinate the work with various Ministries/Departments and agencies in Centre and States.
- To institute an appropriate High Level Advisory Body to look after waste management, advise the Government on short and long term policy on the subject and update the 1976 data regarding locational and seasonal availability. Start R & D through an All India Coordinated Project on specific subjects, generate new technology, put up Pilot Plants, etc.
- Build up awareness regarding the waste collection, disposal, utilisation and employment generation potential.

13

The Costs of Solid Wastes

Introduction

The social costs of solid wastes have been not easy to determine. Open dumps may be responsible for a small amount of disease and incinerators add to the air pollution costs of a community. Most disposal methods involve some aesthetic displeasure. In rare circumtances the costs can be much higher.

Litter along roadsides constitutes a solid waste problem with considerable psychological impact, judging from the interest in antilitter and highway beautification programs. Occasionally there are massive efforts to pick up litter along highways, generally resulting in the recovery of several hundred kilograms per kilometre. Since there are approximately 5 million km of roads in this country, highway litter might amount to roughly one million metric tons at present—a small but important part of the solid waste problem, and one which is generally believed to be growing.

The major cost of solid waste disposal at present comes from the collection, transportation, and transfer of the wastes, not from the final stages of disposal.

ECONOMIC IMPLICATIONS

Labour and transport absorb the major part of the operating cost solid wastes management services. It is usual for a city to employ between two and five workers for each 1000 population. Three workers/1000 population may represent about 1% of the total national workforce. Transport is required on a scale of about 1 heavy vehicle/15000 population. Thus, solid wastes management services can absorb up to 1% of the gross national product, and are one of the most expensive city services. Efficiency of systems and high labour productivity, therefore, and vital importance.

The level of mechanization that should be adopted for solid wastes management systems relates directly to the cost of labour, as compared to that of plant energy. There is not much variation, worldwide, in energy of mechanical plant costs, but there is an astonishing range of labour costs.

Thus there are no universally applicable solid wastes management systems: every country must evolve an indigenous technology based on the quantity and character of the wastes, the level of national wealth, wage rates, equipment manufacturing capacity, energy costs and the availability of foreign exchange for the purchase of imported plant.

Human Resources

A critical factor in the évaluation of alternative systems of solid wastes management and the establishment of efficient organizationis is the quality of management. It is necesary to deploy a complete set of technical skills which derive from several professional disciplines. These include civil and mechanical engineering, chemical engineering, transport organization, land-use planning and economics. Thus, solid wastes management is teamwork, and whatever the original discipline of the top manager, his prime qualification should be his specialized background in this field. A second requirement, no less important, is skilled "man-management", the purpose of which is to obtain high productivity from a contented staff at all levels.

14

Collection and Transportation of Solid Waste

Introduction

Of the total, 90 per cent solid waste is produced from agriculture and mining. The solid wastes could be usually disposed of with little difficulty in the rural areas where they are generated. Solid-waste collection, processing and disposal have been far more of a complex management program in built-up urban communities. Urban solid-wastes sources have been of two principal categories : industrial and municipal. Municipal waste is a collective solid wastes. These subcategories will be described individually.

Residential Wastes

In urban areas, the volume and composition of residential waste may vary considerably in different communities and climates. In general, percapita waste production has increased with per-capita income, although in some localities solid-waste production has actually decreased while presumably income has not. The quantity of residential solid waste in a city also varies with the season of the year. Wastes like garden trimmings, autumn leaves, clean-up campaigns, summer produce, Christmas residues, etc, are all seasonal in their generation.

The type of housing also influences the waste within a city. High-density areas produce less solid waste per capita; apartments produce solid waste with little or no grass and leaves. Areas having large numbers of garbage-disposal units produce much less food wastes. Typical residential solid waste by weight has been found to range from 35 to 45% paper, 10 to 20% food waste, 5 to 10% glass and ceramics, 5 to 10% metals, 5 to 15% grass and leaves, 2 to 6% textiles, and usually less than 4% wood, plastics, rubber, leather, and inert materials.

Commercial Wastes

The aggregate composition of commercial solid waste has been found to be similar to that of residential waste, but individual generation points have tendency to produce a more homogeneous waste stream. That is restaurants obviously produce more food and related packaging wastes, markets produce some food and mostly paper of carton wastes whereas hotels produce considerable textile wastes, as well as more normal trash. The amount of solid waste generated per collection unit has been found to be solid waste generated per collection unit has been found to be much higher for a commercial solid-waste source than for a residential solid-waste source; because of this, collection vehicles in commercial areas fill up rapidly.

In dense urban areas, commercial solid-waste generation has been more than in less-populous land-use zones. Collection routes in which commercial and residential wastes may be picked up by the same organization may be combined so that vehicles could efficiently make both residential and some commercial stops. In many cities, commercial wastes are collected by private contractors with some municipal regulation. It is desirable that local administrators must be able to control the standards for all collections within a community to insure uniform, fair, and sanitary service conditions.

Industrial Wastes

These include food-processing residues, boiler ash, wood, plastic and metal scraps. chemical sludges, packaging wastes, debris, dirt, etc. Increased industrial application of air and water-pollution controls will also cause an increase in the solid wastes which have been destined for land disposal. Industrial waste collection and disposal are usually undertaken by contractors, and less often by the producer. Industrial waste-generation quantities in a given area depends upon the types and sizes of the industries present rather than area population.

Demolition Wastes

Solid wastes due to demolition and construction activities have been including mixed lumber, roofing and sheeting scraps, stone, rubble, broken concrete, asphalt, brick, plaster, wallboard, pipe, wire, and other residual building materials.

Bulky Wastes

These are including trees, telephone poles, bushes, abandoned autos and auto parts, tires, stoves, refrigerators or other large appliances, furniture, crates; mattresses, etc. Conventional collection vehicles fails to transport such materials. Special opentruck vehicles have to be used for collection of bulky wastes which otherwise may be dumped and abandoned on public or private property. Two turcks with dollies could be used to tow away abandoned vehicles. Vehicles collecting shrub and tree branches have been generally more efficient if equipped with volume-reduction equipment such as a chipper of leaf mulcher to reduce the cellulose wastes.

Hazardous Wastes

Wastes having pathological, exposive, radioactive, or toxic materials are regarded hazardous. Dangerous liquids and solid wastes sometimes get containerized and included in the solid-waste stream. Collection disposal crews must have the training to identify all hazardous wastes. Once hazardous wastes are identified, they must be handled separately, with appropriate safety procedures. Fire-fighting equipment and protective clothing must be present on special and regular collection vehicles for handling hazardous materials and coping with emergencies when accidents requiring emergency action take place.

Storage

Storing solid wastes prior to collection disallows offending esthetic tastes, attraction of vectors, and excessive odors. Storage devices must be convenient for the user and facilitate, safe, effcient collection processing and disposal. Types of storage devices must be described as follows :

Containers

Rigid containers must be made from durable metal or heavy plastic, and covers should fit tightly. Pick-up handles should be strong because if a handle may break it could cause in serious injury to the user or collector. Damaged containers having sharp edges should be immediately replaced to avoid cuts and bruises. Containers should be stored on a flat concrete base or elevated above the ground for reducting metal container corrosion and spilling. Some vinylplastic containers may get oxidized and crack with time or cold-exposure ; polyethylene and certain other durable plastics may have a superior useful life.

Bags

Plastic and paper storage bags and holders are having certain advantages and disadvantages. The various advantages have been as follows :

(1) one-way disposal without retrieval :

(2) a universal and standard-size, cleaner container;

(3) faster collection because they do not have to be uncovered and returned after being emptied;

(4) reduced spillage and resultant litter;

(5) easier storage and handling as liners or suspended bag holders;

(6) reduced back strains and similar collector and user injuries because bags are far lighter, will not hold heavy loads, and are easier to handle than rigid cans :

(7) more sanitary and cause less noise, dust, odor, and micro-organism release; and adaptable to mechanized collection by truck equipment.

In summary, plastic bags have been simpler to store, more expandable, less expensive, and easier to tie and handle than comparable paper bags. However, both paper and plastic bags are more easily punctured by sharp objects, and dogs, raccoons, rats and the like can tear these open. Bags have been a continuing cost and contribute to solid waste in themselves. Thirty-two to 40-gallon bags and rigid containers are generally the standard for solid-waste collection. Plastic storage bags (polyethylene) have been commonly 1.5 mil thick while paper bags have been commonly two-ply and have plastic additives for wet strength. Solid-waste generation points that produce large amounts of solid waste between collections should use large bins (1.5-15 cu yd) rather than numerous smaller bags or containers. The larger bins also may use stationary compactors to reduce storage volumes and collection and disposal costs. Vehicles capable of hoisting and dumping bins should service them as required.

Twice-weekly collection has been desirable since fly propagation may be reduced, over once-per-week collection by 75%. The bins should be seamless, watertight, sturdy, and maintained frequently to prevent vermin infestation and odors. Bin collection has been useful for handling large volumes of industrial, agricultural and density residential wastes.

Drop Boxes

Drop boxes have been larger storage bins and are useful for servicing rural, industrial, and demolition sites. Large bins with or without stationary compactors cost less per unit to solid waste than small bags cans or bins. Centrally located boxes of 20- to 40 cu-yd capacity can service a large rural area at less cost, than a small sanitary landfill. Vehicles capable of hoisting these boxes onto their chassis can be used for collection and transport. An empty box is normally dropped off when a full box is collected to avoid leaving the area without a solid-waste storage device. This method needs one vehicle pickup trip for each collection, but could be an economical procedure if properly planned and managed.

Other Storage Devices

Backyard solid-waste shelters that need emptying by a collection crew using hand tools are inefficient, cause odors, support rats, files, and other vermin, and create a fire hazard. Underground pit storage of containers is also undesirable because spilled garbage is difficult to reach and clean out, resulting in odors and rat and fly propagation. Hard-to-reach containers in pits may also need collectors to use dangerous lifting positions.

Compactors

Storage containers equipped with heavy-duty compaction devices can be able to reduce space requirements increasing solidwaste densities severalfold. In communities which have outlawed private on-solid-waste incinerators, compactors can get installed in the former incinerator locations. Household compactor units, a fairly new innovation are currently being used only in more affluent residences. Compactors, shredders, and other volume-reduction devices obviously do not decrease the total weight of generated solid waste.

Source Segregation and Reclamation

Producer-provided segregation of reclaimable materials at the source has been a simple method of solid-waste separation. Segregated materials have minimum contamination, but need special equipment or extra trips for handling and collection.

Resource conservation has been presently the major motivation for waste segregation. For communities incinerating solid-waste, the separation of combustibles and non-combustibles may be desirable. The dominant use of sanitaty landfills for solid-waste disposal reduces the need for source segregation Class I disposal sites having little potential for groundwater contanination require no waste segregation since they have no limitation on acceptable wastes. Class-II disposal sites (located above the groundwater table and protected from surface runoff, with controllable surface drainage) can be able to accept ordinary household and most common scrap metal.

Source segregation of metal cans, paper and other valuable components may be becoming a public responsibility in the future. Many communities are having in the past practiced separate collection of garbage, cans, glass and paper products, but the trend has been toward combined collection to reduce collection and disposal costs. Unlimited collection of all generated solid waste at source should cause a cleaner community; reduced service pickup or restricted costly collection systems result in open dumping litter.

Collection-System Parameters

Many combinations of collection vehicles, routes, crew size, collection frequency, service type, work scheduling, set-out location, etc,. have been used. The most appropriate system will be depending on local conditions. Total costs of refuse collection will be a function of population.

In planning and implementing a solid-waste-collection system for any particular area, collection-system design and operation parameters need to be carefully considered. Once implemented, a collection system must be continually monitored for assuring that maximun cost-effectiveness is being realized. Best known productivity performances of various types of collection systems, have been a useful start for determining the degree to which a designed collection system is reaching its potential.

Crew Size

The number of men in a collection crew is dependent on the system, equipment, and type of service offered. In general, cost savings and high service levels can be realized by employing a minimum-sized crew; there seldom is justification for using more than a two-man crew for curb alley collection or a three man crew for backyard collection. Except for the driver, the crew-member time has been mostly non-productive during disposal trips. Some communities try to also utilize crew members for street or alley clean-up or to set out or set back containers during filled-truck disposal trips. Systems using a single crew member as a driver/loader with a side-loading or front-loading vehicle are shown to be efficient for curb and alley collection routes. Mechanical equipment to eliminate manual pickup and utilization of single-operator trucks have been new and significant developments.

Task Assignment

Crew scheduling and collection-vehicle routing has been interdependent. The type of routes used may be fixed daily routes, weekly routes, single-load routes, or other systems. Routing must be flexible enough to absorb seasonal waste-generation variations and should avoid heavy traffic or lefthand turns. Collection services which need customer set-out allow less flexibility in routing. Crew scheduling basically involves a choice between a fixed working-hour system and an incentive-work system; the former involves a fixed work day (e.g., 8 hours) with overtime if necessary; the latter generally involves an assigned task rather than a specific number of hours worked. An assigned-task system in which early-finishing crews assist other crews until all tasks have been completed has been used successfully in Inglewood, California. Although incentive systems may result in more users serviced per worker day, there are disadvantages. Workers who are hurrying to finish may be less careful and consequently handle containers roughly and create excessive spillage. Good rout-collection management supervision has been always important, but particularly so with incentive systems if high work standards have to be maintained, poor service (such as missed stops) avoided, and customer complaints got minimized.

Type of Service

Frequency of collection. Solid waste must be collected at least once per week. In general, the amount of solid waste collected per stop receiving twic-a-week collection would be greater than the total amount of solid waste per stop receiving once-a-week collection. This phenomenon must be taken into account when determining collection frquency. It may, in fact, get related to more thorough disposal by the waste generators rather than burning on-site or increased littering.

Backyard collection. Collectors remove refuse from the backyards, of premises and return any containers. Collectors often transfer a household container, in the backyard, to a larger receptacle which is carried out and emptied into the vehicle. Burlap bags and plastic"swamper" containers are generally used for such transfers. The disadvantages of this service have been higher costs; invasion of privacy by collectors entering private property (trespassing and robbery complaints have been received with this system), accidents to workers, who often fall down in icy or wet weather or are bitten by dogs; and damage to fences, landscaping and so forth as collectors generally try to use shortcuts to the adjacent lot rather than walk longer distances with a partial load. Backyard, but low productivity and high costs frequently forces a decision to use different collection system.

Modified versions of the backyard collection system have been a set-out service and set back service. In set-out service, containers are carried out to the curb by collectors, generally preceding the regular collection vehicle crew. Customers are responsible for returning their own empty containers. Ideally, full cans would therefore on the street only a short time, thus reducing litter. The lids should be replaced on the empty containers by the crew. Set-back-only service needs customers to bring their containers to the curb. Collection crews return the empty containers to the yard. This method is having less routing flexibility than the other two methods.

A danger involved in carrying bags or containers has been the tendency for collectors to carry several containers at one time to reduce extra trips, resulting in injuries or accidents.

Careful worker training, supervision and good equipment have been necessary to obtain competent collection.

Curb collection. Curb collection has been the most productive type of service. Collectors are not needed to carry waste containers a significant distance. Pickup of the solid waste, however, makes many injuries; solid-waste collection is a work-intensive occupation with high illness risk. Crowded solid-waste truck-loading areas such as have been common with large collection crews should be avoided: they cause collectors to dump while in awkward positions.

Curb collection is occasionally done with pickups by two or more crew members working opposite side of the street. This technique has been generally less efficient than a single collector alternately working one side of the street, then the other. Collecting both sides of a street at the same time should generally get avoided; an exception has been streets with very little traffic and few parked vehicles. Traffic-related accidents have been a danger on more crowded streets.

Alley collection. Solid-waste collection from alleys has been common in many older cities. It has been similar to curb collection except that both sides of the alley are generally picked up simultaneously. Dead-end alleys should not be entered if the vehicle will have to back up to exit, as backing a large vehicle is dangerous. Alley collection generally gets incorporated with backyard service by requiring containers to be stored in proximity to the alley.

Equipment

Equipment maintenance. Proper collection-equipment maintenance reduces annual system oiperating costs. Maintenance yards should be provided by the collection system itself since outside venicle-maintenance facilities are frequently less dependable. Using inefficient equipment past its useful life should be avoided.

Equipment design. Collection vehicles must be enclosed and watertight for all-weather service. Open trucks must be used only for collection of bulky wastes, and must be covered with a tarpaulin during waste transport to prevent littering. A low loading height has been desirable to increase work productivity. Safety devices are essential, including backup lights, noise- making and other warning devices. There must be emergency stop controls on compaction equipment and a safe place for crew members to ride during short trips.

The optimum vehicle capacity has been determined by the routing scheme. Packers involved in slow inefficient collection systems are generally small (less than 20 cu yds); high-density systems with greater productivity generally require larger vehicles (20- to 40-cu yd packers). The turning radius, local highway load weight for trees, overpass clearance, etc.) must be taken into consideration before choosing vehicles.

Operation. Compactor have been the preferred waste-collection equipment. Compactors significantly increase solidwaste density from 100 to as much as 600 lbm per cu yd, thus increasing collection route work capacity (Table 14.1). Many vehicles cannot be driven during the compaction cycle; a frequent cause of this has been compaction equipment driven by the same engine that propels the vehicle. Dual mechanisms for separate compaction and road operation have been becoming popular, as they increase collection productivity.

Table 14.1

Variation of Equipment Capacity with City Size and Number of Weekly Collections

Population class (1000's) From To less than	*No. cities*	*Average capacity equipment once/ week collection (cu yd)*	*No. cities*	*Average capacity equipment twice/ week collection (cu yd)*
10—20	3	17.6	9	18.2
20—30	6	14.7	5	16.5
30—40	3	24.5	3	14.0
40—50	2	15.4	3	17.3
50—60	4	16.5	3	21.7
60—70	2	25.5	1	30.0
70—80	2	16.0	0	—
80—90	1	18.0	1	18.0
90-100	2	17.4	2	23.4
Combined average	25	18.6	27	18.5
100—200	5	24.5	2	18.9
200—300	2	17.0	2	15.8
300—400	2	17.8	3	17.8
400—500	0	—	1	20.5
Combined average	9	21.4	8	17.8
500 and over	5	19.4	3	18.77

The design and location of the loading hopper has been an important vehicular parameter. Small one- or two-man crews in which the driver takes part in the loading operation may employ side-loading vehicles. Front-loading vehicles need the dangerous practice of collectors working in front of the vehicle.

Rear-loading vehicles make communication between the driver and the collector difficult—also a dangerous situation. All loading hoppers must be large enough to avoid crowding while emptying waste containers; and they should be low enough for easy access. Equipping hopper-load vehicles with "dustless" or other hoist equipment to empty bins can be able to reduce overall fleet requirements and eliminate retracting routes.

Collection equipment designed retrieval of solid-waste containers is receiving considerable attention. Under ideal collection conditions, this equipment can be very efficient; however the presence of hindrances like parked cars and low trees can be able to reduce the effectiveness of mechanical collection. Mechanical collection vehicles must be used in conjunction with uniform containers. (One system uses 80 gallon containers: another uses 32-to 40 gallon poly ethylene bags). The community should supply or need these standard containers to achieve uniform operation.

In some communities a relay collection system may be efficient. Smaller vehicles (in some cases, nothing more complex than scooters) having one or more attached bins pick up household waste ; when full, they return and empty their waste into a larger 'mother' vehicle. One large 'mother' vehicle can compact and haul several small 'satellite' vehicles. This system is having the cost and environmental advantages inherent in the use of smaller vehicles.

Other transport systems. Pneumatic-pipeline solid-waste collection systems are being used successfully for short-distance transport from hospital and commercial, industrial and large apartment-building complexes These systems generally use 18-inch-minimum piping, automatic inlet hopper valves, heavy duty vacuum fans, and air-pollution-control devices. Particular care must be taken in the design of pneumatic systems so as to disallow fires. Liquid carriage pipelines for the transportion of solid waste slurries are being studied extensively for long distances, and the preliminary results are promising. Such a system is having a high initial capital cost which could be compensated in part by lower annual operating costs.

Transfer Stations

It order to determine the feasibility of using a transfer station, cost-benefit studies of alternative collection/transport systems must be worked out and trade-offs determined. If sufficient savings have been derived from decreased collection and haul costs, then a transfer station must be practical. Whenever possible, adjacent collection areas must be consolidated into centrally-located regional transfer stations, to take advantage of any economies of scale. Transfer stations would be more likely to be economical if initial conditions have been including traffic congestion, large solid waste quantities, and the disposal sites is located so as to require long hauling distances.

Solid-waste Processing

Volume reduction processing frequently can be carried out at a transfer station. Methods like shredding and high-density baling can also increase the haul-vehicle carrying capacity and thus reduce total operating costs. An economic analysis becomes necessary to determine if waste-densification systems have been justifiable.

Baling. Solid waste at present has been baled in several locations. Baled solid waste could be more easily transported as bales are more easily handled, or fairly uniform size (e.g., 4 ft long. ft in cross-section), and of high density (58-64 1b/cu ft). Higher density can be able to increase the amount of solid waste transported in volume systems such as railroads, and can produce less litter than alternative transport systems. This method is quite cheap. This method allows greater control over the transfer-vehicle load.

Conditions at transfer stations have been typically odorous, noisy, and dusty. If stations are located in urban areas, they preferably should be enclosed. Enclosed transfer stations also eliminate control and automatic weighing facilities must be part of basic transfer-station design.

Long-Distance Transport

Rail Haul, Rail haul generally becomes competitive if the following conditions have been met :

(1) round-trip distances must be greater than 100 miles;

(2) at least 1000 tons per day have been transported;

(3) baling takes place at the transfer station ; and

(4) railroad cars are used exclusively for solid waste.

Barging. It is a special case of long-distance transport which is limited to a few suitable geographical areas. Barging costs have been relatively low.

Semitrailer trucks. Semitrailer costs have been found to vary widely for solid-waste trasportation because factors like labour rates, speed, limits, trucks size, and truck operation and maintenance requirements differ by area.

Waste loads have to be covered during transport. Mechanical unloading or drag. lines placed prior to loading a fixed-bed truck can be employed to unload, using a bulldozer to pull the solid-waste lines. Care should be taken because snapped can bring about serious injuries.

Pipeline transport. Pipelines are mentioned briefly already; their potential for low-cost, long-distance solid-waste transport could not be demonstrated. A possible advantage of a pipeline system has been that it could be used for both solid and liquid wastes transport. Future waste-management systems may combine solid and liquid disposal. Pipeline transport of liquid sludge has been found to be economical.

Street Cleaning

City streets should be kept clean due to the many reasons;

(1) to prevent disease, injury, and annoyance from street dirt ;

(2) to prevent vehicle damage from sharp, metallic objects;

(3) to promote safety in reducing fires from dry leaves and litter, and slips and skids from wet leaves;

(4) to increase community appearance;

(5) to disallow the cloggings of sewers and storm drains; and

(6) to decrease water pollution from street runoff.

A proper solid-waste collection system will help reduce street dirt and litter, but some street cleaning will still be needed.

Manual

Street cleaning in congested areas of large cities may require manual methods. Bright, fluorescent uniforms must be worn by clean-up personnel for their safety. New equipment, like vacuum sweepers and automatic brooms, are replacing the conventional ''white wing' armed with a broom and dust cart. Such mechanical equipment permits one operator to cover a large area at less cost in the more open locations. The collected sweepings are normally stored in bins; solid waste bins along the street are unatractive and become litter sources. Using a small relay vehicle for collecting the sweepings from several street cleaners should be considered. Vehicular or house flushing (usually with fire hydrants) has been costly, generates water pollution, and is not recommended except in a few special situations.

Flushing Machines

Large trucks with pressure nozzles are used for street flushing in many communities. The advantages of this system have been as follows:

(1) rapid and thorough pavement cleaning despite parked cars or other obstructions;

(2) continued operation with wet pavements;

(3) acceptable operation at night ; and

(4) the suppression of dust.

The disadvantages to the system are;

(1) cost has been generally higher than for dry cleaning systems;

(2) it may be increasing drainage and water-pollution problems;

(3) flushing is not a complete method; it moves light dirt to the curb and leaves it there;

(4) splashing of vehicles and pedestrians cannot always be avoided; and

(5) flushing may not be possible during the peak traffic periods.

Well-trained personnel have been necessary for a successful system; sewer or storm drain requirement get increased by the system. The maintenance of catch-basin residue has been away from simple flushing and toward mechanical brush-sweeping and water sprinkling machines.

Sweeping Machines

Conventional sweeping machines are having a rotary broom, vacuum pumps or air-circulating pumps, sometimes watersprinkling tanks, and a storage compartment for the sweepings. As most dirt and litter gets accumulated along the edges of the street, a right-hand-drive vehicle with two passes per street can be quite effective. As parked cars prevent cleaning, temporary parking restrictions are frequently imposed. Signs restricting parking must be clear and well-posted to avoid citizen complaints. Noisy vehicles should not be allowed in residential areas during normal sleeping hours, and vehicles should be well-ighted to avoid accidents.

Street sweepings consists mainly of paper, leaves and inerts, and consequently pose normal disposal requirements. The individual collection vehicles may be able to transport the sweepings directly to the disposal site or transfer may be used. The practice of street dumping loads of sweepings for subsequent loading onto trucks must be avoided as sweepings may be dispersed by wind, traffic, small animals, or pedestrians. Street-sweepings collection has been a far more costly unit process than on-solid-waste collection.

Collection from Communal Sites

The organization of refuse collection is greatly simplified by the use of large communal storage sites. Then city of Delhi, for example, is served by about 1250 sites having capacities from 0.5 to 10 tons for a population which exceeds 4 million. Delhi probably has about 800,000 separate sources of wastes (dwellings and shops) and a frequent collection direct possibly greater expenditure. However, when sites are widely spaced, a great deal of domestic wastes are deposited in the streets by householders too lazy to carry it to the depot or masonry enclosure. It is significant, therefore, that only by employing about 10,000 sweepers in Delhi kept in a tolerable standard of cleanliness.

While the use of large communal sites appear to be a fairly cheap and simple solution, it may transfer much of the burden of refuse collection on the street cleansing service and actually increase total costs, becuase it is cheaper to collect refuse direct from a house than to sweep it up from the street. In Delhi, in the year 1974 Rs. 31,000,000 were spent on sweeping, and only Rs. 7,330,000 on refuse collection.

The use of large, widely spaced communal storage sites is usually a failure because the demand placed on the house-holder goes beyond his willingess to co-operate. Communal storage points should, therefore, be at frequent intervals. Madras, Bangalore and Manila provide fixed concrete receptacles with capacities between 100 and 500 litres on footways or verges at intervals of 50 to 200 metres. The objections to these have been stressed earlier, but they are fairly successful because they place a reasonable and acceptable duty on the residents, thus very little domestic waste is the street.

Both the large masonary enclosure and the smaller concerte bin are inefficient in the use of manpower and vehicles. Wastes have to be removed by rake or shovel and basket; it is a slow process and vehicle waiting time during the loading process is excessive. Vehicles employed on this work tend to interfere with other traffic in the street. The following work performance have been recorded:

Delhi, masonary enclosures	1.4 tons/man/day 7 tons/vehicle/day
Bangalore, concrete pipes	1.2 tons/man/day 6 tons/vehicle/day

Drums of 200-litre capacity are far from an ideal solution, but the fact that two men can usually empty them directly into a vehicle with a low load line greatly increase daily performance: probably to about 5 tons/man/day and 10 tons/vehicle/day.

Block collection

In this system, a collection vehicle travels a regular route at prescribed intervals, usually every two days or every three days, and it stops at every street intersection, where a bell is rung. At this signal the residents of all the streets leading from that intersection bring their wastes containers to the vehicle and hand them to the crew to be emptied. A crew of one or two men is adequate in number as they do not need to leave the vehicle.

Block collection should be operated frequently, as otherwise the weight of wastes to be carried to the vehicle may be beyond the capacity of some of the residents. It has a significant advantage over a kerbside collection in that bins are not left out on the street for long periods.

This system is operated in Mexico City and during a study carried out by the Solid Wastes Division of the Sub-Secretariat for Environmental Protection the following data were collected. They represent the average of four routes:

Collection route, total length 2.74 km	2.74 KM
Number of stops on route 25	25
Average distance between stops 110 metres	110 METRES
Bins emptied per stop 33	33
Total bins emptied on route 858	858
Average load, total weight 3653 kg	3653 KG
Dwellings on route 841	841
Average weight/dwelling 4.34 kg	4.34 KG
Total time for load 2 hours 27 min	2 HOURS 27 MIN
Average travel time point to point 1.15 min	1.15 MIN
Average loading time per point 7.10 min	7.10 MIN

The daily perfomance achieved by this system is about 3.5 tons/man/day and 7.0 tons/vehicle/day.

Kerbside collection

This, like the block collection, requires a regular service and a fairly precise time table. Residents must place their bins on the footway in advance of the collection time and remove them after they have been emptied. It is very important that bins of a standard type should be used, as otherwise it is likely that wastes will be put out in improvised containers, such as cardboard boxes, or even in loose heaps; when this occurs some of the wastes are inevitably scattered by animals and wind, thus increasing the work of street cleansing.

- Kerbside collection is never entirely satisfactory. Problems include:
- bins sorted through by scavengers,

- bins stolen,
- bins turned over by goats or cattle,
- failure of the householder to retrieve the bin quickly.

The worst example of this last problem is when the bins are kept permanently on the footway, this is not uncommon in high income residential areas of certain Asian cities.

However, kerbside collection is unavoidable with some types of house construction, and it is the cheapest method of house-to-house collection. When the rate of wastes generation is high and collection infrequent, very high labour productivity can be achieved. For example, in one city in USA a one-man crew collects up to 10 tons/day—400 dwellings at an average of 25 kg/dwelling. In all developing countries, however, the wagerate to vehicle-cost ratio would be much less than in USA and it would be profitable to employ a crew of at least four men. Productivity would also be reduced as the weight per dwelling would be much less.

Door-to-door collection

This is the system in which the householder does no work: the collector enters the garden or courtyard, carries the bin to the vehicle, empties it, and returns it to its usual place. It is costly in labour because of the high proportion of working time spent walking in and out of premises and from one dwelling to the next, but is the only really satisfactory system. A study in USA showed that this system costs about twice that of kerbside collection, but this ratio would be greatly reduced in countries where labour cost is low.

The problem with door to door collection in developing countries is that vehicle productivity would be very much less than in Europe or USA if collection was at high frequency. Since high vehicle productivity must be the main aim of developing countries, door-to-door collection by the conventional western method of heavy motor vehicle and crew is very unlikely to be a viable system unless the interval between collections is extended to a week. This is unlikely to be acceptable in tropical countries

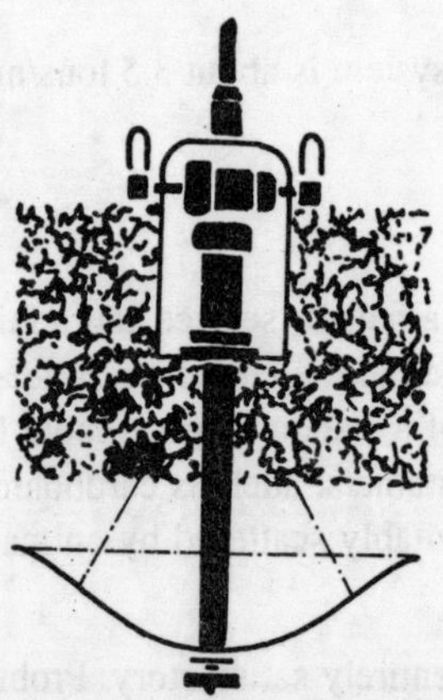

Figure 14.1. The centri-rator.

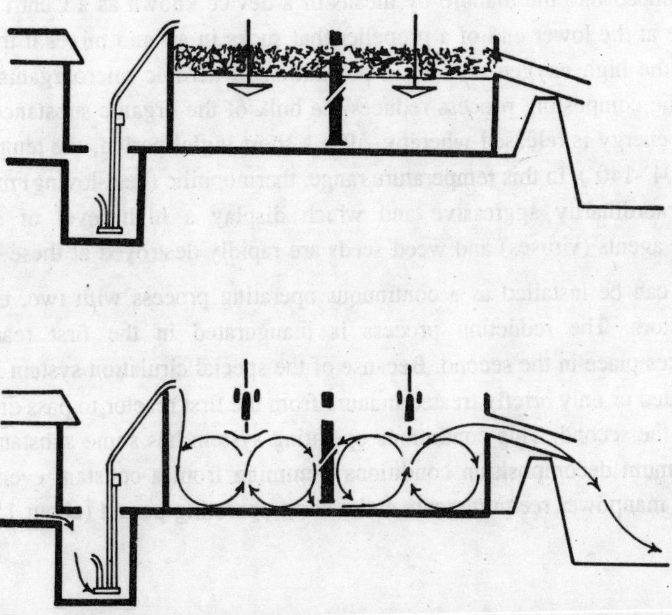

Figure 14.2. The licom system for the composting of manure.

as a priniciple and in any case the majority of dwelling in most cities need a daily collection (or communal storage).

Removal Refuse from Intensive Stock farming—The Licom System

The removal of solid and liquid manure from large-scale stock farms is a serious problem. Collection and removal must be so arranged that groundwater and waterways (streams, rivers, etc.) are not contaminated.

Animal excrement contains plant nutrients and humic substance which contribute to humus and the fertilization of the soil. It manure is supplied in excessive quantities or at the wrong time there is a danger that the nutrients will be washed away or that there will be eutrophication (oversupply of nutrients leading to overabundant growth of plants) in the waterways.

Manure contains a series of strong-smelling compounds such as hydrogen sulfide, mercapton and ammonia, which can cause serious nuisance to the community. Viruses and weed seeds are also contained in manure. If these germs and seeds are not killed off, weed growth will be encouraged and there will be a danger of soil and plant contamination.

In most intensive-stock-farming concerns the manure is in liquid form. In a liquid-manure silo, anaerobic conditions (poor in oxygen) prevail. One result of prevalent anaerobic conditions is the formation of ammonia, hydrogen sulfide and other strong-smelling substances. If oxygen is added, these compounds are converted by oxidation into odorless nitrates and sulfates, carbon dioxide and water. This is the process of aerobic decomposition into compost.

With the aid of the Licom system, aerobic composting can be applied to the liquid manure. The oxygen required is introduced into the manure by means of a device known as a Centri Rator. The air is applied through a pipe at the lower end of a propeller that sucks in air and mixes it thoroughly with the manure. Owing to the high oxygen content, a population of aerobic microorganisms is created, which in the course of the composting process reduces the bulk of the organic substance to inorganic. In the reduction process energy is released whereby, after a short initial period, the temperature of the manure rises to 40-60 (104°-140°). In this temperature range, thermophilic (heat-loving) microorganisms develop which are extraordinarily aggressive and which display a high level of decomposition performance. Pathogenic agents (viruses) and weed seeds are rapidly destroyed at these temperatures.

The Licon system can be installed as a continuous operating process with two, of if necessary more, interlinked reactors. The reduction process is inaugurated in the first reactor, and the decomposition proper takes place in the second. Because of the special cirulation system in the reactors it is impossible for untreated or only briefly treated manure from the first reactor to pass directly through the connecting pipe into the second. This continuous operating system has some substantial economic advantages, such as optimum decomposition conditions stemming from a constant even temperature, reliable disinfection, low manpower reequirements and a short operating period (about 15 days).

15

Disposal of Solid Wastes

Introduction

Nearly 75-80% of all collected residential and commercial solid wastes are sent to open dumps. Less than 10% is buried in sanitary landfills. A small but locally significant quantity is dumped at sea. The remainder is incinerated, partly to obtain energy and the amount of land required for dumps and sanitary landfills.

A number of different collection and disposal methods are in use and will be discussed in this section.

Hog Feeding

At one time, hogs were extensively used for garbage disposal. In colonial hogs wandering in the streets served as scavengers of the wastes in the gutters. As urban areas developed, hog farms nearby found it desirable to collect urban garbage to hogs.

It was known that the feeding of raw garbage to hogs led to high incidence of trichinosis in humans, but the practice continued until the mid-1950s. During the years 1950 to 1955 the rapid spread of a virus disease of hogs, vesicular exanthema, largely through raw garbage, led many countries to institute regulations forbidding the feeding of raw garbage to hogs. Although heat treatment (such as exposing the raw garbage to live steam for a half hour or more) cured the problem, the feeding of garbage to hogs decreased in the late 1950s and 1960s and is presently almost signficant.

Open Dumps

Disposing of solid wastes in open dumps is the most common method. Much of the uncollected refuse is disposed in a similar manner. This method is cheaper than other disposal methods and have the advantage of avoiding the acute pollution problems associated with discharging wastes into waterways or polluting the air from incineration. However, open dumps produce health and air pollution problems and are not an acceptable method of disposal. They can cause public health problems by encouraging the growth of population on flies (which can transmit typhoid fever, cholera, dysentery, tuberculosis, anthrax, and other diseases), rats (which can transmit plague, murine typhus fever, leptospirorsis, rabies, rickettsialpox, and other diseases), cockroaches, mosquitoes (which can transmit malaria, yellow fever, dangue, mosquito-brone encephalitis, and filariasis), and other pests. Air pollution problems arise when the dumped wastes are burned so as to reduce their volume and conserve space or when spontaneous combustion or arson leads to fires.

The blight of open dumps is aesthetically intolerable and depresses land values.

Random Refuse Dump

Many local authorities have no regular refuse disposal outlet and depend on natural areas in which refuse is deposited untreated and uncontrolled. Refuse dumps of this type are to be found also on the outskirts of large urban aggregates, frequently in nearby leisure areas. Sites selected for random refuse drumps include deserted quarries, small valley rifts, and drained dredging ponds. The refuse is normally tipped into the hollow over a ledge at the top and left to roll down the length of slope. As it makes its way down the refuse is "sorted", according to the size and specific gravity of the components: light objects and ones with rough surface tend to remain on the upper slope, while, those of medium size stay in the middle and heavy objects come to rest lower down or at the foot.

The sorting is better defined the longer the slope. In the course of the sorting process a zone is formed in the middle and on the upper slope in which material heats up of its own accord as a result of bacterial decomposition and/or in which, according to the amount of oxygen present, methane gas can develop. In this sector temperatures of up to 55°C (131°F) and above have been recorded.

As a result of the self-heating process and the presence of mathane gas, fires can break out, which, owing to the shortage of oxygen, mostly take the form of low-temperature smoldering accompanied by heavy smoke. This can ignite other articles in the refuse and produce noxious fumes. Fires in rubbish dumps especially those occurring at the heart of them (core fires), are difficult to control. If water is sprayed on them, the activity of the anaerobic bacteria (forming methane in the absence of air) is heightend, giving rise to further danger of spontaneous combustion in the course of weeks or days. Core fires can be controlled with any certainty only when the deposit is broken up with heavy earth-moving equipment.

Next to the fire hazard, the most serious damage to the environment from uncontrolled refuse

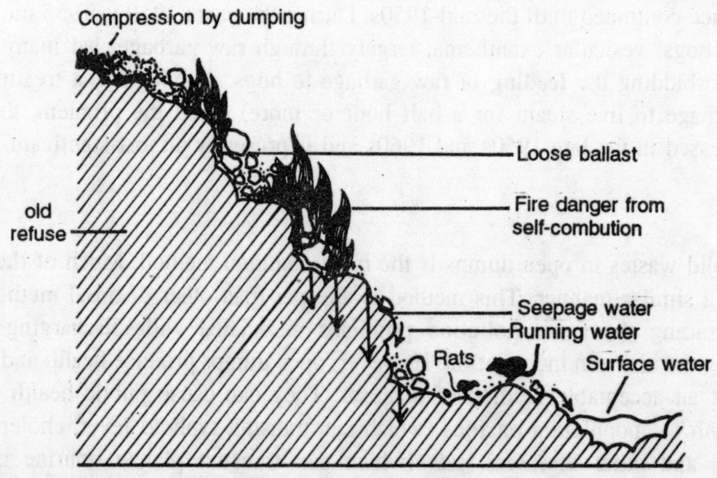

Figure. 15.1. The uncontrolled deposit in outline.

dumps is disturbance of the ground and surface water. Quite frequently small streams are found at badly sites dumps after rainfall which are severely contaminated by organic and mineral waste. If the ground below the dumps is pervious to water, pollution of a bacterial and chemical nature can infect the groundwater and be disseminated with it over areas several miles in extent.

Uncontrolled dumps provide almost ideal breeding grounds and living conditions for rats. If the food supplies are adequate, the self-heating of the dump provides the rats,even in periods of severe frost, with well-tempered habitats for breeding and rearing their offspring. As a result, deaths from freezing are rarc even in severe winter, which would normally lend to a drastic reduction in the population.

Uncontrolled refuse dumps also provide fertile breeding grounds for insect pests and weeds, which spread out from there into adjoining cultivated areas.

Appreciation of the dangers entailed in these uncontrolled dumps should result in a general reform and in the introduction of organized refuse disposal.

Sanitary Landfills

According to the Sanitary Engineers, sanitary landfilling is ''a method of disposing of refuse on land without creating nuisances or hazards to public health or safety, but utilizing the principles of engineering to confine the refuse to the smallest practical area, to reduce it to the smallest practical volume, and to cover it with a layer of earth at the conclusion of each day's operation, or at such more frequent intervals as may be''. When properly run a sanitary landfill is odorless smokeless, and ratless.

In the area method sanitary landfill operation the solid wastes are spread out over an area, compacted, then covered with soil or similar materials on the top end the sides.

In the trench method trench 10-15 feet deep and upto 20 feet wide is dug and the refuse is pushed into it, compacted by a bulldozer, and covered with dirt.

The area and the trench methods need about an acre to handle the trash and garbage of 10,000 people for a year.

In the ramp method gulleys, ravines, or quarries are used, the solid wastes are dumped along the slopes, and covered with earth.

In a well-run sanitary landfill operation the soil covering has been deep enough to keep out rodents and flies, and it prevents the escape of noxious odors. A carefully sited sanitary landfill should not be near a community's groundwater reservoir, because rain-water may leach pollutants into the water supply. Organic material in the fill decays and the internal temperature rises as high as 150-160°F high enough to kill heat-sensitive germs in the wastes. Gases such as carbon dioxide and methane are also generated, with peak production occuring within the first 2 years. As methane is highly combustible, a danger exists to buildings built over such landfills. Sanitary landfills can undergo settling for a long time, which also makes it risky to build too quickly over filled areas. Generally 90% of the ultimate settling happens in the first 5 years; the remaining 10% may take several years longer. Landfill areas are better set aside for playing fields, parks, arboretums, and the like, not used for building dwellings.

Inside the filled region, there could be aerobic reactions to the extent that oxygen is present and anaerobic reactions (with productions of methane, H_2S, CO_2, and water), both involving bacterial action. The inside temperature may rise to 70°C in a few days from the heat produced. The amounts of air and water present influence the speed of decomposition with wet landfills sometimes showing rather complete decomposition of organic matter within a few years, while dry landfills show little decomposition after several decades. Gas production can sometimes be dangerous.

The advatages of a sanitary landfill, as opposed to an open, uncovered dump, have been (1) the public health problems are minimized because flies, rats and other pests are unable to breed in the covered refuse; (2) there is no air pollution from burning and none from dust or odors; and (3) fire hazards are very small; (4) there is danger of ground water, surface water or pollution, however, if the landfill site is improperly chosen or if it is dug out too deep.

Landfills, if properly planned, can be used later for construction sites or recreational facilities such as parks and golf courses.

Land disposal of any kind requires large tracts of land. The increasing shortage of land and the ever-growing mountain of waste make the cost prohibitive to many communities. Leaching by surface water or ground water may contaminate streams and other water supplies. Air pollution may result from the decomposition of wastes, especially in open dumps but also in landfills if the wastes are improperly covered.

Site Selection

Site selection for sanitary-landfill requires, in addition to technical engineering competence, attention to the economic, social and political aspects of land use planning. When several possible landfill sites are being compared, professionals from these various fields can be used to evaluate independently the practicability of a particular location. A rational analytical site-selection methodology can employ a matrix-type analysis based on a weighted scoring system, which should include all criteria that will be discussed in this chapter.

Planning and Land Use

Land-use plants and planning goals have been established by the appropriate local and regional agencies; they obviously, must be consulted as the preliminary step in site selection. A local or regional master plan, for example, may preclude the use of an otherwise technically suitable size. Present and futrue zoning of adjacent land must be taken into consideration in determining the suitability of a site for receiving waste and its ultimate use. For example; the proximity of present or future airports has been a matter of concern since birds may be attracted to an operating landfill thus and endanger low-flying aircraft.

Climate, Topography and Natural Features

The site-selection process must take in consideration all climatic characteristics such as prevailing winds, precipitation, evapo-transpiration and temperature variations as they relate to ordors, dust leachate generation, blowing litter, cover soil, erosion, etc, The effect of snow, ice and rain on the ability of vehicles to traverse access roads has been an important operating factor. Steep roads should be avoided as they also decrease vehicle maneuverability. The frost-penetration depth has been also an important factor in excavating cover soil. Frost-protection measures such as the use of insulating material (e.g., straw or leaves) to cover soil will increase operating costs.

The topography of the site is having important implications for landfill capacity, drainage, ultimate land use, surface and ground water pollution control, site access and related operating factors. Whenever possible, floodplain land must be avoided because of the high cost of preventing water pollution and because costs for protective measures such as dikes and levees have been high. Stormdrainage and floor records should always be consulted when choosing a potential landfill site.

Cost Considerations

The square-foot cost of the land has been secondary to the cost of the land per ton of waste disposed. Possible savings could accrue by leasing land from a private owner with a partial consideration being the ultimate improvement to the land. However, since landfills may need hundreds of years of maintenance due to settlement, methane-gas generation and leachate pollution, it may be desirable provide for perpetual private or public upkeep. A greater tonnage could be disposed at low costs into steep pits or canyons, rather than flat areas.

Development costs will normally include : clearing the timber and brush, diversion of surface drainage runoff or streams, providing the necessary utilities (remote sites may require electrical generators, chemical toilets, water storage, tank, trucks, etc), fencing, access roads, buildings and scales. Haul distance is another major economic factor in site selection. Landfill sites should be as close as possible to the collection area.

Soil characteristics and cover-soil availability have been important, practical considerations. Importing cover-soil material may be hard to realize, except at great cost.

Public Acceptance

Public acceptance must be a constant key consideration both during and after the site-selection process. A site having minimal potential for public opposition, all other things being equal, should be chosen over a site generating considerable local controversy.

Site Planning and Design

Many of the basic engineering data requirements for good landfill planning have been complementary to those for proper site selection. The optimum development plan, however, should include comprehensive site data, upon which would be developed following facility plans :

(1) Detailed drainage facilities, all-weather access roads. fencing, buildings, sanitation, water lighting, telephone, scales, equipment-maintenance yard and all other site improvements;

(2) A series of Out-and-fill plans presenting the landfill at various stages of completion ;

(3) A detailed description of daily operating procedures;

(4) A phased analysis of equipment and manpower requirements for the life of the site;

(5) An estimate of projected capital and operating costs including final finished landfill landscaping improvements; and

(6) An ultimate land-use plan for the site.

The engineers preparing the site-development plan should be knowledgeable of all applicable regulatory requirements. Necessary, statutory permits should be obtained early in the planning stage to avoid costly delays.

Site Improvements

The access roadway leading into the gate-house should have at least two paved lanes of sufficient width to accommodate heavy vehicles ; grade of the access road should never exceed 8%. Lime, portland cement, or asphalt and oil binders are desirable road-surface materials that can help control dust in dry weather and prevent heavy machinery from being mired in wet weather.

Permanent or portable chain-link fences have been found to be suitable for protecting the site and catching windblown papers. An attractive entrance having a lockable gate should be provided along with appropriate signs identifying the site and presenting pertinent information.

Drainage ditches to divert surface waters away from the site should be fully designed channels and appurtenant structures. Surface slopes of the completed landfill should be less than 2% grade. Cut earthen slopes should not exceed 2:1, and local soil characteristics and rainfall quantities may dictate flatter slopes. Debris basins outside the fill area should be constructed to collect and pond drainage waters, thereby allowing suspended material to settle.

Appropriate sanitary facilities should be provided for landfill and other employees : toilets, sinks, showers, lockers, and lunchroom facilities are desirable amenities. Equipment storage and maintenance facilities should also be provided.

At some sites, truck-scale facilities may be installed. The scale size and type will depend on local requirements. The scale size and type will depend on local requirements. The scale should be capable of weighing the largest vehicle that may use the landfill; a 50- to 60 ft platform is usually sufficient. Automatic printouts and computer data control have been at large sophisticated installations. Scale-house installations can be made portable for ready removal to another location.

Progress Plans

A series of plans should be undertaken for the phased landfilling program and include the following elements :

(1) a base map which shows the original topography, and intial site improvements ;

(2) several interim plans depicting the site topography, excavations and improvements at various construction stages ; and

(3) a final plan illustrating the ultimate contours.

Appropriate survey control monuments must be established to reference the horizontal and vertical locations. A time schedule should accompany the plans.

Equipment and Manpower Requirement

A minimum and two persons is desirable to work at a sanitary landfill for safety reasons.

Equipment requirements will vary with landfill characteristics. Typical solid waste-handling equipment used at landfills includes crawlers or rubber-tired tractors, together with accessories including dozer blades, buckets, multipurpose buckets, and landfillblades. More specialized heavy vehicles often used at larger landfilling include draglines, scrapers, steel wheel compactors, water trucks and other conservation equipment.

Cost

The site-planning and design process must consider capital, operating, and maintenance costs. In many ways, unit costs are affected by the volume of the landfilling operation. Obviously, with a greater population serviced, theré have been increased weights of solid waste disposed daily. The economic ability to provide the site improvements listed above is also size-dependent in that small operations typically cannot afford more than minimum essentials.

The sanitary landfill has been generally a relatively economical waste-treatment and disposal process when compared with other acceptable processes and high standards of operation should be maintained.

Dry versus Wet, Warm versus Cold

The planning and design factors mentioned above are on a general level. Specific planning will also take into account a variety of local site conditions. Of particular concern are the operating requirements for extremely cold and wet areas. Extended low-temperature operations will need the following procedures;

(1) Excavation and stockpiling of covet fill before frost conditions prevail;

(2) Placement on prepared ground of insulating material such as leaves, hay or straw to a suitable depth, as much as 3 ft;

(3) Excavation of trenches or cut/fill areas as needed, including insulated cover material;

(4) Storage of equipment in a heated shed;

(5) Equipping tractors and other landfill equipment with grousers if work on ice is necessary; and

(6) Provision of a heated cab for the operator,

Extremely wet areas should have paved access road and suitable dumping area.

Environmental Factors

The critical environmental factors which must be considered in sanitary-landfill design have been vehicle traffic, leachate gas production, odor, noise. aesthetics, air pollution, dust, fires, vectors and birds. Each of these environmental factors and procedures for their abatement has been discussed in this section.

Leachate

Though some leachate has been produced by organic solidwaste decomposition and compaction of wet refuse, the main factor causing leachate generation has been the inflow of water. Excessive rainfall or other water-caused saturation of solid waste typically results in leachate generation and possible water pollution. Leachatc may contain dissolved and suspended materials of all sort, within the

Table 15.1

Range of Pollutant Concentrations in Leachate from Typical Domestic Refuse

Ion	*Concentration-mg/litre Range*
Iron	200-1700
Zinc	1-135
Phosphate	5-130
Sulfate	25-500
Chloride	100-2400
Sodium	100-3800
Nitrogen	20-500
Hardness (as $CaCO_3$)	200-5250
Chemical oxygen demand	100-51,000
Total residue	1000-45,000
Nickel	0.01-0.8
Copper	0.10-9.0
pH	4.00-8.5

limits given in Table 15.2. Table 15.2 shows leachate constituents derived from organics and ash fill under several types of environmental parameters. Water inflow to a land fill can be reduced by the following measures :

(1) locate the landfill so as avoid direct ground-water or surface water contact ;

(2) use proper surface-cover grading to increasing rainwater runoff;

(3) plant final fill site-cover with vegetation which will avoid erosion while enhancing runoff and evapotranspiration;

(4) use a non-permeable cover material with maximum lateral-flow characteristics ;

(5) delay leachate generation by placing solid waste and soil cover with low moisture contents and high compaction densities; and

(6) install drainage facilities to divert off-site surface and subsurface water. It is possible to determine the local rainfall and solid-waste parameters in order to calculate the estimated field capacity which will cause leachate generation.

If fill soil and solid waste receive enough water to reach their respective field capacities, leachate may start to be discharged. However, some localized leachate may get discharged due to concentrated channeling effects. If underlying soil or trock strata are impervious, this leachate will "rise" and discharge on the ground surface either at the landfill toe or somewhere down the slopes.

Gas

During the decomposition process, landfills generate several types of gases. During the first several days the decomposition process is aerobic, considerable heat gets generated, and the principal gas produced is carbon dioxide. Once all the landfill's free oxygen has been depleted, organic

Table 15.2

Constituents in Leachate from Refuse and Incinerator Ash
(Based on weight of refuse as received)

	Per cent leached					
Material leached	*1*	*1*	*2*	*3*	*3*	*3*
Permanganate						
30 min	0.039					
4 hr	0.060	0.037				
Chloride	0.105	0.127		0.11	0.087	
Ammoniacal nitrogen	0.055	0.037		0.036		
Biologic oxygen demand	0.515	0.249		1.27		
Organic carbon	0.285	0.163				
Sulfate	0.130	0.084 (as SO_4)		0.011	0.22	0.30
Sulfide	0.011					
Albuminoid nitrogen	0.005					
Alkalinity (as $CaCO_3$)				0.39	0.042	
Calcium				0.08	0.021	2.57
Magnesium				0.015	0.014	0.24
Sodium			0.260	0.075	0.078	0.29
Potassium			0.135	0.09	0.049	0.38
Total iron				0.01		
Inorganic phosphate				0.0007		
Nitrate					0.0025	
Organic nitrogen	0.0075	0.0072		0.016		

decomposition then becomes facultative and finally anaerobic so that the principal gases generated are methane and carbon dioxide. Some hydrogensulfide generation has been noted in landfills which have substantial quantities of sulfate salts.

The impacts of carbon dioxide and methane upon the environment differ. Carbon dioxide is water soluble and can be adorbed into the leachate increasing its corrosiveness and acidity. Methane diffuses readily and is exposive at atmospheric pressure in concentrations from 5 to 15% by volume. Table shows the physical properties of methane and carbon dioxide.

Gas control can be both difficult and expensive. Two common techniques used to control gas movement are clay barriers and gas vents. Gas vents have been necessary whenever the ultimate use of the landfill site involves covering with an impervious material.

Odors, Fires and Vectors

The problems of odors, fires, and vector can all be minimized by proper soil cover. Earth mover-compactors should remove burning solid waste from the fill area and cover it with dirt immediately. Rats and other mammalian vectors are discouraged from feeding on or propagating in solid waste which receives a well-compacted 6 in. layer of low clay soil cover daily. This type of soil cover also discourages bird foraging; crows and seagulls are especially likely to feed on uncovered solid waste.

Flies, another potential vector, have difficulty emerging from well-compacted soil cover. Continuous grading of soil cover to fill in low spots is essential to prevent the development of stagnant pools of water in which mosquitoes may be able to breed.

Noise and Aesthetics

Routing vehicle traffic through industrial, commercial or low-density-population areas decreases the adverse noise impacts of landfill-related vehicle traffic. Construction of noise barriers surrounding the landfill, such as earth berms or walls, is good practice, and these with trees and planting can also improve landfill aesthetics.

Dust and Air Pollution

Both main and access roads should be of all-weather type, which should be sprinkled with water or another suitable dust-control agent. Cover soils should be of a type not susceptible to wind erosion; be planted. Wetting to control dust should be limited to the working zones. To control noise and air pollution all landfill equipment should be fitted with appropriate abatement devices.

Public Acceptance

Engineers must be prepared for the possibility of negative local public response to a sanitary landfill. A programme designed to increase public acceptance for the landfill must be included in the early planning stages. An unnecessarily defensive attitude should be avoided, however, opposition to locating and operating a landfill can come from any of the population groups listed below :

(1) Property owners who believe a nearby landfill will lower their land values and social status.

(2) Citizens opposed to the expense.

(3) Citizens who resist change.

(4) Citizens concerned about conserving (not wasting) resources and natural areas.

(5) Neighbours concerned about traffic, safety and health blowing paper, insects, rats, noise, unaesthetic landfilling and other environmental effects.

(6) Politically opposed voters.

(7) Citizens who believe a prior agency record of poor dump operations will continue.

Points to be stressed in any public-relations program should be geared toward technically responsible answers to the questioners. Tools that may be used to gain public acceptance are listed below.

(1) *Accurate information.* Sanitary landfills should be well run.

(2) *Comprehensive plans for regulating truck traffic.* The prospect of heavy truck traffic raises genuine and justifiable concern. Plan for regulating speed, weight loading, noise and exhaust odor should be given.

(3) *Increase citizen involvement.* Citizens advisory groups, workshop and civic group involvement will develop positive attitudes among residents.

(4) *Use of media.* Films on solid-waste management (available from federal state and local agencies, as well as private industry), radio and television spots and news releases can be used to spread information to a large number of concerned citizens. Attractive landfill models appropriate symbols and slogans can be skillfully used.

Operating Procedures

Methods

Landfilling procedures are of two basic types (see Fig. 15.1 and 15.2) : the area fill and its modifications, and the trench fill and its modifications. The trench fill has been common in flat or rolling areas. Solid waste is placed in excavated trenches, and the exacavated earth is used for cover soil. The area method has been common in ravines, canyons, open pits or similarly concave areas. Soil cover is obtained by cutting down the fill wall areas. The solid waste is confined by equipment compaction, and solid waste or cover soil admixture. In both mehods, distance from the fill working face to cover-soil material must be minimized to conserve time and fuel. The new working face created each day must be kept to a minimum (as little as 20 to 30 ft) to conserve cover soil and reduce litter. The pattern of lift construction must be carefully scheduled to route vehicles over previously filled areas for additional compaction of the solid waste and soil (such routing should be avoided in wet weather, however).

Collection vehicles must place their loads of solid waste as close as possible to the fill working face. Solid waste should be spread and compacted on an inclined working face, with a slope of 20° to 30° in layers no more than 2 ft thick after spreading and compaction by more than 2, and preferably 5, passes of the dozer or compacting must be directed toward maintaining thin, even waste and soil layers and realizing uniform cell heights, slopes and widths.

As the fill cell progresses a minimum of 6 in. of compacted soil cover should be continuously placed so that only a minimum working face remains uncovered at any time. At the end of the working day, the remaining working face should be covered and the soil compacted with 2 or more passes of the equipment.

Indiscriminate salvaging and storage of solid waste (scavenging) should not be allowed at the landfill site, primarily due to health, safety and aesthetic considerations.

Special or Processed Wastes

Special Wastes

Crushable, bulky items, such as automobile bodies, can be deposited at the toe of the working face. Demolition debris is often durected to a separate storage area, used for access subbase road construction, or placed directly at the toe of the working face.

Reinforcing bars, pipe, rolls of cable, and coils of wire often may become entangled in the landfill equipment and cause mechanical damage, or injure personnel. These entrapping and sharp objects should be covered immediately with dirt and solid waste. Dead animals should be kept in excavated holes and covered immediately ; when they are received in large numbers, they may be buried in a specially prepared trench or pit. Rendering plants should be utilized to reclaim animal products whenever possible.

Processed Wastes

Landfills have been tending to utilize more shredded waste, baled waste, and residues from reclaimed processes. These processed wastes can greatly influence landfill practices, Mechanical shredding pretreatment includes cutting , tearing, ripping, and impact-shattering processes which reduce wastes to particles to a dimensional range of 2 to 6 in. Shredding may improve landfill aesthetics. Reclaimed solid-waste residues usually have most of the ferrous and other metal removed. Compost residue has a reduced biodegradable content.

Shredded wastes and other process residues should be handled in the same manner as in a normal sanitary landfill. However, aerobic decomposition can proceed rapidly in shredded wastes giving little methane and rapid settlement. Incinerator and pyrolysis residues must be handled carefully to prevent landfill fires due to hot embers in the residue.

High-density baled solid waste is produced by compressing solid waste into large blocks. Advantages to landfills receiving baled solid waste may include; reduced transfer, haul, and handfilling costs ; improved aesthetics; greater landfill solid-waste capacity due to higher in-place density lower cover-soil requirements ; reduced litter and vector problems ; lower cover-soil requirements ; reduced litter and vector problems ; and considerably less fill settlement.

At a landfill disposing baled solid waste, bales are placed in uniform layers 2 or 3 bales high per lift using a forklift or front-end loader. Plots of iso-concentration levels may assist in determining gas-migration patterns.

Gas collection devices vary from a simple "dry well" to an inverted gas capturing container. The most accurate instrument for analyiss of landfill gases has been a gas chromatograph. Portable measuring device, such as explosimeters.are used for rough in situ gas detection.

Other

In settlement monitoring, it is advisable to run a survey control benchmark-elevation circuit around and with the site to provide necessary references for measuring fill settlement. A good method for masuring differential settlement is to use plates set within the fill connected to a vertical pipe or bar which extends to the surface. First-year settlement may be as much as 20% of the initial fill depth and may continue at reduced rates logarithmically for many years.

Cover-soil erosion should be monitored if the ultimate use of the site will require vegetation cover. The burial locations of specific problematic or hazardous wastes within the landfill should be recorded as the landfill is being constructed.

Cost

Monitoring is relatively low in costs compared to the total solid-waste expenditures.

Final Land Use

The best final use of a landfill has been one that creatively takes full advantage of existing site conditions and landfill operating procedures to produce an optimum design which satisfies community needs, Ideally, the final land use should be planned prior to designing the landfill.

Construction of buildings on completed landfills is usually avoided due to potential problems from gas migration, corrosion, foundation bearing inadequacies and settlement. Areas of undisturbed ground must be reserved for permanent stuctures, Extra cover-soil depth becomes necessary in areas in which utility lines, trees, or deep-rooted plants are to be located. Existing trees on a site are a valuable asset if the fill can be designed to preserve them. The completed site must be attractively landscaped with native plants whenever feasible. Special assistance is usually needed when preparing a detailed landscaping plan.

Sanitary Landfill Disposal

Land disposal, or more accurately, the burying of wastes, is the only approved method of disposal which is performed at a single site. Incineration composting, and salvage are either a form of refuse handling or processing. They are complete methods of disposal, and they require disposal residue or the removal for sale or internal consumption of compost.

Landfilling is not really a self-contained method of disposal, because matter can be leached by water moving through the system. When rainfall infiltrates the landfill surface or when the ground water table saturates a part of the landfill, a return to equilibrium moisture conditions requires that water in excess of field capacity will drain from the landfill. As the water (leachate) drains from the landfill, it carries mineral and organic matter in suspension or as climate and geology of each landfill. In some geological formations water drainage is primarily vertical, and unless a nearby ground water source is affected, it is difficult to horizontally trace pollution from sanitary landfills. In other formations the impervious layers interrupt the vertical drainage and convey the leachate horizontally until the impervious aquifer outcrops and causes contamination of surface waters.

Sanitary landfill can be defined very simply as the use of solid wastes for land-reclamation, a typical example being the restoration, by filling to the original level, of man-made surface dereliction such as a disused surface mineral excavation. Solid wastes may also be used to improve natural features by raising the level of low-lying land to enable it to be used to improve natural features by raising the level of low-lying land to enable it to be used for cultivation or industrial development. Described in this way, sanitary landfilling has the virtue of being a method of refuse disposal which confers environmental improvement by restoring dereliction or improving natural contours.

Most solid wastes, however, are very offensive materials which provide an attractive habitat for such disease vectors as flies and rodents, and which pass through a lengthy process of decomposition the products of which can cause serious water pollution. Thus sanitary landfilling has two essential features which differentiate it from crude dumping :

- only sites that will be improved, not degraded, by a change of level, are selected ;
- simple engineering techniques are used to control the manner in which the wastes are deposited, so that dangers to public health and the.environment are avoided.

The methods of control include two classical requirements:

- wastes must be deposited and levelled in layers not exceeding about 2 metres in depth;
- the whole of the surface of each layer of wastes is covered with soil or other suitable material to a depth of 15-25 cm on the same day as the wastes are delivered to the site.

Limiting the depth of a layer is required partly because of the instabilty of newly deposited wastes over which vehicles must pass; also to limit the fire risks which arise from various causes. Covering the wastes serves several purposes ; it is no longer accessible to flies; attempts by rats to penetrate the covering material are immediately apparent; odours are avoided; and most important, the heat resulting from decomposition of organic matter is retained, and assists in the destruction of fly larvae and pathogenic organisms in the wastes.

Sanitary landfill is usually by far the cheapest method of refuse disposal. Because of this, and its comparative simplicity of operation, one would expect that it would have been adopted the world over, except for the rare case of a city having no suitable site within reasonable distance, but this is not the case.

Unfortunately most of the world's solid wastes are disposed of by uncontrolled dumping which blights the land for any further use and causes serious risks of water pollution and vector breeding. In some cases, wastes are levelled by a bulldozer so that ultimately the land can be used for another purpose, but this does not prevent pollution occurring throughout the filling operations and for some time afterwards. Very few cities operate sanitary landfilling to standards which totally control health and evironmental dangers; most of those that do are in the industrialized countries.

The record of the weather nations, however, is by no means a good one. Namerous landfill sites have been so badly operated as to cause devaluation of property in their vicinity, and bitter criticism from the public.

"Bad practices of the past have often created a climate of public opinion that leads property owners to oppose the location of any kind of treatment plant or disposal site in their vicinity, and renders acquisition of sites extremely difficult, even when high standards of operation are promised. Even professional urban planners are often reluctant to allocate and for solid wastes disposal."

Paradoxically, many municipalities in the industrialized countries which avoided paving that extra Rs.10/ton needed to provide good management and operation of a sanitary landfill site have ultimately found themselves paying Rs. 150/ton for incineration not through any lack of suitable landfill sites, but because public opinion has been outraged by poor landfill standards.

The record of the developing countries is even worse; crude dumping is still the most common method of disposal and causes the following hazards :

- fly generation,
- encouragement of rodents,
- static water pollution and aerial nuisance.
- river pollution.
- sea pollution,
- fire and smoke pollution,

Decomposition in Sanitary Landfill

Decomposition in a sanitary landfill arises from chemical changes which are brought about by bacteria which are present in the organic content of the wastes when they are collected, and which multiply rapidly, particularly when the wastes are effectively sealed. There are three main types of bacteria which are found in a landfill : the saprogenic, the agent of decomposition; the zymogenic, which are agents of fermentation (they are used in commercial processes such as brewing); and the pathogenic, which are associated with human diseases.

Both the saprogenic and the zymogenic bacteria are found in forms which vary according to the supply of oxygen. There are aerobic types for which oxygen is a necessity. There are aerobic types for which oxygen is a necessity. The are obligate anaerobes for which oxygen is a disadvantage, and facultative anaerobes which can adapt themselves to varying supplies of oxygen.

During the first stage of decomposition, aerobic bacteria are dominant, but as the available oxygen declines their work is taken over by anerobic bacteria; facultative anaerobes will be present at all times.

The bacteria have the ability to create certain colonial solutions known as enzymes which have the power to break up stable compounds. The enzymes are used by the bacteria to reduce organic matter to a form of nourishment for them and it is this process of breaking down, and making soluble, relatively stable compounds that we know as decomposition.

The consequences of all these complex changes include:

- temperature changes within the landfill;
- the production of gases: hydrogen, oxygen, nitrogen, carbon dioxide, carbon monoxide, methane, and sulphuretted hydrogen;
- the production of soluble and particulate decomposition products which can be carried away from the site as a leachate when passes through the landfill.

Temperature irrespective of ambient temperture, the internal temperature of a landfill rises soon after the deposit of the wastes as a result of heat emitted by the activities of aerobic bacteria.

The main significance of this rise in temperature is its possible significance in the destruction of pathogens. The conclusions reached by two British investigators, Jones and Owen ("Some Notes on the scientific Aspects of Controlled Tipping", City of Manchester, 1934) were as follows:

"...it is highly improbable that pathogenic bacteria can multiply and continue to live in a tip." The factors to be taken into account when considering the likelihood of danger arising from pathogenic bacteria are as follows:

"The temperature inside a tip rises during the early period of decomposition to a point; in most cases, higher than the thermal death points of pathogenic germs."

"Pathogenic germs are difficult to keep alive even in the laboratory."

"The saprogenic organisms will tend to outnumber and exterminate them."

"Unless the organisms are spore formers they will perish rapidly under ordinary conditions of earth burial; all the common pathogenic likely to be found in a controlled tip are non-spore formers."

Gases. Four gases are of significance: sulphuretted hydrogen, carbon monoxide, carbon dioxide and methane. Sulphuretted hydrogen is toxic, but only small traces have been found, and any large concentration would give ample warning of its presence by its extremely offensive odour. Carbon monoxide, which is persent in large amounts, is diffused very slowly and offers no risk on the surface; in excavations on sanitary landfills it has been found at a level of one per cent which would cause unconsciousness and death if a person were to be trapped.

Because methane is diffused very slowly from the site, in concentrations very much below the level at which ignition would occur, it rarely presents any risk. Conditions have arisen, however, where methane has seeped through fissures in the soil into building adjoining the landfill site; in such cases a dangerous concentration could occur.

Leachate. Organic and inorganic matter present in crude refuse, and the products of decompostions, can be leached by water; either rainfall or ground water, passing through the wastes. This is a potential cause of serious water pollution. In a series of experiments it was found that a layer of refuse 1.5 metres deep, compacted and covered with soil, would absorb rainfall for several weeks and thereafter produce a leachate equivalent to about one third the annual rainfall in that area, which was 635 mm. This leachate was 20-30 times as strong as settled sewage, having a BOD of 6000-7000 mg/litre, and contained organic carbon, ammoniacal and organic nitrogen, ammonia, chloride and sulphate. The leachate continued for several years, but within less than three years pollutants had declined to a very low level except for the sulphate.

It follows that wastes should not be deposited in such a way that water can pass through them to a stream, not to an underground water supply.

Control of Hazards

There are a number of hazards—such as pathogenic organisms, insects, rodents, birds, airborne litter, fire and water pollution—which have to be controlled in a sanitary landfill.

Pathogenic Organisms. Control over pathogens is dependent upon a rigorous policy of covering the wastes, soon after deposit, with a layer of soil or other suitable material at least 15 cm in thickness. This serves both to isolate the wastes and to retain the heat which is quickly generated during aerobic decompostion.

It is probable, however, that workers at the site will be exposed to the wastes before they are covered. In developing countries the risks to workers by skin contact wastes may be greater than in the industrialized countries because street sweepings may contain human faecal matter. Protective clothing is necessary, therefore, and should include rubber boots, gloves, and overalls which are changed and washed regularly. The provision of personal hygiene facilities at the site is also important.

Insects. The main source of insects will be the eggs of flies which have been deposited in the wastes before they arrive at the site. Most of these will be buried deep in the wastes and will succumb to the temperature increase. Near the surface, however the rise in temperature will not be so great and the eggs will hatch into larvae ; only the provision of an adequate layer of suitable covering material, well compacted, will prevent their emergence to recommend a covering thickness of 25 cm.

Another obvious advantage of covering is that it denies access to the wastes to flies that are already at the site and thus avoids a major problem of open dumping; continuous cycles of breeding in exposed wastes at the site.

One of the recommended precaution in sanitary landfill is not to permit cans and motor tyres to lie about on the surface. They not only spoil the appearance but when filled with water they encourage the breeding of mosquitoes. Depressions in the surface of the site which permit ponding after rain should be avoided for the same reason.

If for any reason these methods of control are not adequate it will be necessary to have recourse to insecticides. They should be used daily as long as is necesary. The tate of application recommended in "Sanitary Landfill Design and Operation" (U.S. EPA. 1972) Is as follows :

Insecticide	*Approximate rate of application (g/ha)*
Malathion	672
Dichlorvos	336
Naled	111-222
Dimethoate	111-222
Diazinon	336
Fenthion	450
Ronnel	450

It is writer's experience that the need for insecticides arises only because of some failure of the normal control measures such as allowing a large surface area of wastes to remain uncovered, or the use of a very porous or thin cover which allows larvae to reach the surface.

Rodents. Rodents may be attracted from surrounding farms, or they may be delivered to the site in a load of wastes. A clean, unbroken and relatively smooth surface of covering material allows a burrow to be instantly detected and recognized. Regular inspection should be made and if the beginning of rate infestation is observed, a baiting program should be carried out by an experience exterminator. It is normal practice to use an anti-coagulant poison.

Airborne dust. During long dry periods the surface of a sanitary landfill can become very dusty, causing unpleasant working conditions for the men, and excessive wear of vehicles and plant. Watering carts are frequently used to reduce this nuisance; and waste oils are suggested in "Sanitary Landfill Design and Operation ":

"Periodic treatment or multiple spraying at a rate of 1-4 litres per sq. metre may be necessary. After several treatment, a packed, oily, soil crust usually develops that has good resistance to traffic abrasion and is moderately resistant to water. Good penetration by the oil can be expected in more permeable soils. Clayey soils or tightly knit surface may resist penetration in which case it may be desirable to scarify the surface lightly, apply-oil-and compact the surface".

Airborne litter. Airborne litter is often a major nuisance in the industrialized countries, and during gales at desert sites in countries having a significant paper and plastic content in the wastes. In the latter case environmental degradation of this kind has been observed extending over many sq. km. It is helpful to keep the size of the working face to a minimun, but this does not prevent litter being carried away during the period when the wastes are being discharged from a vehicle. It is necessary, therefore, to employ a movable barrier of perforated screens to prevent litter being blown off the site. Convenient dimensions for a screen are 1/2 metres long and about the same in height. The frame can be made from timber or light steel angle. To enable the screen to stand unsupported horizontal transverse members are attached by brackets to each bottom corner, these should be about 1 metre long. The whole area of the screen is covered with wire netting (chicken wire) of 20 mm to 40 mm gauge. Enough screens are necessary to form an area sufficiently long to catch everything blown from the working area. They must be cleaned regularly if they are to remain effective, and in very high winds they may require to be temporarily supported by guy ropes secured to pegs driven into the ground.

Fire. Fire at a sanitary landfill can arise from innumerable causes: hot ashes in a vehicle delivering wastes; a cigarette thrown down by a worker; the sun's rays through a fragment of glass on the surface. With some kinds of wastes the consequence of fire may be very serious and underground fires have been known that ultimately caused the collapse of the surface into voids caused by the fire. The high density, high vegetable, wastes of developing countries are unlikely to give rise to risks of this kind but is important to try to avoid fires and to be prepared to deal with them if they arise. Workers should not be allowed to start fires to dispose of bulky waste such as tree loppings, and private scavengers, who may use open fires for cooking, should be prevented from obtaining access, especially outside normal working hours.

It is useful to have a supply of water on the site: the water cart used for laying dust could serve a double purpose. Once a fire takes hold, however, it is usually necessary to dig out the material which is burning before water can be applied effectively. In the early stages this can be done by long rakes but once a fire becomes deep-seated the excavation of a large area by bulldozer may become necessary.

It is important to add that the writer has never seen a deep-seated fire outside Europe. The wastes of tropical and developing countries have often been on fire, but the fires have invariably been on the surface and have never presented the problems of control that occur in Europe where wastes sometimes contain a high proportion of solid fuel such as cinder and coal fragments.

Birds. Birds are normally attracted to landfill sites and may cause risk to aircraft if there is an airport in the vicinity. This problem is exacerbated if the birds have routing flight path between the sanitary landfall site and another source of attraction, such as a body of water, and their flight path crosses the approaches to a landing ground. Many methods of discouraging birds have been tried, from cannon to falcons, but none has been entirely successful.

This probelm is one that does not appear to be controllable and the solution lies in the selection of sites which avoid creating the risk of bird-strikes on aircraft.

Surface-water pollution. The pollution of static water, ditches, river, or the sea can occur when a sanitary landfill adjoins a body of water. The normal source of the leachate causing this pollution is rain falling on the surface of the fill, percolating through it, and passing over an impermeable base to water at a lower level. Only a proportion of total precipitation emerges as leachate; some is lost by evaporation and transpiration. In the British experiments of 1961 an annual rainfall of 635 mm produced a leachate equivalent to 200 mm. In arid areas there may be insufficient rainfall to create a leachate, but in all other cases the production of leachate from rainfall is unavoidable during the early stages of filling. The quantity of leachate can be substantially increased when upland water drains across the site of the landfill, but the worst case is when a stream crosses the site. The solution to these problems lie in appropriate site engineering such as:

- diversion or culverting of all watercourses which flow across the site,
- diversion of upland water by means of drainage ditches along appropriate contours.
- containment of leachate arising from precipitation by the construction of an impermeable barrier where necessary, such as a clay embankment adjoining a river,
- grading the final level of the site so that part of the precipitation is drained across the surface, thereby reducing percolation below the level needed to produce a leachate.

Works of this nature will obviously add to the cost of a sanitary landfill project, but it is usual to find that when capital expenditure is spread over the life of the project the cost/ton is still very much cheaper than for any alternative method of disposal. Furthermore, some of these forms of expenditure, such as culverts or river walls, represent capital assets of continuing value when the reclaimed land is handed over for its final use, perhaps agriculture or recreation.

Ground-water pollution. Every proposed landfill site should be the subject of survey and evaluation by a hydrogeologist, and where there is risk of leachate passing directly to underground water which is the source of a public supply, or to wells, the site must be rejected. The following comments are designed only to indicate the nature of this problem and do not pre-empt the judgement of the local hydrogeolodist.

Potential sites can be divided into three main groups :

- impermeable sites,
- sites on which such as sand gravel may provided a measure of filtration,
- fissured sites from which a leachate may be able to travel many kilometres.

The classic impermeable site is a clay pit from which there are no pollution risks because the leachate cannot escape.

On sites which provide filtration the Britsh Technical Committee reported as follows :

"Experiment filtration at a rate of about 30 cm per day of the polluted percolates through anaerobic filters of graded sand and gravel produced remarkable results. In a distance of about 7.3 metres practically all bacterial pollution and most of the organic pollution was removed. This was confirmed in full scale tipping.."

"The experiments.. indicated that sub soils such as sand and gravel would act as an effective filter for bacterial and organic contaminates. For such contaminates as chloride and sulphate, reliance would mainly have to be placed on dilution but generally it is considered this will be sufficient in many cases when the quantity of water abstracted from an underground source of water supply though to be at risk is related to the quantity of percolate......"

On this question the following observations are made in "Sanitary Landfill Design and Operation" (EPA,1972):

"It is a basic premise that groundwater and the deposited solid waste not be allowed to interact. It is unwise to assume that a leachate will be diluted in groundwater because very little mixing occurs in an aquifer since the groundwater flow there is usually laminar".

"When issuing permits... many States require that ground–water and deposited solid wastes be 0.6 to 10 metres apart. Generally a 1.5 metre separation will remove enough readily decomposed organics and coliform bacteria to make the liquid bacteriologically safe. On the other hand, mineral pollutants can travel long distances through soil or rock formations. In addition to other considerations, the sanitary landfill designer must evaluate the ; (1) current and projected use of the water resources of the area; (2) effect of the leachate on groundwater quality; (3) direction of groundwater movement; (4) interrelationship of this aquifer with other aquifers and surface waters."

Where there is doubt as to the charateristics of a proposed site the Bristish Technical Committee suggests the following precautions:

(*a*) Waterproofing the base of pit prior to tipping, i.e. with puddled clay (or chalk), collection of the percolate and its disposal...

(*b*) Rapid tipping of successive layers of refuse in order to utilize its capacity to absorb rainfall and to cover the final layer with a waterproof earth cover with a good cross-fall to provide a ready run-off of rainfall".

Similar advice is given in "Sanitary Landfill Design and Operation."

"An impermeable liner may be used to control the movement of fluids. One of the most commonly used is a wellcompacted natural clay soil, usually constructed as a membrane 30 to 90 cm thick. It must, however, be kept moist.

"The use of an impermeable barrier requires that some method be provided of the contained fluid."

Many developing countireis may find the collection and treatment of leachate a difficult and costly proccdure and one which has no residual value in relation to the ultimate use of the land.

The recommended approach is to reject sites at which problems are revealed by hydrogeological survey. The major pollution problem in the developing countries appears to be the contamination of

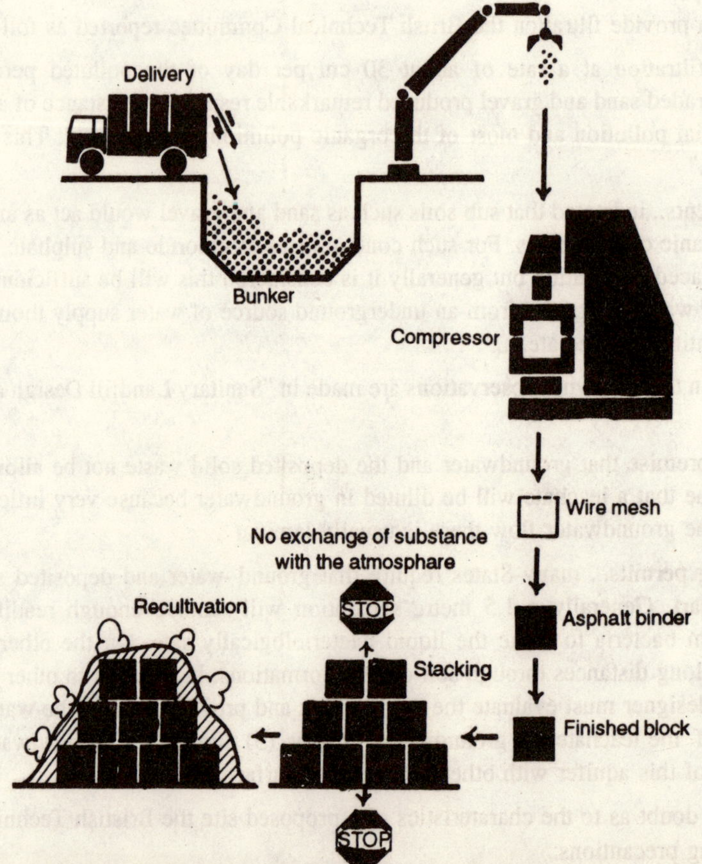

Figure 15.2. Deposit of refuse Blocks.

surface water by run-off and this can be controlled by relatively simple site engineering.

Disposal in "Refuse Blocks"

For many years a procedure has been in use by which the refuse is tamped with special wheels or some similar heavy deposit device. A machine has recently been developed in Japan whereby the tamping is effected much more efficiently.

Under the new system the refuse is fed into a bunker and, by means of a hopper, to a press. On arrival the waste is compressed from all sides by a tamping machine exerting a pressure of 200 kg per square meter. A cube of about 1 cubic meter is formed, and from the upper plate of the press smaller cylinders effect further compression inside the cube, with the result that the whole block forms a solid mass that will not disintegrate of its own accord. By this means any ''spring return'' effect is obviated. and the compressed material retains its existing condition from any subsequent expansion. After removal of the pressure plates the cube is so firm that it can be removed by a crane .

In order to bind the smaller, loose external in the block, the whole is enclosed in a wire mesh and subsequently in an asphalt mastic. After pressing, a block of 1 cubic meter should have a weight of about 1.2 tons. This represents a compression ratio of between 1 : 6 and 1 : 10, assuming an initial density of 150-200 kg per cubic meter.

According to a report from Japan, as this block continues to season, a gradual conversion should occur into a peatlike substance. Provided this happens it is to be regarded as an environmentally acceptable method of waste disposal. The procedure also offers considerable possibilities for the transmission of large quantities of refuse from densely populated congested areas to outlying dumps. The blocks can be loaded by cranes without difficulty onto standard open trucks or railway cars. They

Figure 15.3. Dump compressor.

can thus be carried long distances to refuse dumps without causing nuisance from smell, dust and the like in transit. At the dump the blocks can be stacked on the ground without the need for installing expensive large-scale handling machinery and providing covering material, which is often difficult to secure and can take up a lot of space.

A further advantage of this deposit method is that no substances from the virtually gas and water-proof refuse blocks can come into contact with the environment. There will consequently be no pollution of groundwater by seepage from the deposit, nor any spontaneous combustion of the material by heat and gas generation from bacterial decomposition processes. In this way the recultivation of the enclosed deposits and/or their employment in stacks is considerably eased. Furthermore, as is normally the case with an organized deposit no settlement will occur. The refuse is excluded for long periods from the biosphere and from circulation in it.

For regions in which depositing as a process of refuse disposal is standard, this method developed in Japan is deserving of closer attention.

The Dump Compressor

For many small and medium-sized municipal authorities that are responsible for disposing of the refuse, combustion and composting may prove too costly on account of the heavy investment and incidental expenses involved, especially if insufficient refuse is available to justify the costs of large-scale plants. It is not always feasible for adjoining districts to combine in the erection of a central disposal platn, if only because transport costs are increased sharply as road distances are extended. For many municipal districts this leaves the deposit method as the only viable alternative.

Straightforward dumping domestic refuse presents serious dangers of both air and water contamination, which can be avoided only by the workmanlike incorporation of the refuse in compact deposits. For the purpose a refuse compressor has been developed to distribute, crush and compress the refuse spread over the dump. The refuse compresor is provided with an earth-moving blade, which in some models takes the form of a loading scoop, which serves to produce an even distribution of the refuse. The earth-moving blade is wider than the compressor, and so keeps the machine free from loose refuse. The second functional feature of the compressor is the tamper or lacerating wheel. By means of the tampers high pressures are exerted on small surfaces, whereby the refuse is crushed. Tamper wheel compress the refuse better than lacerating wheels, but they have by no means as good a reduction performance. If large quantities of long—fiber special refuse are deposited—plastic sheets, for instance— it is better to fit the compressor with lacerating wheels than with tamper wheels, as the latter get tangled up more easily. To provide a solution for special problems, combinations of tamper wheels and lacerating wheels are also possible. The compressor also has a high climbing performance, so it is unimportant whether the organized dump is sited on slope or in a hollow.

By the constant application, crushing and compression of the refuse better use can be made of the dump site. The smoldering which can be caused by the formation of methane gas in loosely packed refuse is obviated by the compression. Plagues of rats and the growth of weeds are kept in check, because the rats cannot build breeding grounds and weeds cannot germinate. Settlements that occur in deposits (owing to the rotting of organic substances when the refuse is not compressed) are not completely eliminated by compression with tamper and lacerating wheels, but the uniform consistency of the material keeps them at a controllable level. Dumps treated in this way can be planted to woodland once the operations are complete and used as parkland or for leisure purposes, without any disadvantageous effects being likely on the plant stands.

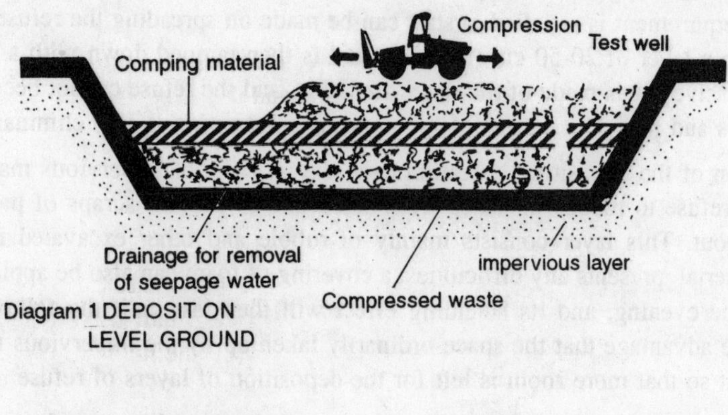

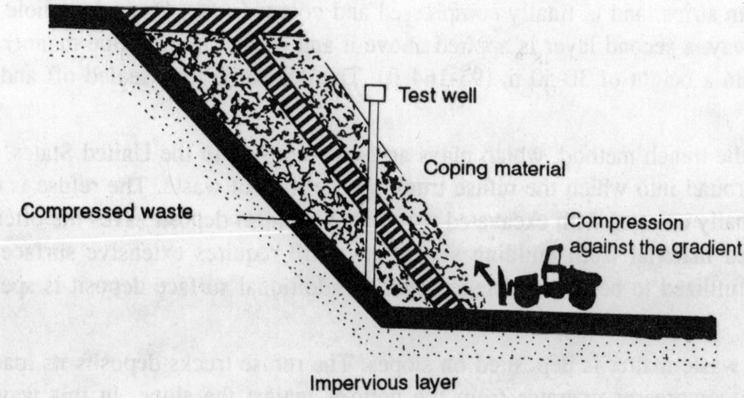

Figure 15.4. Deposit on sloping ground.

The refuse compressor has a very high output performance in relation to the manpower required for its use. One man with one machine can distribute and compress between 300 and 500 cubic meters of refuse per hour according to engine performance. As a rule the same worker can also handle without difficulty the other aspects of the operation, such as supervision and bookkeeping. It is unusual today for a modern organized dump to be run without mechanical equipment. With its aid all the tasks of distribution, clearance, crushing, homogenization, loading and recultivation can be carried out at relatively low investment and incidental costs.

Organized Dumps

The predominant part of household and industrial waste will continue to be disposed of in organized dumps. This is mainly due to the fact that depositing it in this way is the cheapest method. Investment, labour and incidental costs together amount to roughly one-tenth of the costs incurred in other disposal processes. Organized dumps have the further advantage that problematic refuse (special waste) can also be disposed of in this way.

When refuse is disposed of by this method it is important to ensure that the body of the deposit does not come into contact with groundwater, that heavy metals and organic pollutants do not get washed away. Provided this requirement is satisfied, a start can be made on spreading the refuse in layers. The refuse is spread out in a layer of 20-50 cm (8-20 in.) and is then tamped down with a compressor. In this way the most intensive use is made of the space available, and the refuse cannot become a breeding ground for insect pests and rats. The danger of spontaneous combustion is also eliminated.

Upon completion of the depositing operation each day, a layer of impervious material is spread over the compressed refuse to reduce nuisance from smell and to prevent scraps of paper and plastic from being blown about. This layer consists mainly of rubble and other excavated material. If the application of this material presents any difficulties, a covering of foam can also be applied by spraying over the deposit in the evening, and its shielding effect will then last until the following day. This procedure also has the advantage that the space ordinarily taken up by the impervious material can be saved to a large extent so that more room is left for the deposition of layers of refuse.

In the case of surface deposits, which are normally applied on level terrain, the refuse coming in each day is laid out in strips, and is finally compressed and covered over. Once the whole surface has been treated in this way, a second layer is applied above it and treated in the same manner. The dump so produced can attain a height of 30-50 m (95-164 ft). The deposit is then sealed off and the land is cultivated again.

A variation is the trench method, which plays an important role in the United States. A trench is excavated on level ground into which the refuse truck tips its load of waste. The refuse is compressed by machine and is finally covered with excavated material. This form deposit saves the often expensive transport of excavated material from building sites, but it still requires extensive surface areas. The surface areas can be utilized to better advantage when an additional surface deposit is spead over the trenches.

In hill country waste matter is deposited on slopes. The refuse trucks deposits its load at the top of the slope and the compressor operates from the bottom against the slope. In this way a uniform consistency is attained. The covering material is applied from the bottom upward.

A relatively neat process is achieved by what are known as "rot" deposits. In this case refuse is pulverized in suitable machines and left to rot in stacks on the ground. When the rotting is complete, the waste is leveled off and compressed as previously described. The rotted deposit has the advantage that hardly any further decomposition will occur, so that subsidence is largely avoided. Owing to the settling by rotting, furthermore, considerable dumping space is saved.

Pyrolysis

Certain wastes can be pyrolyzed to provide chemicals. Pyrolysis has been an operation using intense heat to cause chemical changes but not combustion. Trash rich in paper (as most trash is) and other organic materials yields combustible gases when pyrolyzed and the gases can be burned for fuel. About 180 million rubber tires have been discarded annually. Firestone Rubber Company has found that the pyrolysis of rubber tires can yield two marketable products, a petroleum-like mixture of gases and liquid (55%) and a carbonized residue (45%)

Incineration

There have been several hundred municipal incinerators in use accounting for about half the tonnage burned. True incineration involves the burning of solid wastes at high temperatures, to leave ashes, glass, metal, and unburned combustibles amounting to perhaps one-fourth the original weight, which must then be disposed of in a landfill or other dump. Air pollution has been often a problem, because of their emissions of pollutants into the atmosphere. Incinerator technology has been developing in recent years, with air pollution control a particular concern.

The important factors to be observed have been the maintenance of a minimum temperature of 850°C in the combustion chamber to burn up the smoke produced, and the presence of grit extractors to remove finely divided dust before it gets discharged into the atmosphere.

Incireation reduces the volume of waste by 60 to 80 per cent and reduces the public health problems associated with the accumulation of refuse. This method is adaptable to a wide range of capacities from small domestic incinerators to large centralized municipal plants. It can handle a mixture of garbage and rubbish, and the physical nature of the residue—called clinker—aids in the disposal of the waste. Much of the heat that is generated can be recovered with modern equipment. However, other forms of pollution are created. Air pollution from the products of fuel combustion must be taken into account, but the principal air pollutant is fly ash. Several methods entails the use of water with devices called scrubbers; but the large amount of water needed introduces a water pollution problem. The increasing quantities of plastic products and plastic packaging materials aggravate the air pollution problem due to the formation of gaseous decomposition products that will require research on methods for special treatment of the stack gas. The ultimate requirements have been difficult to predict.

Use of small incinerators

This method of disposing of refuse is widely used but has not been found wholly satisfactory. The reason is that many of the plants used are poorly designed and cause serious local air pollution. Many states have now laid down stringent design criteria for incinerators and some ban small-scale incinerators absolutely. There is however, one reason why small-scale incinerators should be inefficient, and there are many on the market now which are completely satisfactory. The important factors to be observed are the maintenance of a minimum temperature of 850°C in the combustion chamber to burn up the smoke produced and the presence of grit extractors to remove finely divided dust before it is discharged into the atmosphere.

The use of waste disposal units

Waste disposal units in kitchens can normally deal only with some 10 per cent by weight of refuse generated in a dwelling. There are considerable limitations regarding the types of materials that can be placed into them. Plastics, textiles and even fibrous food remains can clog them up—in fact these are potentially more harmful than glass bottles which are crushed quite satisfactorily. The groundup remains are then carried by the sewage system to the sewage works where they are removed in standard thickener equipment.

Controlled Tipping

By this technique, which is still by far the cheapest in many places, refuse has been permanently deposited on land, sealed in cells isolated from the surface with layers of earth, Controlled tipping is generally practised in conjunction with land reclamation schemes, where it serves the usefull purpose of reclaiming marshes, exhausted quarries and gravel pits for agricultural, industrial and recreational

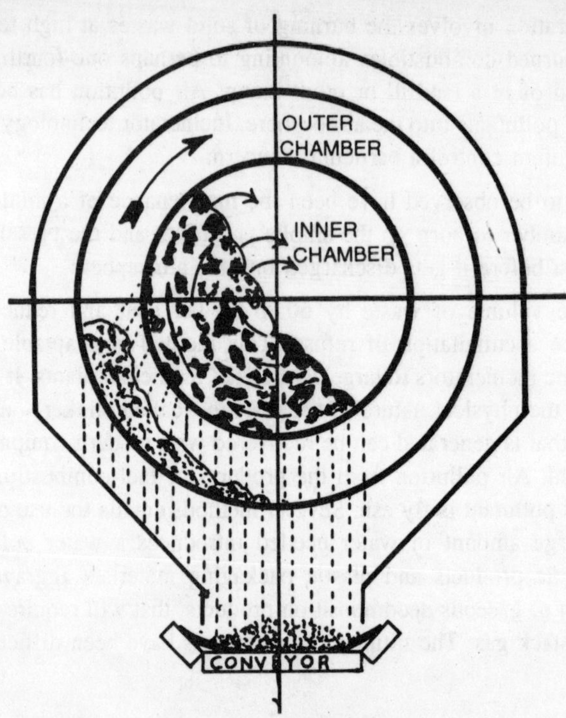

Figure 15.5 Rotating drum pulverizer.

uses. But, on the whole the availability of areas suitable for such controlled tipping close to the big cities, has been diminishing. Furthermore, controlled tipping, unless very carefully carried out can soon create a considerable nuisance and even a danger to health. Some local authorities demand that refuse is pulverized prior to disposal and also specify the minimum thickness of earth cover to be used.

Today controlled tipping is no longer regarded as good solution to the problem of refuse for a number of reasons. There is far less heavy material like cinder and ash in refuse today there used to be. Instead, refuse has more lightweight materials such as paper and, particularly, polythene film. In addition, plastics in the refuse do not rot and therefore remain unchanged in the soil for ever. This implies that the subsequent usefulness of the soil for agricultural purposes has been less than it used to be when all the components of refuse were able to rot fully.

Local authorities with the ratepayers looking over their shoulders, have to decide which method of waste disposal they are going to select. From a very short-term point of view tipping usually costs somewhat less than incineration, but with the rapidly increasing cost of land this is unlikely to be for very much longer. Unfortunately. once the decision has been made to look at the problem from the short-term point of view; only the damage has been done and much of the land has been ruined for agriculture.

Pulverization of Refuse Combined with Controlled Tipping

Pulverization of refuse prior to controlled tipping obviously adds to the cost of refuse disposal but has many advantages :

(*a*) There is far less likelihood that pieces of rubbish blow away, as happens with normal tipping practice;

(*b*) Less earth cover is required than with unpulverized waste ;

(*c*) Rats, mice and insect larvae get destroyed ;

(*d*) The rate of rotting is very much accelerated so that less time elapses before the material has been converted into humus.

Three types of pulverizing machine are in common use:

Rotating Drum Machines

These pulverize the refuse by attrition. Water is added to the crude refuse and the mix is then passed into the revolving drums which have either a circular, an octagonal or a hexagonal cross-section. The drums operate either on a batch or on a continuous basis, but the number of revolutions and the time it is necessary to have the refuse in the revolving drum to obtain good disintegration vary according to the detailed internal design of the plant.

There has been usually a separating device for materials which cannot be broken down—mainly plastics, rubbers and metals. The final material that comes out of the drums normally has a moisture content of 40-50 per cent and a density of 0.5-0.6 kg/dm^3 much higher than the density of the refuse originally treated. Such material can be tipped more easily and more satisfactorily than typical low density town refuse with its large quantities of bulky and loose materials. The disadvantage of wet grinding is the fact that the large quantity of added water increases transportation costs. Drumtype pulverizers usually have capacities of about 10-12 tonnes per hour and require 35 kw of power to operate. Magnetic separators have been always provided to remove iron and tin cans.

Hammer Mills

Hammer milling is operated either as a dry process or as a wet one, water being added to produce a denser product which is more suitable for tipping. The mills themselves vary considerably ; there are both fixed hammer and swing-hammer types of mills, and either single or double rotor models are available. The biggest hammer mills are able to process up to 30 tonnes of refuse per hour.

Hammer mills consist basically of a horizontal rotor, which carries a number of swinging or fixed hammers. There are impact plates and grates through which the processed refuse materials is finally ejected. For refuse treatment the grate bars, which determine the final size of the pulverized particles obtained, are rather far apart so that on average the maximum size of the pulverized refuse is about 5.8 cm in diameter. Feed openings with many of the plants used for refuse pulverization are up to 1.2 m × 2.3 km, so that even such large items as old chairs and old sofas can be conveniently torn up into sizes suitable for controlled tipping. Metals are also fed into the pulverizers and shredded without difficulty. Iron and steel components are recovered on the discharge conveyor belt by magnetic separators, Hammer mills require very powerful motors, to cover instantaneous peaks when they are fed with large awkward units such as refrigerators or old gas cookers. Motors with a capacity of 150-200 kW are recommended. On average the power consumption is fairly modest, amounting to about 4 kW per tonne processed per hour.

The Dutch Rasp

This consists of a steel drum which has a perforated end, which is fitted with hardened steel pins.

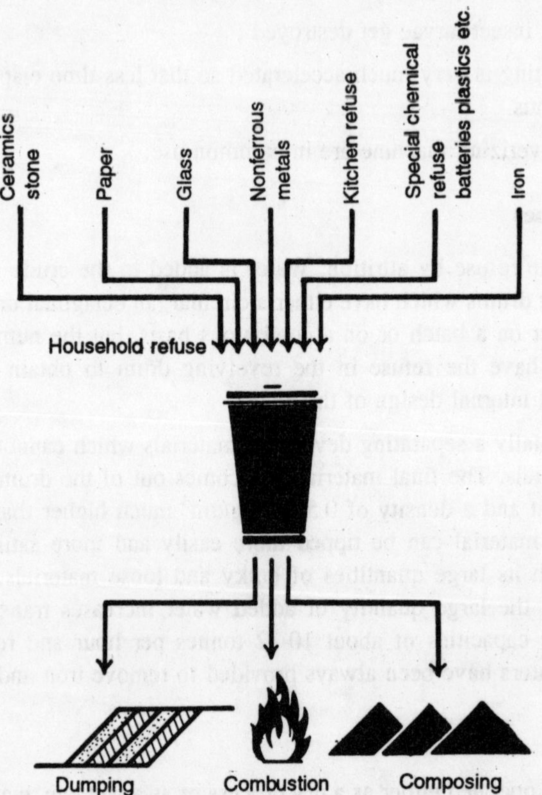

Figure 15.6. Modern Waste Disposal.

The refuse is led past these steel pins and the ground-up components drop through on to a conveyor belt beneath. Hard and bulky objects which cannot be reduced in size in this way finally come out at the side. Plants are made capable of dealing with 10 tonnes of refuse per hour; they have a power requirement of about 60 kW and operate at about 5 revolutions per minute. The disadvantage of the Dutch rasp is that the plant is unable to deal with harder substances and is consequence there is a fairly high proportion of material which does not pass through the size reduction process.

Preliminary Sorting of Domestic Refuse

The three main methods employed at present for the treatment of refuse from settled areas–dumping, combustion and composting—call for no preliminary sorting of the refuse which can be treated as it stands. It must be recognized, however, that without preliminary sorting, many valuable raw materials such as paper, iron and nonferrous metals, and glass are destroyed, and their value is thus lost to the economy. Without preliminarly sorting, moreover, there is a danger that noxious substances may be left in the dumps or in compost and cause considerable damage.

The elimination of particular refuse elements in the treatment process is also not without its

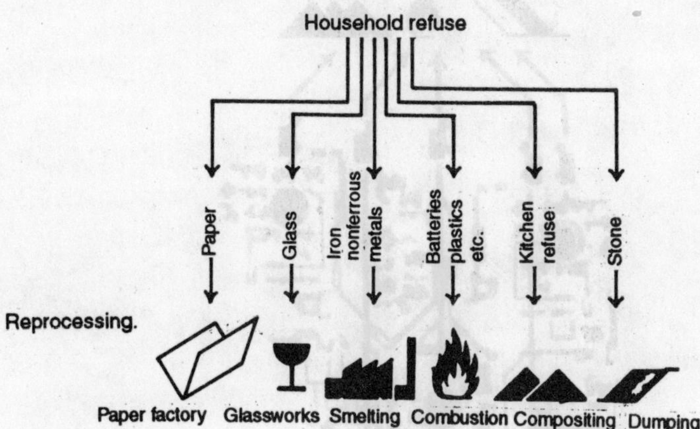

Figure 15.7. Preliminary Sorting of Household Refuse.

difficulties and calls for heavy expenditure on machinery and labour. It thus has to be considered whether separate treatment of the individual elements in domestic refuse is of advantage in relation to the practice follwed so far.

A separate disposal bin for each type of domestic refuse would as a role, be out of the question as it would be far too difficult to supervise and would be uneconomic . If the constituents of domestic refuse are assembled in a few well-defined subgroups, however, and provided reasonable control and discipline are exercised, this solution should present no difficulties.

The most important subgroups for waste from settled areas would include organic kitchen refuse; paper and board; glass ; stone, ceramic ash and other noncombustible material groups ; plastics ; and iron and nonferrous metals. Altogether there are six subgroups. the separate assembly of which is of some economic importance. First organic kitchen refuse, because of its susceptibility to putrefaction and the nuisance it can cause for the population through smell, should be collected more often and if possible converted into compost. All other waste is, for the most part, chemcically and biologically noncorrosive and consequently need be collected less frequently. Second, paper can be reused as a raw material in paper production or, if it is not required for that purpose can be used for energy production. Third glass is similarly reused as a raw material in the manufacture of low-quality glassware (glass bottles). Fourth, iron and nonferrous metals can be handled together, as their magnetic separation presents no technical difficulties. These, too, represent a valuable raw material which can be used again in engineering processes. Material in the ceramic stone and ash group can be left in dumps without hesitation as no chemical or biological reactions are to be expected in this case. Finally plastics that cannot be reutilized for technical production purposes can be used for energy production.

The separate assortment of refuse in groups certainly entails difficulties in the home and for collecting and disposal organizations, but these can be largely overcome by recent advances in the construction of suitable collecting bins. In any case, this form of refuse disposal and the related

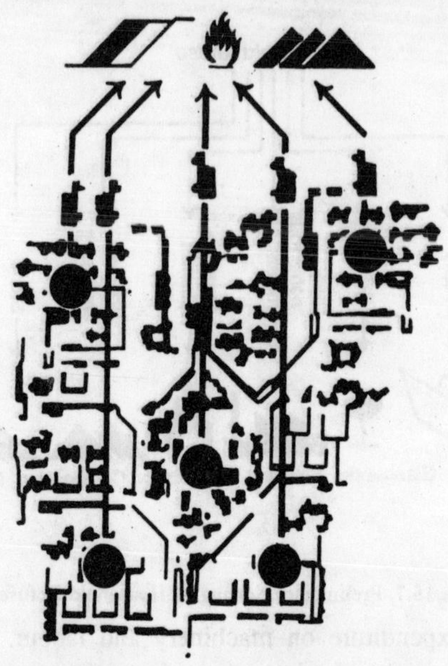

Figure 15.8. Centralized Refuse Treatment.

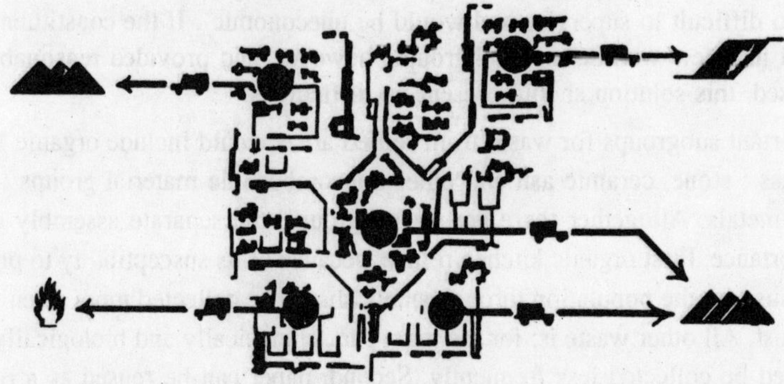

Figure 15.9. Decentralized Utilization of Refuse.

reallocation of parts of the residue to production processes considerably eases the load carried by disposal plants. It is therefore desirable that the steps already taken for the separate collection of old papers and aluminium cans should be continued and extended to other groups of materials.

Disposal of Domestic and Industrial Refuse in Industrial Agglomerations

Social changes in the past twenty years have resulted in the inhabitants of big cities moving out to surrounding areas, Population centers have consequently been extended, and the cost of waste removal has risen accordingly, owing to the greater transport distances involved.

Local authorities responsible for refuse disposal have three main pocedures to choose from. The cheapest of these is to deposit the refuse in a dump. As dumps may no longer be opened up at random and left unattended because of the threats they pose to the environment, big centralized dumps are now provided to take over the duties previously shared by a number of smaller ones. The refuse consequently has to be transported over long distances, which not only entails heavy expenses but has other diadvantages connected with increases in traffic. It is not uncommon for a city to have to handle well over a hundred truckloads of rubbish a day, for which the distance has to be covered twice (there and back). The burden could be eased only by installing a number of smaller dumps around the area, but this comes up against the difficulty that sufficient sites are not usually to be found that comply with the legal requirements. Furthermore the technical and mechanical equipment required costs as much to run and maintain on a small dump as on a large one Staffing expenses are also disproportionately high on a small dump compared with a large one. If waste disposal has to be decentralized, the dump is thus not a favourable solution.

By decentralizing waste disposal, the long distances to be covered from the point of origin to the treatment point can be avoided and the costs of transport and staffing can be reduced. Two other methods of waste disposal merit consideration. Disposal of the refuse by burning (see later) is a practical solution provided the sites are well removed from residential dwellings and all necessary safety measures are complied with. Composting the refuse (see also later) meets the reequirements of decentralization almost ideally. To judge by experience so far, compost plants operate to the best advantage in communities on the order of 60,000-80,000 inhabitants (the smaller handicraft and manufacturing undertaking in the area being assessed, as regards their refuse output as equivalent to a particualar number of inhabitants). In addition to refuse sewage sludge (see also later) can also be treated by the composting process, thereby solving a particular problem facing every city authority. The compost, which is the end product of the process, is best disposed of in the vicinity of the compost plant. In the case of a decentarlized plant (provided it is not faced with too heavy transport costs), the compost may be disposed of even more easily because of the extensive areas that should be available in the neighborhood.

Ocean Dumping

Ocean dumping of wastes, from garbage to radioactive wastes and discarded chemical warfare agents, has been, in the past, a favourite disposal method under the prevalent notion that the ocean is an inexhaustible sink. We are now learning that the ocean does not have an infinite capacity for covering, holding, absorbing, or decomposing materials, nor the capacity for anywhere near the load of pollution that would result if the rate of ocean disposal were continued as in the past.

A study made in 1968 identified 126 ocean disposal sites used by twenty coastal cities. There were seven major categories of waste: harbour dredgings, industrial wastes, municipal sewage sludge, refuse and garbage, construction and demolition debris, military explosives and chemicals, and miscellaneous wastes—a total of 38 million tons per year in coastal and estuarine areas. Other sources of waste include agricultural land run-off pesticides,. fertilizers, sewage and garbage from vessels and oil spills. Often overlooked is the fact that a large part of the air-pollutants eventually setlle out and end up in the ocean. A major concern is that ocean dumping has increased fourfold during a period of twenty years and is expected to greatly increase relative to other means of disposal in the future. Based on a 1970 study, the Council on Environmental Quality recommended a series of drastic actions to control ocean dumping.

Little is known about either the immediate results or the cumulative effects of ocean disposal. By the time adverse effects become evident, it may be too late to discontinue the disposal practices in time to prevent long-term pollution problems. We have a preview of what could happen in the case of DDT, plasticizers such as the poly chlorinated biphenyls (PCBs), and more long-lasting and potentially dangerous substances such as lead, mercury, cadmium and selenium. The amount of wastes transported and dumped into the ocean is now relatively small compared to the total volume of pollutants that reach the oceąn, Industrial and municipal liquid wastes are the largest source of pollutants.

The occans cover 70 per cent of the earth's surface. The 140 million square miles of water surface are crucial to the equilibrium of the world's atmospheric environment—the oxygen and carbon dioxide balance and the global climate. The oceans are a major part of the earth's hydrologic system. They are an important part of the biosphere, supporting a rich biota of marine plant and animal life, and are economically economically valuable to man for the production of food, minerals and other necessities.

Some coastal cities dump solid wastes into the occean. New York formerly dumped wastes off the New Jersey short but the State of New Jersey in 1933 won a Supreme Court decision forbidding this practice. New York still dumps wastes over an area several square kilometeres in size 20 km out into the Atlantic Ocean ; this area has been described by critics as a "dead sea" of muck and black goo, largely because of the sewage sludge disposed there.

The solid and liquid wastes are dumped into the ocean by means of barges and pipes from shore. Raw sewage and other wastes pumped through underwater pipes to offshore locations are included among materials dumped into the ocean, but only in a broad sense of the term.

Dredge spoils make up the largest single category of wastes dumped into the ocean, amounting to 6-7 times as much tonnage as all categories combined. Dredge spoils are material removed from a channel to maintain or improve navigation, and they consist of rocks, detritus, clay, silt, sand, and municipal or industrial sludges.

In 1968, nearly 90% of 4.5 million tons of sewage sludge deposited in the Atlantic came from the New York area, and nearly all the rest came from Philadelphia. Sewage sludge is a by-product of the treatment of waste water. Most sludge is either incinerated or dumped on land.

Industrial wastes include many heavily polluted liquids such as waste acids, refinery wastes, paper mill wastes, pharmaceutical wastes, pesticides, and many others. Those that are especially toxic are packed into drums that are dumped well offshore into deeper zones. The U.S. Army Chemical Corps once disposed of some poison gases by loading them into an old cargo ship and sinking the vessel into a deep ocean trench, Some radioactive wastes used to be dumped into the ocean, but that no longer is done by the United States. Some industrial wastes are burned in special incinerator ships cruising offshore.

Construction and demolition debris, a minor fraction of all wastes dumped into the ocean, consists of building stone, masonry material, pipe, wood, and cellar dirt for which on land disposal sites may not be economically accessible.

Solid wastes—garbage, trash and other refuse picked up by trash removal operations in coastal cities—are not disposed of in the ocean to any great extent.

Problem of Ocean Dumping

Very little is comfidently known about the effects of ocean dumping on marine life, but much is suspected and feared. Because the wastes include organic matter that is degraded by organisms requiring oxygen, they cause some oxygen depletion in the waters at the dumps sites. The water column at the dump site may uninhabitable by fish or shellfish, but natural mixing action and surface aeration eventually restore the oxygen. Solid wastes that are not biodegradable change the habitats for fish in the dump areas. The most serious threat to fish and shellfish is the presence of a number of toxic metals, pesticides, and other poisons such as the polychlorinated biphenyls (PCBs). Sediments in the New York Bight, for example, contain abnormally high concentrations of copper, chromium, lead, zinc and nickel. Little is known of the long-range effects on marine life of exposure to toxic substances.

No part of the New York Bight where ocean dumping occurs is a biological "desert", a region devoid of marine life. Plankton, bottom-feeding organisms, and commercial fish and shellfish are present, but fin rot disease occurs among several species. Fin rot disease is a progressive erosion of the fin rays and overlying tissue starting at the outer edge and working to the base of the fin. The disease itself is a symptom that something is wrong with the water.

Sewage sludge and polluted dredge spoils usually contain bacteria, viruses and intestinal parasites. These are taken up by shellfish and consequently parts of the shellfish-producing wastes of the American coastline—about one fifth of the 10 million acres-are closed to shellfishing. Residents along Long Island Sound fear that bacterial contamination of waters off their beaches is occurring. Ocean currents circulate dumped material, but they do not carry the wastes in huge globs. Some of the Long Lsland beaches have been fouled, but the origin of the pollution has not been proved.

In 1972, a multi-government conference agreed to an International Convention on the Prevention of Marine Pollution by Dumping Wastes.

The EPA assumes that toxic wastes must be mixed and diluted, and the establishment of criteria on levels of toxic substances allowed in the ocean-dumped wastes has been intended to prevent changes in the ocean beyond those that might be expected from natural causes. For synthetic pollutants the criteria are designed to limit the concentration in the waste to 0.0001—0.01 of the lethal concentrations and in the ocean these levels would fall to much lower levels because of mixing.

A total prohibition has been placed in dumping biological, chemical, and radiological warfare agents, high-level radioactive wastes and substances that are persistent, nonbiodegradable, or naturally floating (e.g., plastics). Solid wastes may not contain more than 0.75 milligram per kilogram of mercury or 0.6 milligram per kilogram of cadmium; and liquid waste are limited to 1.5 and 3.0 milligrams per kilogram of mercury and cadmium, respectively. Oils and greases may not produce a visible seen of the surface at a dilution of 1 : 100. Organohalogens (*e.g.*, certain pesticides and the PCBs) are also sharply limited. Other strictly regulated wastes requiring special handling are those containing arsenic, beryllium, chromium copper, lead, nickel, selenium. Since many of the larger cities and metropolitan areas of the world are on near the coastlines, the ocean offers a ready disposal site for certain types of wastes when inland or coastal resources are difficult to obtain. We are therefore concerned with finding a methodology for solid waste disposal at sea which meets the following criteria:

1. The port facilities must efficiently handle a large number of trucks and also provide docking services for the sizable barges; material storage, reduction by grinding, baling, and encapsulation at these facilities must be done in an efficient, nuisance-free manner.
2. The transport of the compacted wastes through harbor waters must be in a secure manner: refuse shall not fall off even during stormy periods; and

3. The disposal at sea must be in an approved area; the bales must sink to the bottom and stay there, with no breaking off and flotation of particles; adequate electronic fix and photographic records of location and depth, as well as date and time of day, must be kept. Provision must be made for frequent or constant monitoring.

Disposal costs at sea, conforming to the above criteria, have been higher than for sanitary landfill but less for incineration. The method is therefore attractive.

The situation as it now stands with regard to disposal at sea could be summarized as follows:

1. Specific guidelines, criteria, or standards governing disposal at sea are not available at this time. Interim guidelines announced by several agencies specify a moratorium to develop pilot programmes and monitoring systems.

2. Dumping will probably be restricted to designated areas beyond the continental shelf. The water depth will be at least 1000 fathoms (6000 feet) and the site would be at least 25 miles from the nearest land.

3. Disposal of solid wastes at sea would be restricted to inert material which have a bulk density greater than sea water. (At 1000 fathoms this is about 70 lb/cu ft.) All other solid wastes having organic materials shall be packaged or containerized, transported, and deposited at an approved site using methods designed to protect the public health and the marine environment

Ocean as a Resource

The ocean is a ready dumping found for waste materials from the large cities located on or near the coastlines. Sewage sludge, so-called acid and industrial wastes, and explosives get dumped for years and are continuing to be disposed of at designated and approved locations (Table 15.3)

Table 15.3

Number of Marine Wastes Disposal Areas
(By Region and Waste Type)

Type of Waste	*Disposal, Areas*			
	Pacific	*Atlantic*	*Culf*	*Total*
Dredge Spoil	10	11	17	38
Industrial Wastes	7	9	6	22
Sewage Sludge		2		2
Refuse	3			3
Radioactive Wastes	6	11	1	18
Explosive Wastes	14	16	10	40
Total	40	49	34	123

However, the concern with the ecology of the sea and of the coastal zone and the recent activities of aroused citizens are bringing these practices under sharp scrutiny. Their is a gradual diminution of the life activity as the depth increases; the temperature at 6000 feet is steady at about 4°C, and the pressure is over 2500 lb/sq in. Since they are the bottom of the food chain, they are of vital concern, particularly in the long-range aspects. There must be no doubt that ocean dumping will have no harmful effect on the ocean ecology.

Conversation with biologists working in the marine environment disclosed some interesting facts of solid-wastes dumping. The submarine ''Alvin'' got accidentally sunk a few years back in a depth of about 4500 feet.When it was recovered about one year later, it was found that the apple and the bologna sandwich on the mess table had not deteriorated. This points up the theory that bacteria functioning under these temperatures and pressures are not amendable to such land-based foods. We expect that the methane-producing bacteria already in the solid waste will continue to function, but at a greatly reduced rate.

The ocean has been a resource which should be used. Of course, this use cannot imperil the ecology of the marine environment.

Possibility of Dumping Inside the Continental Shelf

Considertion has been given to disposing of baled refuse at locations closer in shore. The possibilities of contamination and damage to sea life are greater at such locations. Nevertheless, properly anchored bales could provide artificial reefs so necessary to the inprovement of fish habitats.

The moratorium previously mentioned relates not only to dumping on the continental shelf but also to dumping in the deep ocean (depths of more than 6000 feet). It is interesting to note that the distance from Boston, Massachusetts, to the nearest possible disposal area off the continental shelf is about 230 miles. The distance to the continental shelf from other East Coast metropolitan areas in much shorter—in most cases it is under 100 miles.

Disposal by Burning

Refuse combustion plants should be kept running without interruption over long periods, as otherwise erosion of constantly reheated and cooled-down parts by thermal expansion and contraction is too great. To ensure continuous operation, combustible material must be constantly available. This is achieved by means of a refuse bunker, into which the refuse vehicles empty their loads directly. The capacious refuse bunker also helps to ensure that when refuse comes in large quantities the plant will not be overloaded.

The refuse is transferred from the refuse bunker to the feed hopper of the furnace by means of a grab crane, without previous sorting, sifting or crushing. The removal from the bunker can be effected automatically or by hand control. The grab operator sits in a glass turret and does not come into contact with the refuse. Air to assist combustion is drawn from the refuse bunker, which lowers the air-pressure there and prevents dust or fumes from being driven out into the open.

The refuse is then loaded by a mechanical charger onto a grating, where it is burned into sterile slag. Throughout the burning process the refuse is turned over and raked about so that variations in its composition and fluctuations in water content and calorific value are evened out. The mechanical charger and the rake mechanism are driven by a pressure oil unit or an electric motor. Air is injected into the burning refuse from below to ensure that enough oxygen is present. A secondary airstream is provided at high pressure above the grating to make sure that any gases that form will burn off completely.

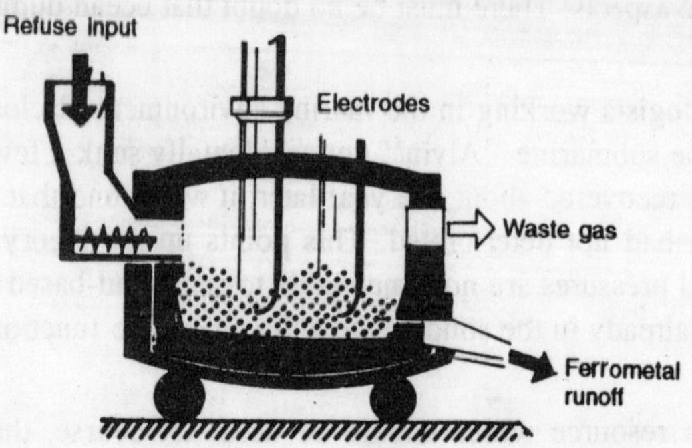

Figure 15.10. Electronic arc furnace for refuse smelting.

When the noncombustible elements leave the grating as slag, they are cooled in the water bath of the slag-removing plant at about 80°C (176°F). They are then ejected with a low residue of water in a sterile condition, free from dust and fumes. The slag can then be dumped or can be utilized on roadworks. If the operation has been carried out correctly, the slag will assume no more than 10% of the volume of the original refuse. Reduction of the refuse by combustion thus saves considerable deposit space.

Although the fluctuations in calorific value are considerable, combustion takes place normally at temperatures in excess of 800°C (1472°F). All the organic components of the combustion gased are burned at this level and the gases are virtually odorless.

A combustion problem arises, however, from the ever-increasing proportion of plastics in the refuse, especially in the form of PVC. Upon combustion of PVC, chlorine is released, which combines with water vapor in the outer air to form hydrochloric acid. This can only be counteracted by the installation of a gas purification plant. After leaving the furnace, the fumes are cooled, cleared of dust (up of 80%—90%) and discharged through a chimney into atmosphere.

To recover some of the cost of the operation, the heat generated by combustion can be utilized

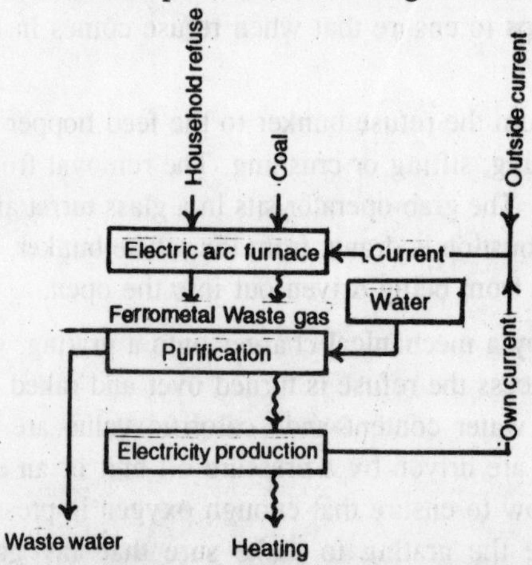

Figure 15.11. Substance and energy flow in refuse smelting.

to produce steam or hot water. Steam turbines can be used to generate electricity, and the hot water or steam can be used for heating.

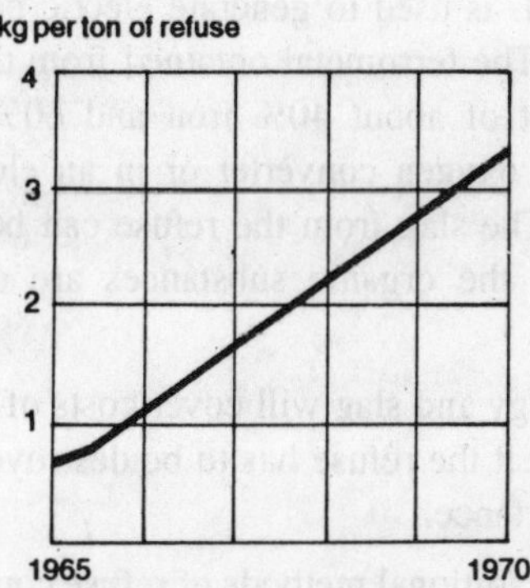

Figure 15.12. Specific hydrogen chloride emission from household refuse with an increasing

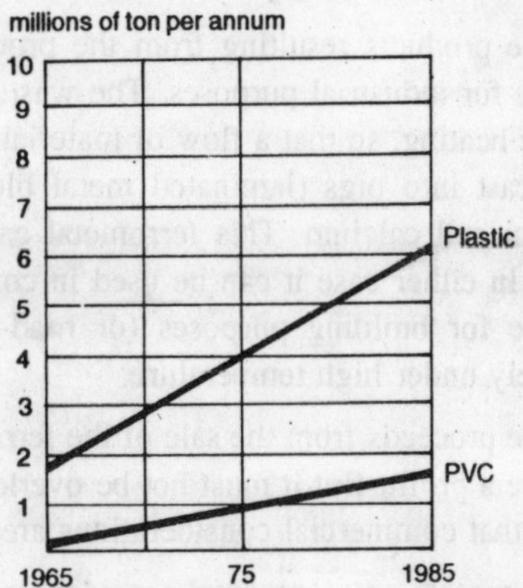

Figure 15.13 Consumption of plastics and PVC in West Germany

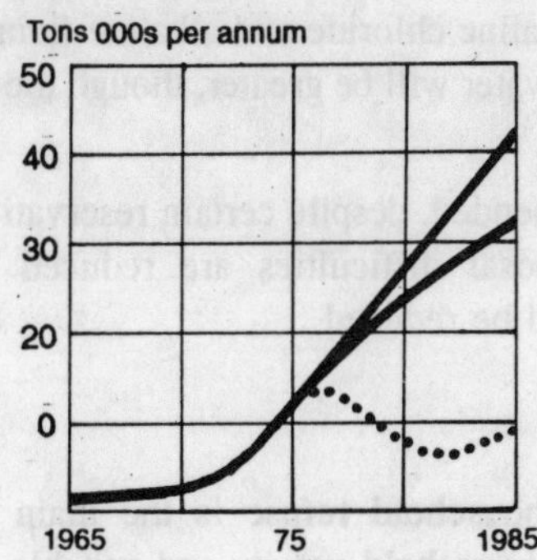

Figure 15.14 Emissions of hydrogen chloride from refuse combustion with and without waste-gas purification.

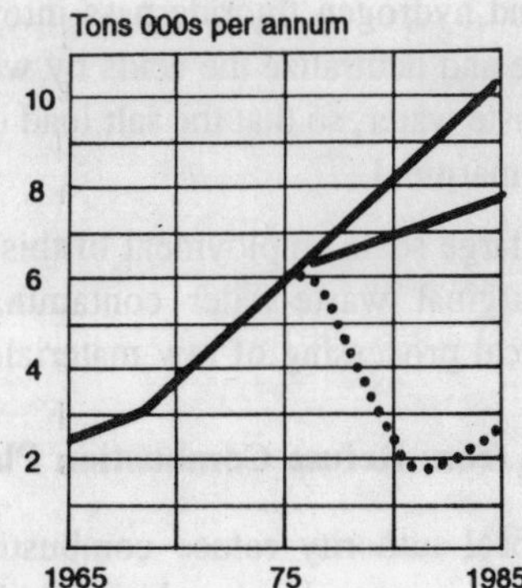

Figure 15.15 Emissions of sulfur dioxide from refuse combustion with and without waste-gas purification.

By burning the refuse, waste from settlement areas is decontaminated, the volume of dumped material is reduced, and the energy contained in the refuse is exploited in the interests of the community,

Refuse Smelting in a Electric Arc Furnace

Refuse smelting in a electric arc furnace is a process that contains the chemical cycle within more narrow limit than other types of refuse combustion. With this process the slag left behind by the refuse can also be utilized, whereas otherwise is usually has to be discarded.

Refuse that is to be smelted is treated in an electric arc furnace at temperatures 1500°C—1700°C (2732-3192°F), while directly beneath the electric arc, temperatures of 3000°—3500°C (5432°—6332°F) prevail. A metallurgical reduction is effected in the same way as in a blast furnace, so that it is fair to describe it as a smelting process.

The products resulting from the process are waste gas, ferrometals and slag, all of which are utilizable for industrial purposes. The waste gas in purified and is used to generate elecric power and domestic heating, so that a flow of material and enegy results. The ferrometal obtained from the runoff can be cast into pigs (laminated metal blocks). These consist of about 40% iron and 60% silicon, aluminum and calcium. This ferrometal can be cast either an oxygen converter or in an electric arc furnace. In either case it can be used in concrete production. The slag from the refuse can be used as aggregate for building purposes (or road-building), since all the organic substances are destroyed completely under high temperature.

The proceeds from the sale of the ferrometal, electric energy and slag will cover costs of the plant will leave a profit. But it must not be overlooked in principle that the refuse has to be destroyed in any case, so that commercial considerations are of secondary importance.

As regards environmental considerations, compared with traditional methods of refuse combustion, smelting in an electric are furnace presents few problems. All the organic compounds and agents of disease are destroyed completely. There are no solid residues to be deposited at considerable expense. The ferrometal and the slag are made use of. Dust, fumes and acid are eliminated from the waste gas. The clan gas can be used for electric energy production. In the refuse smelting process, hydrogen chloride and hydrogen fluoride pass into the waste gas, and at the same time alkalis are evaporated in the furnace and neutralize the acids by washing. In this way alkaline chloride and alkaline fluoride pass into the waste water, so that the salt load of settlement in waste water will be greater, though the increase should be marginal.

The large scale employment of this process is to be recomended, despite certain reservation (such as the marginal waste-water contamination), as refuse disposal difficulties are reduced and the unmethodical processing of raw materials (iron ores, etc.) could be reduced.

Emissions from Refuse Combustion Plants

In local-authority refuse combustion plants, in which household refuse is the main element consumed at present—but in which rubble, business waste, the household variety and suitable types of industrial waste will be consumed to an ever-increasing extent in future—the waste products should be reduced to a sterile condition no longer capable of putrefaction and, as far as possible, impervious to water. In the process both volume and weight are substantially reduced. During combustion, atmospheric contaminants are emitted in the form of dust and gases.

Almost all local combustion plants are equipped with electric filters to effect dust separation, the efficiency of which is over 95%. The emission of dust in this way from even large scale municipal plants, compared with the total amount of dust emitted from other sources, is very low.

Owing to the high firebox temperatures of at least 800°C (1472°F), combined with the normally high air surplus and the length of time the combustion gases remain in the firebox (where by organic compounds are largely consumed), the emission of organic compounds from refuse combustion plants is not of great importance, either now or for the future, The emission of fumes from refuse bunker can

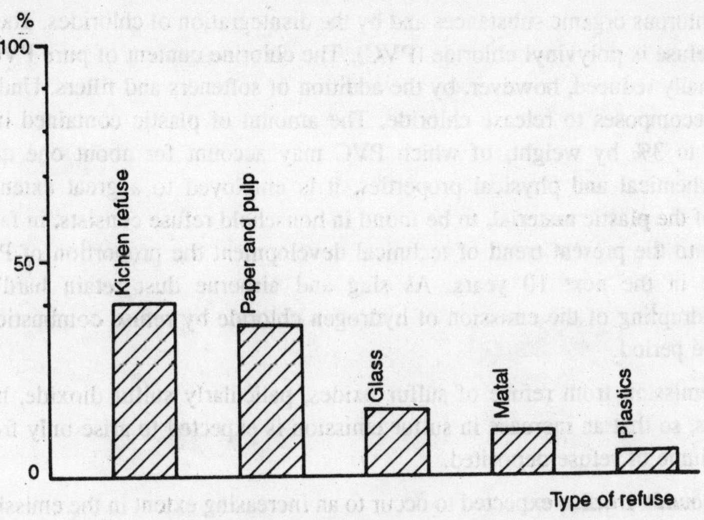

Figure 15.16. Composition of domestic refuse.

be sharply reduced by cutting down the retention period of the waste and by extraction of the combustion air by suction from the bunker.

The emission of gaseous inorganic contaminants, on the other hand, will present imcreasing difficulties in the future. Special importance attaches to chloride compounds, which are formed by the

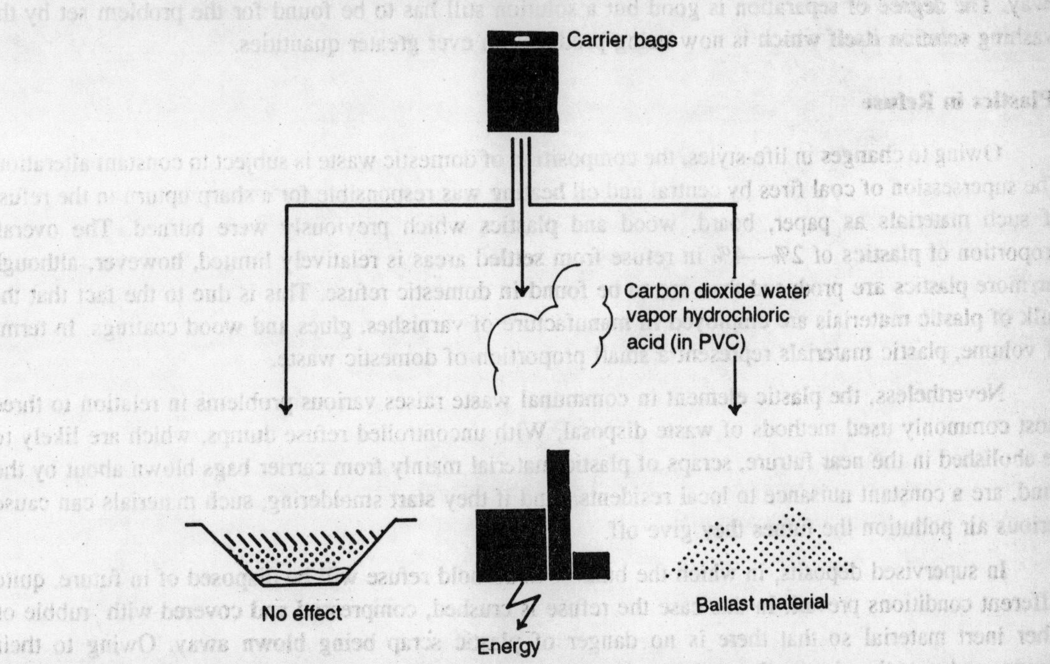

Figure 15.17. Effects of plastics in refuse in various manufacturing processes.

combustion of the chlorous organic substances and by the disntegration of chlorides. The most important chloride carrier in refuse is polyvinyl chloride (PVC). The chlorine content of pure PVC, amounting to about 56%, is normally reduced, however, by the addition of softeners and fillers. Under the effects of the heating, PVC decomposes to release chloride. The amount of plastic contained in refuse usually averages about 2% to 3% by weight, of which PVC may account for about one quarter. As PVC possesses suitable chemical and physical properties, it is employed to a great extent for packaging material. The bulk of the plastic material, to be found in household refuse consists, in fact, of packaging material, According to the present trend of technical development the proportion of PVC in refuse is likely to quadruple in the next 10 years. As slag and airborne dust retain hardly any gaseous contaminants, a quadrupling of the emission of hydrogen chloride by refuse combustion plants is also likely to occur in the period.

The specific emission from refuse of sulfur oxides, paticularly sulfur dioxide, has not changed much in recent years, so that an increase in sulfur emission is expected to arise only from the constant expansion in the volume of refuse deposited.

Fluoride compounds are also expected to occur to an increasing extent in the emissions from refuse combustion. These arise essentially from plastics containing fluorides and from fluoride motive fluids in sprays, As the use both of plastics and of spray guns is on the increase, a substantial expansion in the emission of fluoride can be expected here too.

In order to separate chloride, sulfur and fluoride compounds from refuse combustion plants, two methods have been tried : (1) The dry waste-gas purification process, by which alkaline reagents are added to the acid contaminants. The degree of the seperation is very low, however, and bears no relation to the cost. (2) The wet waste-gas purification process, by which dust and contaminants are washed away. The degree of separation is good but a solution still has to be found for thc problcm set by the washing solution itself which is now being produced in ever greater quantities.

Plastics in Refuse

Owing to changes in life-styles, the composition of domestic waste is subject to constant alteration. The supersession of coal fires by central and oil heating was responsible for a sharp upturn in the refuse of such materials as paper, board, wood and plastics which previously were burned. The overall proportion of plastics of 2%—4% in refuse from settled areas is relatively limited, however, although far more plastics are produced than are to be found in domestic refuse. This is due to the fact that the bulk of plastic materials are employed in manufacture of varnishes, glues and wood coatings. In terms of volume, plastic materials represent a small proportion of domestic waste.

Nevertheless, the plastic element in communal waste raises various problems in relation to three most commonly used methods of waste disposal. With uncontrolled refuse dumps, which are likely to be abolished in the near futrure, scraps of plastic material mainly from carrier bags blown about by the wind, are a constant nuisance to local residents. And if they start smoldering, such materials can cause serious air pollution the fumes they give off.

In supervised deposits, in which the bulk of household refuse will be disposed of in future, quite different conditions prevail. In this case the refuse is crushed, compressed and covered with rubble or other inert material so that there is no danger of plastic scrap being blown away. Owing to their compressed state the plastics themselves raise no sort of difficulties in the deposit. They give off harmful substance into the atmosphere or into groundwater, either immediately or in the longer run. They are also easy to compress and crush and do not interfere with the decomposition processes. Long-fiber plastic waste can get tangled up in the cylinders of the compressor, however and if large quantities of such waste have to be dealt with, this should be taken in account in selecting the compressor.

In the composting of refuse from settled areas, plastic material, owing to its high resistance to corrosion, represents a ballast element, which has to be extracted or burned. Dificulties still arise, however, in the case of long-fiber waste, such as women's nylon stockings, which get wound around the axles of the crushing and conveyor appliances and so reduce their performance or bring them to standstill. During the actual composting process the plastics suffer hardly any change. At the most, some phosphorus-containing softners may be released from them so that the plastic parts become brittle. Provided they are pulverized sufficiently, plastics cause no problems either in compost is usually sifted.

The possibility of developing and producing plastics that are reducible biologically or physically (by ultraviolet rays) has long been under consideration. These plastics are not suitable for use in a number of sectors, e.g., for durable goods, however, and could best be used for packing purposes,

When refuse is burned, the bulk plastics polyethylene polystyrene and polypropylene are reduced to innocuous carbon dioxide and water. Less favorable are the combustion products of polyvinyl chloride (PVC), which in addition contain hydrochloric acid. This gaseous hydrochloric acid, on the one hand, produces corrosion in the combustion plant and, on the other, has harmful effects on the environment. Both these dangers can be overcome with present technology, however, so that unless it is worthwhile from an economic point of view to restore them to the industrial cycle, plastics can be burned without hesitation.

16

Mining Wastes

The Mole (*Talpa europea*) has been described as the first miner. Mole hills are composed of its waste, the spoil derived from its tunnellings underground being voided on to the surface. Mole hills have a considerable ecological effect locally. In some upland areas in Wales mole hills may appear to cover up to half the area of some pasture, thus considerably reducing its value for grazing. The freshly broken-up soil also provides sites for germinating seeds, and in this way many weeds invade well-developed pasture of mainly economic grass species. The mole hills may even become sites for invasion by hawthorn and other shrubs, as the sheep or cattle tend to avoid the hills when grazing, although hawthorn seedlings in really well-grazed grass seldom survive. They also bring worms and soil insects to the surface, and one way of detecing whether moles are active underground is to observe when blackbirds and thrushes approach the hills where new soil and fresh food appear. It is difficult to estimate the total weight of the mole's mining waste, but throughout Britain it is likely to amount to at least a million tonnes in a year, so it can hardly be described as a trivial problem.

Some of these, as those produced in stone quarries, are valuable raw materials easily absorbed in road building, but much waste is dumped in massive heaps, disfiguring the landscape. Today some of this is being removed, and many heaps are being imaginatively vegetated, in ways described by A.D. Bradshaw and M.J. Chadwick in their outstanding book *The Restoration of Land*. Restoration is usually of a utilitarian nature, for amenity or for some form of recreation, but this could be (and occasionally is) amended to produce some form of vegetation valuable for wildlife conservation. It is perhaps ironical that there has recently been some local opposition to the restoration or removal of large spoil heaps. Coal tips, china clay deposits and even hillsides covered with slabs of broken slate from quarries are sometimes considered to be part of our 'industrial heritage' and therefore candidates for preservation in their starkest form. It is perhaps fortunate, though the process may be very slow, that they are likely eventually to develop some concealing vegetation, and they may in the very long term become covered by shrubs and trees. But, in most cases, some process to accelerate vegetation is clearly desirable.

Coal mining

Considerable areas of ground are required when deep coal mines are established. In Britain in recent years this has given rise to controversy in such places as the Vale of Belvoir, which is quite a distance from earlier mines. In this case the objection has mainly been to the change in a pleasant agricultural landscape, but wildlife will clearly be affected, particularly by the spoil which will have to be excavated in addition to the coal. In some mines the environmental damage done by the spoil is only temporary, for there are plans to put it back underground. As this is costly, however, it is by no means a universal practice, and as a rule the spoil is dumped in small mountains all around the area of the pit.

These characteristic black 'mountains' are a particularly familiar sight in Country Durham, parts of Yorkshire, Nottinghamshire, South Wales and several areas in Scotland. In the past they were generally left untouched, and for many years almost no plants appeared on the surface. Seeds often germinated, and growth started, but the plant suffered the fate of the seed that fell 'on stony ground' in the Parable of the Sower. The spoil heaps were often steep-sided, and therefore unstable, and small 'landslides' were common. Very occasionally these gave rise to disasters such as occurred at Aberfan in South Wales on 21 October 1966 when 144 individuals including 116 school children were fatally buried by a collapsing spoil heap.

Plant growth is difficult on coal mine spoil, because of the nature of the material. It consists almost entirely of finely crushed rock and coal, with little or no soil and almost no nutrients. Furthermore, it often contains pyrite (iron sulphide) in quite large quantities. This discourages growth, by producing very acid conditions which may be between pH 2 and 4. The acid material, and the associated metal, may also be leached out of the spoil and may pollute neighbouring watercourses. In one such case an upolluted stream, with slightly alkaline water (pH 7.9) and very low concentrations of metal, received leachate from a mine. Below the spot where this occurred the pH remained in the region of 3.3 for over 10 km, and iron levels were raised to 40 ppm.

In recent years, there have been many very successful attempts to vegetate coal spoil. This requires treatment of the surface layer with lime-possibly as much as 50 tonnes per ha in very pyritic soil—and fertiliser. In some cases deep layers of expensive topsoil have been put over the surface, in other cases only subsoil has been used. The addition of any kind of soil makes restoration easier, but is not strictly necessary. The lime is generally well retained, and has a more or less permanent effect so far as it penetrates. Inorganic fertilisers need to be applied lightly and often because heavy dressings before the vegetation is established tend to be washed away. As well as soil, organic dressings such as sewage sludge may be applied successfully. The trouble here is that sewage is liable to contain high levels of metals, particularly lead, which may make the ground unsuitable for food crops or gardens where vegetables for human consumption are grown. If only amenity grass is required, sewage can generally be used. Actually, even fairly high levels of lead are seldom taken up by animals living on grass grown on such soils, and wild mammals generally show no sign of being affected.

Trees planted directly into coal tips die, even if fertilised. If planted after a grass cover has been established, they may survive, although they will only grow slowly. The best results have been obtained by digging holes and filling these with a couple of wheelbarrows full of soil. Such trees will grow quite well at first, but will only continue to flourish if the surrounding spoil has been limed and continues to be fertilised for some years after restoration begins.

Whereas well-limed and fertilised tips will grow most grasses used in agriculture and found on rough grazing areas in Britain, sometimes the best results are obtained, with minimum trouble and expense, will strains of grasses which are tolerant of high levels of any existing heavy metals. Collieries in Britain differ considerably from one another. Some tips are reasonably low in metals, others contain large quantities. It is always wise to analyse the spoil at any site where revegetation is planned. In most cases, it is essential to introduce some permanent drainage system, particularly as the lower layers may be very impervious.

Open-cast or strip mining is the cheapest, quickest and most economical method of winning coal from the earth. It is widely practised in many countries, particularly in the USA and Australia, where

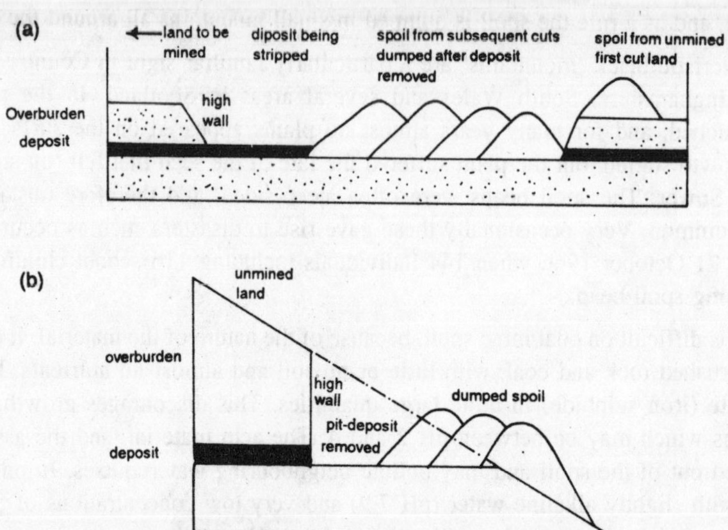

Figure 16.1. Open cast mining: a, Shallow deposit in flat country; b, In hilly country.

the environmental impact is often horrendous. This was at one time also true of Britain. Today open-cast mining is governed by stringent planning regulations. It is normally only allowed where it will help in the restoration of land already mined, although some lower grade agricultural land has also been used. As a result of these restrictions, open-cast mining in Britain has little effect on areas of conservation interest or on wildlife. It is permissible to dig up only areas of minor importance, and it is also always stipulated that the land should be left in an improved condition, generally for farming.

Figure 16.1 shows how open-cast mining operates. There is usually only a comparatively thin deposit of coal, although it must be thick enough to be economically worth excavating. To get at it, the overburden of 'ordinary' soil and rock has to be removed—this overburden is generally 20 to 30 times the volume of the coal. A hole is dug down to the coal layer, this is removed, and the first lot of spoil is dumped on to the surface. This is the only serious excursion outside the excavated area. The rest of the spoil is piled within the excavation after the coal has been removed. In some cases, the topsoil for the whole area to be mined is ripped off and stored, but this may be an unnecessary complication, as the topsoil soon deteriorates unless spread comparatively thinly, and useful soil fauna like earthworms do not generally survive.When restoration is to take place, after all the coal has been removed, the spoil is spread out evenly over the cavity. Although the volume of the spoil is less (by the volume of the coal) than that of the material removed, it usually fills the hole even when compacted as much as possible; the subsidence in succeeding years is generally minimal. When such sites have been efficiently drained and appropriately fertilised, they have often been restored to full agricultural productivity. The depth of soil has proved sufficient to grow good crops of cereals. Where hedges and trees have been planted, many former open-cast sites are difficult to distinguish from neighbouring fields which have not been so exploited. Incidentally, on open-cast restoration areas, as elsewhere, any trees planted will grow best if grass is discouraged for a circle of up to a metre all around, so allowing them to obtain better supplies of water and fertiliser. Successful rehabilifation of open-cast areas depends on good planning and careful after-use, and proper application of fertiliser is crucial. Some of the best results have been obtained where pasture with a good complement of clover has been grazed for a number of years before returning the fields to arable cultivation. As mentioned above, if required for wildlife conservation, such restored land has great potentialities, but may be too expensive to be acquired for such purposes.

Iron ore

The world's riches iron ore deposits are in Australia, and other important sources exist in America. However, iron ore has been mined for many years in Brtain, and some of the environmental effects still remain. Strata of iron ore, generally much thicker than those of coal, are found quite near to the surface. In the early days of mining, the ore was removed with minimum damage to the countryside. The overburden was removed in wheelbarrows, the topsoil generally being kept separate. Comparatively small areas were excavated at a time, so that even with few mechanical aids the topsoil was only stored for a brief period. When the ore had been removed, the subsoil and then the topsoil was replaced, and the field system, some depth below the original surface, was restored. As a rule, the fields were immediately able to yield reasonable crops.

At the beginning of the 20th century, 'progress' struck the industry in England. Immense steam-powered excavators were employed, profitably in Leicestershire and Northamptonshire, to remove iron ore at greater depths than had been previously exploited, so that a much greater amount of overburden was removed. As this was a period of agricultural depression, there was little economic inducement to rehabilitate the land. The mining companies could buy agricultural land with underlying iron ore so cheaply that they were prepared to abandon it at the end of their operations. The result was that, by 1945, there were over a thousand hectares of totally derelict land in what was otherwise a rich agricultural region. The overburden had been replaced with the least possible trouble, mixing any topsoil with the mass of the material, and the result was 'hill and dale' land, which was difficult even for walking. Much was simply neglected. The soil was comparatively lacking in nutrients, particularly phosphorus, but nevertheless some vegetation started to appear. The main species found included Couch Grass (*Agropyron repens*), Bush Grass (*Calamagrostis epigejos*), Creeping Thistle (*Cirsium arvense*), Spear Thistle (*Cirsium vulgare*), Beaked Hawksbeared (*Crepis vesicaria*), and Rosebay Willow Herb (*Chamaenerion angustifolium*). Seedlings of shrubs including Hawthorn (*Crataegus oxycanthoides*), and Goat Willow (*Salix caprea*), and trees such as Silver Birch (*Betula verrucosa*), and Oak (*Quercus robur*) appeared but seldom survived. Legumes, which might have done much to improve the levels of nitrogen, were generally absent, presumably because of the low phosphorus levels. The growth was generally so poor that it could not even be said to produce some semblance of a 'wilderness', and the sites did not prove attractive to birds or small mammals. This was presumably in part due to the poverty of the soil fauna— earthworms in particular were virtually absent.

The only remotely economic use made of the unimproved hill and dale land was to plant trees. Hand planting of 2-3 year transplants, especially of Larch (*Larix decidua*) and Sycamore (*Acer pseudoplatanus*) was carried out. Where fertiliser, particularly phosphate, was added, the growth has been reasonable, and quite respectable 'woods', with all the appropriate wildlife, have developed. Even the soil fauna has slowly returned. These areas, with appropriate management, would be particularly suited to wildlife conservation, as the forestry, although aesthetically more pleasing than the derelict ground, is not very productive.

Since the Second World War planning regulations have insisted on full restoration of ironstone workings. In 1951 the Ironstone Restoration Fund was set up, with a levy based on every tonne of ore extracted, to cover the costs of restoration. The sites are carefully levelled, usually (as in the 19th century) lower than originally, but sometimes extra material has been brought in to restore the original ground level. The area is cross-ripped, and any large stones near the surface are removed. Disking and harrowing to produce a satisfactory tilth then follows, fertiliser (based on soil analysis) is applied, and a good grass ley generally results. It has been found that these soils need substantially more fertiliser for some ten years than do normal agricultural soils, and these extra applications are paid for by the Ironstone Restoration Fund. At the end of ten years the restored land is hard to distinguish from the surrounding countryside where no mining has taken place.

In the past, ironstone mining has destroyed areas of ancient seminatural wood and other features of conservation importance. Today such developments would not be permitted. The restored land, like that derived from open-cast coal mining, could clearly be used for many purposes in addition to agriculture. It could be returned to forest, and this might be desirable in situations where mining areas are still adjacent to ancient woodland areas, from which emigration of plants and animals would be possible. Under these circumstances, something very similar to genuine woodland should be produced.

Sand and Gravel

The building industry has an almost insatiable appetite for sand and gravel. In Britain some 100 million tonnes of these materials are dug from 1,000 ha of generally good agricultural land every year. Although there is very little actual waste from sand and gravel digging, some holes made by the extraction may be used as waste tips, and the gravel workings often provide valuable wildlife sites.

Many deposits of sand and gravel are in river valleys, and the deposits are below the level of the water-table. Thus when the material is removed, the holes fill with water. The layer underlying the excavations usually has a comparatively high clay or silt fraction, and therefore is a satisfactory medium for plant growth. The nutrient levels are frequently adequate for reasonable plant growth. So, just left to itself, a wet gravel pit soon develops its characteristic vegetation, and requires comparatively little management, particularly in the early stages. If good beds of reed (*Phragmites australis*) are required, some judicious planting may be useful, and may accelerate the process, but many species, including Reed Mace (*Typha angustifolia*), Yellow Iris (*Iris pseudacorus*), Great Willow Herb (*Epilobium hirsutum*) and Water Mint (*Mentha aquatica*) will soon grow on the margins, and various pond weeds in the open water. Before long Alder (*Alnus glutinosa*), Willow (*Salix alba*) and sal low (*Salix cinerea*) may appear and flourish on the banks surrounding the water.

Almost anywhere in Britain where open areas of water are created, duck and other birds move in. Mallard (*Anas platyrynchos*), Moorhen (*Gallinula chloropus*) and Coot (*Fulica atra*) are almost inevitable arrivals, and many less common species will almost certainly accompany them. The growth of gravel pits in the last century has allowed the great Crested Grebe (*Podiceps cristatus*), which was an endangered species a century ago with only 32 pairs in England, to increase to a healthy population of well over 5,000 pairs. Gravel pits are also quickly taken over by many aquatic insects and other invertebrates, and fish, including Perch (*Perca fluviatms*) and Roach (*Rutilus rutilus*) flourish—it is often uncertain how they find their way into these locations.

While almost all wet gravel pits are potential wildlife habitats, only a proportion remain available for this purpose. Some are used for noisy sports such as water skiing, which discourage the animals, others are restored to agricultural and recreational use. This is not easy, as suitable filling material may be hard to find. As the water communicates with that outside, domestic waste, which would cause pollution, is unsuitable, although it has been used when the material is sealed off by a thick butyl rubber lining. As such a lining is unlikely to last for many years, only refuse which is relatively free from toxic substances and organic matter can safely be used. Also, particularly when used for recreation, the areas of open water may have a higher cash value than the same area of agricultural land. Thus in many parts of England, e.g. the Cotswolds and the valley of the Great Ouse, the whole face of the countryside is being changed, with increasingly large areas of open water being created as the gravel is dug out and transported to building sites. Those areas employed for boating and other sports will encourage some forms of wildlife, even where they are far from being sanctuaries, so the contribution to conservation is generally positive.

Dry gravel pits are usually on good agricultural land, grade 2 at least, and there is generally pressure to restore them to the farmers who previously worked the land. In such areas the topsoil is normally preserved while the gravel is removed. These pits can be filled with relatively inert materials, such as pulverised fly ash from power stations, which is then covered with at least a metre depth of subsoil and topsoil. With suitable management, and the correct application of fertilisers, the land is soon brought back almost to the state it was in before gravel extraction began.

Clay for Brickmaking

All over Britain where there were deposits of clay which, when baked, produced a hard mass, brickworks sprang up. Certain small enterprises produced very attractive products, some of which continue to be made and have a luxury value. However, most small brickworks have closed, and their sites may be used for other purposes. One at Ramsey Heights in Huntingdonshire has been made into a centre for the study of wildlife by the Bedfordshire and Huntingdonshire Wildlife Trust. There are similar developments in other parts of the country.

Today almost all the major brickworks are in Buckinghamshire, Bedfordshire and around Peterborough, on the strip of Oxford clay which runs diagonally across England. The works are all immediately adjacent to the pits from which the clay is dug. The clay is near to the surface, so only a small amount of surface soil has to be scraped off. After that, only brickmaking clay is dug out, work being stopped as soon as the best material is exhausted. As a result there is virtually no waste to dispose of.

Although the material at the bottom of the pit may not be top-class brickmaking clay, it is still relatively impervious material, and the pits are ideal places for the safe disposal of waste. Many are used for this purpose, and trains and lorries bring a continual stream of municipal waste to Buckinghamshire from London and other industrial towns. The sites are well managed, and the waste is always quickly covered with soil, although in the past these operations were not always so efficiently performed. The intention is to restore all brick pits for agricultural, recreational or building purposes. As mentioned earlier, where domestic refuse containing much cellulose is used, the continuing generation of methane makes it dangerous to build houses, but grass and even crops have been successfully grown.

One very large enterprise has been the use of the brick pits near Peterborough to contain pulverised fuel ash (PFA). Modern, coal-fired electricity generating stations utilise coal which has been ground to a fine powder and is then blown in an airstream into the furnace. The carbon is burned off, and produces the heat which generates the power. The residual ash melts to produce PFA, which is made up of tiny, colourless, glassy spherical particks. Some 95 per cent of PFA is, in size, similar to sand and silt. It contains only I per cent of clay-sized particles, but nevertheless it has fairly good water-holding properties. It contains no organic matter at all, may be very alkaline (pH 11), and is relatively free from nitrogenous salts, but otherwise it contains at least some of the more important plant nutrients. PFA by itself tends to set hard, like cement.

When this type of power generation was first introduced, it was thought that disposal of the PFA would be a major problem, but in fact it has been less serious than was expected. Nearly half the PFA has been used to produce building blocks which are light and have good insulating properties. In fact, it has been suggested that efforts should be made to utilise more of the PFA for this purpose, instead of digging holes in the ground in order to extract clay to make bricks, and then filling up the holes with the PFA which could itself have produced the 'bricks'. However, the London Brick Company, which operated the Peterborough Brick Works, entered into a contract with the Central Electricity Generating Board, to allow the CEGB to deposit their PFA and then to restore the land before returning it to the brick company.

The process called for a number of tanker trains to convey the PFA which contained enough water to make a slurry that could be pipe into the brick pits. These were then filled to near the surface level required. The advantage of the PFA slurry over some wastes used to fill holes was that it did not contract after deposition so there was no risk of subsidence. Then the PFA surface received one of several possible treatments. When only some 5-10 cm of sewage sludge or other organic matter was worked into the surface of the PFA, reasonable grass growth was possible. Poor crops of wheat could be obtained under the same circumstances, if nitrogenous fertilisers were applied. The best results were obtained with a deeper layer of soil worked into the top of the PFA and fertilised. Fairly good tree growth was obtained when planting was effected in pits 30 cm in diameter and 40 cm deep, filled with good soil.

One difficulty with the PFA is that, when fresh, it contains a quantity of the metal boron, which is toxic to many plants, particularly such crops as barley, peas and beans. If the PFA is allowed to weather before planting, much of the boron is leached out. Boron levels also fall considerably during the first years of any sort of cultivation, so it does not seem to present a long-term problem.

At Peterborough, the restored brick pits filled with PFA have been used to grow various crops, and as recreational grass. They have been proposed as sites for house-building, but some authorities are not yet convinced that the ground will be sufficiently stable. However, recent tests have been reassuring, and it is likely that building will in fact go ahead.

The brick pits, whether in course of excavation, or after filling with PFA, do not appear likely locations for wildlife conservation. But at Peterborough part of the area of the old pits has been turned into a nature reserve. One pit has not been filled, but has been allowed to collect rainwater to produce a lake. Trees have sprung up or been planted on areas between excavations, and a management plan has been devised with wildlife conservation as its main objective. The reserve is still in an early stage, and its flora and fauna will in time be enriched, but it is already proving a valuable educational facility.

China Clay

Koolin or china clay is mined in Devon and Cornwall, and has gone far to destroy nearly 100 sq km of the countryside. As well as being used in the British fine porcelain industry, and the manufacture of paper, it is also a major export. Kaolin deposits.were produced by the hydrothermal degradation of granite from below, extending to greate depths, so digging continues deeper and deeper, and the waste cannot be put back into the pits. The amount of waste is considerable—one tonne of kaolin is accompanied by 1 tonne of mica and 7 tonnes of mainly quartz sand. The sand is built up into mountains, the mica goes into lagoons in nearly valleys. The result, until recently, was a veritable moonscape.

Natural revegetation does take place, but very slowly, and may be overwhelmed if more material is used to cover existing tips. One large heap, after 30 years, had a growth of grasses and shrubs tolerant of the existing acidity on its lower slopes, but the rest was still virtually uncovered. Where old clay working becomes colonised by leguminous species, thus raising the nitrogen levels, trees may appear and grow reasonably well, but this process is likely to take at least 100 years. The industry only began in 1770 (previously rougher clays were used by the pottery industry, and kaolin was chosen to compete with the fine porcelain from China, hence 'china clay') and some of the oldest heaps are today still not completely covered with dense vegetation.

Recent work, again mainly by A.D. Bradshaw and his colleagues, has shown that the process of vegetation can be greatly accelerated. It has been usual to leave the steep mountains of spoil, as with plant cover these do not get badly eroded by the weather. The structure of the sand mountains is poor, and they would dry out badly in the east of England, but in the much wetter west country this seldom occurs. Fertilisers and lime must be applied, and hydroseeding (i.e. spraying the seed mixed into a jet of water) is generally used on the steeper slopes. A suitable seeding mixture has been devised, containing Rye Grass (*Lolium perenne*) to establish a quick cover, and Red Fescue (*Festucar rubra*) and Common Bent (*Agrostis tenius*), appropriate local grasses, to build up a permanent cover. Legumes, Red Clover(*Trifolium pratense*), White Clover (T. repens) and Bird's-foot Trefoil (*Lotus corniclatus*), needing adequate lime and phosphate, are required to accumulate nitrogen in the sward. Tree lupins (*Lupinus arboreus*) and gorse (*Ulex europaeus*) grow well and increase the build-up of nitrogen. Eventually a wide selection of trees can be introduced.

The mica waste is usually deposited in ponds. If these are drained good 'soil' results. This retains fertilisers well, and can be used to grow good crops of vegetables such as potatoes. The mica waste may be utilised for cropping the same year as that in which it is produced. Obviously many types of plant can be grown in it.

As the original'moonscape' is, in most areas, left intact but covered with a good vegetation of herbs and trees, the site will always be unsuitable for normal agriculture or ball games requiring flat land. It has therefore obvious potential for wildlife management and for education.

Quarries

Immense quantities of rock and stone are today extracted from quarries for use as building materials, for roadmaking, and for the production of cement and lime. In the modern quarry there is little waste, as almost all the rock can be utilised in one way or another. In some old quarries the overburden remains as a mound which is easily covered with vegetation; but in most cases it is used to cover the scars produced by quarrying.

There is generally very vocal opposition to opening up any new quarry, particularly, when it is in a National Park or area of scenic beauty. Sometimes the people who are most vocal in their condemnation of the electrical industry for polluting the air with sulphur dioxide also lead the opposition to quarrying for the limestone necessary for operating the flue gas desulphurisation (FGD) which removes the gas.

The slate industry of North Wales has produced the greatest amount of waste, and today there are heaps containing perhaps 500 million tonnes in the areas where quarrying was important. These quarries produced slates for roofing, in various sizes, and for tombstones and memorial plaques. The accuracy involved in this work inevitably resulted in large amounts of useless offcuts. The mounds of this acidic rock are well described by Bradshaw and Chadwick as 'a miserable environment for plants'. A few mosses and the occasional birch may be all that relieves their surface. Fortunately, many slate heaps are now being removed for use in roadbuilding. There is then little loss of any wildlife habitat.

Although there is increasing pressure to reuse waste from limestone and chalk quarrying, any such material left over is rapidly colonised by a rich vegetation, and mammals and birds find shelter among the stones. Indeed, the main interest in quarries today is that they so often turn into valuable wildlife habitats. B.N.K. Davis has described what has happened in a series of chalk and limestone quarries, some of which have become the best sites for orchids and other interesting plants in their respective countries. This development of rare vegetation in quarries, particularly those in the chalk,or in limestone country, can be a source of embarrassment to the authorities. There have been many instances where a quarry has been excavated on the understanding that the site will be restored to its original contours and to use for agriculture. Where restoration has been delayed, the quarry may have developed into a site of conservation importance, and may even have been scheduled as an SSSI. As a result, there has been strong opposition on environmental grounds to carrying out the original plans for restoration.

17

Hazardous Wastes

The public, and the media, are very much concerned about the danger to themselves and to the environment from toxic wastes. This was the reason for the excitement generated by the *Karin B*, a ship loaded with leaking barrels of unknown substances, which tried unsuccessfully to land its cargo in Britain in the summer of 1988. Waste cargoes are in fact landed and processed in this country, which has caused us to be stigmatised in some quarters as 'the dustbin of Europe'. In the *Daily Telegraph* of 3 November 1988 we read: 'Grave concern at the Government's "very lax" attitude to the safe disposal of dangerous chemicals and industrial waste was expressed by Sir Hugh Rossi, chairman of the Commons Select Committee on the environment.' We need to know how real this danger is, and what damage is done by hazardous or toxic waste to the environment and to wildlife.

There is no clear definition as to what is meant by 'hazardous waste'. The size of the problem may be inferred from the eleventh report of the Royal Commission on Environmental Pollution which estimated the annual production of hazardous wastes in England and Wales at 4.4 million tonnes Some types of waste are covered by legislation. The Deposit of Poisonous Waste Act 1972 was one of the speediest pieces of legislation ever introduced in Britain. Before 1972 there were few regulations and, as will be shown, poisonous substances were deposited in many sites with little control and few special precautions. The 1972 Act was passed within days when drums containing cyanide were found on we ground used as a children's playground. The Act introduced the concept of 'notifiable' waste, which was a blanket category comprising all toxic or dangerous waste which was not specifically excluded. It became a offence 'to deposit on land poisonous, noxious or polluted waste in circumstances which could give rise to an environmental hazard'. The Act also made it obligatory on those disposing of waste to notify the appropriate authority of the composition, quantity and ultimate destination of any notifiable waste.

The 1972 Act was effective, but cumbersome to operate because of the paper work involved. The Control of Pollution Act 1974 (COPA) was an attempt to deal with pollution and waste disposal in a comprehensive way. Under this Act, Control of Pollution (Special Waste) Regulations 1980 were introduced, and the 1972 Act was eventually repealed. The new regulations introduced a new category of 'special waste' to cover the more dangerous substances. These included medical products available only on prescription, or listed substances in dangerous concentrations. These are described, in somewhat macabre language, as having:

1. The ability to be likely to cause death or serious damage to a tissue if a single dose of not more than 5 cc were to be ingested by a child of 20 kg bodyweight, or

2. The ability to be likely to cause damage to human tissue by inhalation, skin contact or eye contact on exposure to the substance for 15 minutes or less, or

3. A flash point of 21°C or less.

The various categories of wastes are detailed in Appendix 3. It will be seen that the situation is far from simple.

However great the improvement after the passage of the Deposit of Poisonous Waste Act 1972, the country was, at that time, left with the legacy of the period when such regulations did not exist. The situation was not quite as bad as some people imagined, because since 1947 planning controls over all developments had been enforced, so that no new largescale uncontrolled dumps had been established. Nevertheless there were many sites where all manner of toxic substances had been deposited during the previous hundred years, and little was known of the effects on the environment. The situation was studied by D.C.Wilson, E.T.Smith and K.W.Pearce of the Hazardous Materials Service of the Harwell Laboratory, and their findings were published in 1981.

The results of this survey were generally reassuring. The scientists looked into open dumps which had been abandoned in what might well be a dangerous condition, landfill sites which had been partially restored, and land contaminated by past industrial use.

Perhaps the most horrendous example of an open dump was at Malkins Bank in Cheshire, described as 30 ha of 'aesthetic insult'. This dump was developed in the late 19th century and used until 1932 as a salt and chemical works, pumping brine from the ground (where salt deposits were present) and using water from the stream for manufacturing processes. This left the stream culverted with a considerable area of the site overlain with lime sludge in three large lagoons. When manufacturing ceased in 1932, indiscriminate tipping of toxic and noxious waste from several different industries covered some 15 ha of the site to a depth of up to 15 m.

When examined in 1970, the tip was squalid and what was described as 'the odour of rotten eggs and tomcats' could be smelt miles away. Thousands of drums, containing a wide variety of chemical wastes, including reactive substances such as metallic sodium, lay in the sludge. There were bales of paper and stinking tannery wastes. Where tipping had not occurred, the lime sludge had dried out. The whole site was hazardous. Minor explosions were common, and there were open shafts, falling into which would almost certainly have proved fatal.

What might have been a very dangerous situation was greatly mitigated by the presence of a deep layer of lime sludge which covered most of the site, something which was also true of other sites in north-west England. This lime sludge, which neutralised many of the toxic materials, was used to bury the drums, which when possible were crushed before burial. The stream was seriously polluted, though much of this resulted from developments upstream of Malkins Bank. The stream was culverted to avoid further pollution. After levelling, the whole area was buried in at least 0.5 m of soil brought in from outside, and it was then grassed and developed as a golf course. Considering what remains underground, the site will have no further development of the kind likely to bring toxic material up to the surface, but as grassland it should offer no problems. The lime sludge and other materials will, in time, become incorporated in the soil. Obviously this site, used as a golf course, will have wildlife potential, particularly if this is borne in mind when the course is laid out.

Another site investigated by the Harwell team was on flat marshy land on any estuary subject to flooding at high tide. Tipping took place between 1945 and 1972. Large quantities of asbestos- cement, rubble and flux glasses were dumped along the river bank, in an attempt to protect the site from the river. Heaps of multi-coloured chemical wastes, and a considerable amount of domestic waste, raised the fertility so that lush vegetation, including apple and other fruit trees, had grown up spontaneously. Although described as a 'moonscape', the site was used by motorcyclists and fruit-pickers. There is no suggestion that eating the fruit did any damage, although the soil proved to be seriously contaminated with asbestos fibre and heavy metals. The total for zinc, copper and lead reached 37 per cent, and the recorded maximum for cadmium was 0.5 per cent. This site has been reclaimed as a public open space, precautions being taken to minimise pollution of the adjacent estuary. It has been covered by a deep layer of soil, amounting to at least 1.5 m. Sadly the productive fruit trees have been removed.

A third site had been used to deposit calcium sulphide waste from the old Leblanc process for the production of calcium carbonate. There are many such dumps in Lancashire and, as described below, some have developed wonderful wild flower gardens. Others, where wet uncompacted refuse is also present, present problems. Any disturbance releases hydrogen sulphide, a very poisonous gas, smelling of rotten eggs when dilute, but generally undetected when in stronger and more dangerous concentrations. The pH here was 12-14. This site was buried deeply and grassed over successfully.

The most widespread British problem has been that of contaminated land arising from industrial practices which were normal a hundred years ago, or from waste disposal. A notorious case is the Lower Swansea Valley, where Victorian metal smelters have left their legacy.* In almost every town the old town-gas works has left its area of sites contaminated with coal tars, phenols and spent iron oxide (used for gas cleaning) containing up to 50 per cent free sulphur, together with sulphide, sulphate and various cyanide compounds. In the worst cases, the highly contaminated soil has been excavated and reinterred in a safe site. Otherwise the material has been buried. As a rule, care has been taken to avoid developments which would release the material from its otherwise safe location.

The most serious problems in the UK have been caused by the release of methane in housing estates. There have been cases where gardens of houses built on former tips contain unacceptable levels of lead, arsenic, cadmium, copper or zinc. If grassed over, there would probably be no danger, but the soil could contaminate vegetables and might be ingested by small children who are notoriously prone to ingest 'dirt'. Where such contaminated gardens have been identified, the soil has been removed and replaced.

Although we must continue to be vigilant and on the look-out for dangerous old dumps, and although we have no reason to be complacent about our success in dealing with this problem, conditions in Britain seem to be much better than those which obtain in the United States of America. The USA has a far greater area of land available, much of it well away from inhabited zones. In thousands of such places dangerous wastes have been dumped without any remedial treatment, so the dumps are still very dangerous. The most notorious is that at Love Canal, near Niagara. During the 1940s and 1950s, a chemical company disposed of its industrial wastes by sealing them in drums and burying them on this site, as was perfectly legal and according to accepted practice. The drums were then covered with soil, and a great many houses were erected on top of the waste. At first all went well, but then the drums corroded and the contents leaked into nearby water and percolated to the ground. On 7 August 1978, a generation after the area had ceased to be used for chemical waste disposal, conditions were so

*Those who despair of having a clean environment with developing industry should note that the masive modern smelters, even when badly operated, only liberate a minute fraction of the pollution produced by the much smaller Victorian industries.

bad that Love Canal was declared a disaster area, and 239 families living close by were evacuated from their homes. Later, many more—almost the total population—had to be moved. There was obviously great local anger, and concern that people had been seriously injured or even killed by the pollution. The most careful investigations in fact showed that no one had suffered serious harm, but this was due more to good luck than to careful management. It is perhaps not surprising that some extreme environmental groups refuse to accept the official findings, and still insist that deaths and serious illness occurred.

According to the Eleventh Report of the Royal Commission on Environmental Pollution, there are in the USA possibly as many as 20,000 abandoned hazardous waste disposal sites and a 'Superfund' has been set up to clean up those for whom no responsible organisation can be traced. The Commission says: 'There is no evidence that in the UK waste disposal sites have, either in the past, or at present, been operated in a way likely to cause similar problems—certainly not on such a scale—and the need for emergency action is therefore much less likely. This is not to suggest, however, that there have been no examples of bad practice.'In the UK, the emphasis has been in the importance of reusing land for economic and social reasons, because 'new' land for all developments is scarce, particularly in and around urban areas. Such developments generate the funds to clear up the pollution. Also, notwithstanding the comments of Sir Hugh Rossi reported above, we do now have suffcient legislation to prevent further abuse of the environment, and the technical skills to put the regulations into force.

The question we have to ask is what were the effects on the environment, and on wildlife, of the uncontrolled dumping which went on before 1972, and which was even less controlled before the planning acts were passed in 1947. Clearly 'moonscapes' like Malkins Bank were virtually sterile, and the same was largely true of heavily contaminated land such as existed around gas works and in the Lower Swansea Valley. Some of the less heavily polluted soils have acted as outdoor laboratories, becoming spontaneously covered by vegetation. In many cases, as when the soil contains so much copper that the growth of normal commercial strains of Rye Grass is inhibited, the process of natural selection appears to have operated rapidly, and plants which are naturally resistant to copper have been able to grow and to flourish. These resistant strains, when identified, have proved very useful for growing on similar sites where selection has not occurred. Observations made on these polluted sites have, in fact, yielded a whole series of resistant species for such purposes.

Where toxic dumps have polluted rivers, the results have been devastating. Many animal deaths must have taken place when poisonous materials were dumped, or when animals visited the contaminated sites. however, although there may have been local effects, there is little evidence that toxic wastes have had any serious effects on the national population of any forms of British wildlife.

Britain, in fact, has probably gained by being a small, densely populated country. Sites for waste have never been easy to obtain, and large, responsible companies have done much to reduce the waste problem. Ideally, a chemical manufacturer should have no toxic waste to dispose of; in many cases the amount that leaves the factory is quite small. Valuable substances like gold and silver, which might be included in 'waste', are almost completely recovered and reused—this form of recycling makes economic sense. Moreover, many poisonous substances may be treated within the works and rendered harmless, and today this is happening increasingly. Where a firm has developed the means of detoxifying and destroying its own poisons, it may use the same process to treat other people's waste, often providing the basis of a very profitable industry.

The fuss that was made about the importation of drums of polychlor biphenyls (PCBs) from

Canada into Britain in 1989 illustrates the conflicts which may arise. PCBs, used widely in industry until the nineteen seventies, have been shown to be longlasting toxic substances which, like organochlorine pesticides may be accumulated in the tissues particularly of marine mammal like seals, and in marine birds. where they cause serious metabolic upsets including sterility. When this was discovered, industry stopped making PCBs, but considerable dumps remain in many parts of the world. There is no difficulty in destroying PCBs, and transforming them into harmless residues, but creation of suitable plant is expensive. Several firms in Britain have this expertise, which if properly used could soon get rid of all residues. However, when a Canadian firm attempted to import drums of PCBs into Britain, the organisation Greenpeace went to great lengths to stop this, including hampering the ships before they could land, and also by persuading dockers' trade unions to ''black'' such cargoes. In my opinion this was purely political action, and showed a great lack of understanding of environmental problems. Had the PCBs been landed they would by now have ceased to exist. Refusal to allow the cargoes to land may well result in their being dumped at sea or being improperly treated so that the environment is seriously put at risk.

Nevertheless at the end of the day there are many waste products of industry which are potential dangers to the environment. These may be relatively non-toxic chemicals that only need to be diluted sufficiently to be harmless, or they may be substances which break down quite quickly in a containment site; but they all require appropriate treatment. Within Britain, our present laws make it obligatory on any manufacturer producing waste to state its exact chemical constitution and what toxic substances it contains. Given this information, there are responsible firms in Britain which can deal successfully with virtually any poisonous substance. The EEC has regulations similar to those operating in Britain, making it obligatory to state the contents of waste for disposal. This information did not exist in the case of the *Karin B's* cargo. A British firm offered to analyse the drums, and to deal with those substances for which it had facilities. All reputable firms in this field would take the same attitude. Unfortunately there have been instances of small, so-called 'cowboy' firms accepting undescribed toxic waste from abroad and perhaps not dealing with it properly. Provided this is stopped, the importation and destruction of toxic wastes, which is at present widely criticised on environmental grounds, should be encouraged.

The important safeguard of the environment is the site licence, which must be obtained wherever wastes of any kind are disposed of or treated. Responsible contractors, who may be local authorities or commerical companies, adhere very strictly to the conditions of the site licence, otherwise the site may be closed down. Incidentally, these licences are only granted when the operators have the facilities (e.g. analytical laboratories) necessary to show that only approved substances receive the appropriate treatment.

In Britain most of the less dangerous chemical and industrial waste is eliminated by the process known as 'co-disposal', a term which describes the deposition of municipal and certain types of industrial waste in a single landfill site. The procedure has been greatly criticised, and is less commonly used in other European countries. However, the Royal Commission on Environmental Pollution, after careful study of the subject, and after visiting important sites where the operation was being practised, said: 'We conclude that, for many industrial wastes, properly managed co-disposal is an environmentally acceptable option in containment sites.' The same conclusion was reached by parliamentary committees concerned with the environment. Among the various wastes considered unsuitable for such treatment were acid tars, volatile and inflammable organic liquids and drums of toxic substances.

All responsible people involved in managing co-disposal sites will only accept materials of which they know the exact constitution. I myself have been present at such sites when lorries containing unacceptable material, or lorries lacking a proper description of their contents, have been turned away. On one occasion I left the site soon after a load had been rejected, and observed that the driver, unwilling to go the extra miles to a site with the facilities to deal with his load, was tipping it on to the roadside verge. In this case the police were summoned and the culprit was heavily fined as well as being made to remove the material from the roadside.

Co-disposal proceeds as follows. Domestic waste is deposited, compacted, capped with soil and allowed to mature for some time, preferably as long as five years. Then trenches are ripped through the capping material into the body of the now largely decomposed waste. Liquid waste is brought by tanker trucks, and run into the trenches. The liquid should not completely fill the trenches and flood the surface, but even if this happens it is quickly absorbed. After an agreed amount of liquid waste has been absorbed, the capping is restored. Care must be taken to avoid putting liquids which react with one another into the same trench; occasionally the wrong type of mixing has caused the production of poisonous carbon disulphide gas.

The Royal Commission rightly insists that research on co-disposal should continue. It records that when layers of cyanide and phenolic wastes, substances which if liberated could have had seriously damaging effects, were deposited in household waste, only minute amounts were leached out, and that after five years' exposure to rainfall and anaerobic decomposition, more than 95 per cent of the original toxic ingredients had disappeared, mainly by biological conversion into harmless substances.

Critics of co-disposal are concerned that the leachate from such sites will contain large concentrations of heavy metals or other substances contained in the industrial wastes. As these are sites where containment of the leachate occurs, this need not be important. However, many analyses have shown that the municipal refuse, which has acted as a 'sponge' and absorbed the liquid waste, tends to bind metals and many other substances within its mass, so the leachate is generally little different from that produced by simple municipal waste. Nevertheless, all leachate is a potential source of pollution, and care must be taken to see that it does not escape in an untreated state. As already stated, it may be sprayed time and time again on to the surface of the refuse, when evaporation takes place and so the risk of overflowing the banks of the site is reduced. It is also found that when the leachate percolates the refuse for a second time, metals and other substances continue to be absorbed and the leachate becomes weaker. Where large volumes of leachate need to be disposed of, treatment plant is installed, and this will deal with any residues of material co-disposed on the site.

Co-disposal appear to have no harmful effect on the ecology of the waste disposal site, as was indicated when I first compared parts of the Pitsea site, where co-disposal had taken place, with other areas which had only received municipal waste. Two years after the surface had been restored when co-disposal was concluded, the plant cover of the two areas could not be distinguished. Both areas also had similar populations of earthworms or other soil fauna. Furthermore, the extensive lists of plants and animals set out in the appendices to this book were mostly derived from this same co-disposal site; they do not indicate that co-disposal caused any environmental damage. Open water immediately adjacent to the site was examined, and was found to contain a normal and flourishing aquatic fauna. Had pollution outside the site occurred, this freshwater ecosystem would have been the first to react. The conclusion, therefore, is that properly managed co-disposal of appropriate substances does not endanger wildlife.

Some toxic wastes are treated by solidification. The waste liquid is mixed with various lime-containing cements and alumino-silicate compounds called pozzolans to produce a slurry. This sets in a few days, and within six months is rock hard. The process is very much more expensive than co-disposal, and although it is effective it is generally uneconomic, expect with dangerous substances unsuitable for deposition along with municipal waste. It is a safe technique, the solidified material may be easily stored (though its volume may be considerable), and there is little risk of toxic materials escaping. But it lacks the great advantage of co-disposal, where most dangerous wastes are not just stored away, but transformed into harmless substances.

It is clear that most toxic waste in Britain today is dealt with successfully, causing little damage to the environment. Most of the concern expressed by the public, and even by parliamentary committees, is therefore unfounded. We have safe methods of disposal, and sites where this can be performed. Damage can only occur if the law is not observed, or if carelessness leads to unnecessary accidents.

I have described some of the old sites which were used before present laws and planning regulations were enforced. The results were horrific and the fact that there was so little damage outside the sites was largely a matter of luck. Even so, there are instances where the most deplorable practices, such as would not be permitted today, have produced some of the most interesting natural conditions in the country. So, from the conservation viewpoint, improved technology, as in methods in sewage disposal (see later) is not all gain.

Until the beginning of the 20th century, washing soda (sodium carbonate) for the chemical industry and for the housewife was produced from sodium sulphate, coal and limestone by the Leblanc process. This process, widely used throughout Europe, produced serious air pollution such as is almost unknown today in Lancashire (where much of the industry was concentrated), generating many unpleasant and toxic alkaline wastes. These were generally deposited in considerable quantities, sometimes covering an area of some hectares to a depth of several metres. If the dump was adjacent to a stream, this was deprived of all life, assuming there was anything left to kill off as a result of sewage and other industries upstream. On the ground the wastes, which had a pH approaching 14, were far too alkaline to support any form of plant life, and remained bare and unsightly for many years after they were deposited.

Then gradually, washed by the rain and inluenced by the mildly acid carbon dioxide, and possibly affected even more by acid rain containing sulphur dioxide from house chimneys and factories, the free lime (calcium hydroxide) disappeared from the surface layer. This left a very limy soil, containing few nutrients, which allowed calcicole plants to grow with comparatively little competition from other, possibly more aggressive species. These sites have been studied by several botanists, in particular by R.P.Gemmel.

After some 70 years the Leblanc process waste developed the profile described in Table 17.1.

The lime wastes are rich in those species which do well under conditions of high lime content and low fertility. The low nitrogen and phosphorus levels restrict the growth of grasses and clevers which would otherwise suppress the more interesting and uncommon plants. The vegetation is very open or sparse, and allows slow-growing plants, such as orchids, to colonise and form large populations.

Orchids do indeed form the most spectacular feature of these sites. Species which are otherwise only found near the sea, or in other places many miles away, unexpectedly appear, presumably through airborne seeds. Many orchids take up to 15 years from the germination of the seed to flowering, and for most of the period the plants are tiny and obscure, so the appearance of the large flower spike is a pleasant surprise.

The most characteristic species are the Marsh Orchids *Dactylorhiza incarnata, D.majalis* subsp. *praetermissz* and subsp. *purpurella*); but several others occur, including the Common Spotted Orchid (*D.fuchsii*), the Fragrant Orchid (*Gymnadenia conopsea*), the Early Purple Orchid (*Orchis macula*), the Marsh Helleborine (*Epipactis pulustris*) and the Pyramidal orchid (*Anacamptis pyramidalis*). Many orchid hybrids, rare elsewhere, are common. The following hybrids have been identified: *D.fuchsii* × *D.purpurella, D.fuchsii* × *D. praetermissa, D.fuchsii* × *D.incarnata, D.incarnata* × *D.praetermissa, D.incarnata* × *D.purpurella* and *D.fuchsii* × *Gymnadenia conopsea*. There hybrids are sometimes much larger than the normal British orchid, in some cases reaching a height of 60 cm.

Although the orchids are the most spectacular, other plants which can tolerate alkaline soil are present. These include some common grasses, for instance Sheep's Fescue (*Festuca ovina*), Red Fescue, Cock's-foot (*Dactylis glomerata*) and Creeping Bent (*Agrostis stolonifera*). As a rule these grow quite well, but are not sufficiently rampant to form a mat and compete serously with the orchids. Other herbs commonly found are Common Knapweed (*Centaurea nigra*), Wild Angelica (*Angelica sylvestris*), Centaury (*Centaurium erythrea*), the exotic introduced Blue-eyed Grass (*Sisyrinchium bermudiana*), Mouse-ear Hawkweed (*Hieracium pilosella*), Common Hawkweed (*H. vulgatum*), Eyebright (*Euphrasia nemorosa*) and Devil's-bit Scabious (*Succisa practensis*). The long-term prospect of these sites is somewhat doubtful: they will require suitable management, which mainly consists of ensuring that there is no application of nitrogenous, and particularly of phosphate, fertiliser. Even one application allows grasses and legumes to grow so vigorously that the orchids in particular are seriously diminished. The other risk is that shrubs such as Hawthorn, Goat Willow (*Salix caprea*), Creeping Willow (*S.repens*), Grey Willows (*S. cinerea*) and Silver Birch will appear and unless controlled may turn the site into dense scrub which, in time, will revert to forest. Indeed, several fairly dense Birch woods already exist.

The Leblanc process ceased to be used in 1919, and was replaced by the Solvay or Ammonia Soda process which produces sodium carbonate from salt, ammonia and limestone. This also generates large amounts of lime waste which has formed some recent lime-rich habitats, but because greater care has been taken in their disposal, these are not as spectacular as those produced in earlier periods from the Leblanc waste.

Although the old Leblanc waste sites are highly alkaline, they may also contain relatively small patches of very acidic soil, as a result of the way in which waste disposal was formerly carried out. Acidic boiler ash was dumped on to the top of the Leblanc waste, and since little mixing took place, the characteristic vegetation of acid soils developed locally. This supports only a few species, typically

Table 17.1

Description of the soil profile of the Leblanc process colonised by a lime flora after 70 years' exposure (after Gemmell).

Depth (cm)	*pH*	*Description of 'soil'*
0-5	7.7	Black with significant humus content.
5-15	7.7	Dark with low humus content.
15-25	7.8}	Yellowish-brown, markedly stained with ferric deposits.
25-35	8.0}	
35-45	9.4}	Yellowish, slightly stained with ferric deposits.
45-55	9.7}	
55-65	12.2}	While with dark lumps of unburnt coal; calcium hydroxide
65-75	12.1}	present.
Below 80	12.6	Bluish-white with dark lumps of unburnt coal. Calcium hydroxide present, permanently n permanently moist.

Wavy Hair Grass. (*Deschampsia flexuosa*), Mat-grass (*Nardus stricta*), Common Bent and Sheep's Sorrel (*Rumex acetosella*). The poverty of the fauna, as compared with that on the alkaline areas, is probably explained by the comparatively small areas of acid soil, and the fact that no species such as orchids, with tiny seed easily dispersed by wind, were involved.

Other lime wastes, generally less alkaline than the Leblanc material, arose from lime kilns, limestone tailings and rubble, calcareous sands and limestone ballast. Blast furnace slag, from the smelting of iron ore with limestone to make pig iron, creates a material which weathers to produce something resembling natural limestone. These support typical calcicole plants. However, they are even more liable to be invaded with Hawthorn and Willow than are the Leblanc wastes, and the improved fertility caused by leaf fall and humus enrichment encourages further scrub and grass, and gradually eliminates the pioneer calcicole plants.

I have already described how poweer station ash (PFA) is treated to produce grass fields and even agricultural land. In its raw state, the PFA generally contains toxic levels of boron and other metals. Nevertheless, ash with a higher than usual lime content is occasionally produced, and here interesting plants may appear. At one ash site in north-west England dense populations of several orchids, including Marsh Orchids, Common Spotted Orchids and Marsh Helleborines, appeared within a period of 15 years from the original tipping. As willow scrub also flourished, this site, unless skilfully managed, will quite soon lose its interest.

Inland salt marshes have occasionally developed around natural brine springs, but the salt industry, as a result of brine spillage and subsidence caused by underground pumping of brine has brought about the local development of brine pools and salt marshes. As these are some distance inland, the seed of most salt-tolerant species (halophytes) has not reached them. Species which have been found in these situations include Orache (*Atriplex patuda*), Salt marsh- grass (*Purccinella distans*), Sea Spurrey (*Spergularia marina*), Sea Aster (*Aster tripolium*), Sea Arrowgrass (*Triglochin maritima*) and Sea Club-rush (*Scirpus maritimus*). Other areas rich in salt are roadside verges where the carriageway has been heavily treated with de-icing materials. There is evidence of halophytic grasses colonising these verges, starting at the seaside and progressing inland. It is thought that moving vehicles transport the seed.

Colliery spoil is not well colonised, particularly where it is very acidic, because of the content of pyrites, which release sulphuric acid over periods of more than 100 years. In some cases the only obvious coloniser is a poor stand of Wavy Hair Grass (*Deschampsia flexuosa*). Less acidic heaps may grow Mat-grass, Common Bent, Yorkshire Fog (Holcus lanatus) and Creeping Soft-grass. In time some herbs, including Sheep's Sorrel, appear. Eventually willows and birches may establish themselves. to be followed by Sessile Oak (*Quercus petraea*) and Aspen (*Populus pendula*). with Bramble (*Rubus fruitcosus*) both under the trees and in open ground. This shows that given time, even the most inhospitable areas develop some interesting vegetation. It is perhaps a pity that today, when we are so much more careful not to have large quantities of toxic wastes lying about the countryside, we will never again produce anything comparable to the chid-rich area that resulted from the almost criminally careless dumping of Leblanc waste.

18

Pesticides

There is considerable public concern about the way in which toxic pesticides, particularly those used in agriculture, may get into the environment. We have many alarming reports of their presence in drinking water and food, and of their effects on wild birds and other wildlife.

At first I had some doubts as to whether this was a suitable topic to be discussed in this book. Pesticides—and here I mean insecticides, herbicides and fungicides in particular—are poisonous substances used to kill organisms which we wish to eliminate because they damage our crops and reduce their yield. If all the pesticide reached its target and none was 'lost', there would not be a problem of wastes. But, unfortunately, this sort of efficiency seldom exists. Where an insecticide is sprayed to control a comparatively rare insect pest which is present in small numbers, less than one per cent of the spray reaches the target and over 99 per cent is wasted, and this waste material may have important ecological effects. Not all pesticide applications are anything like as inefficient. Thus when the herbicide paraquat is sprayed on grass which the farmer wishes to kill, very little falls anywhere except on the grass where it is rapidly absorbed and where its biological activity is eliminated. There are the two extremes—with most pesticide applications there is a substantial waste of the chemical, and for this reason the subject will be treated here.

The subject is one of considerable magnitude. Table 18.1 shows he areas of farmland in England and Wales treated with the different pesticides. Almost all arable land, and even some grassland, is being treated. This costs the farmers a considerable amount, for in 1987 they spent some £195,000,000 on herbicides, £1,000,000 on insecticides and £99,000,000 on fungicides (the British pesticide industry also sold £563,000,000 worth of their products abroad).

There are those who consider that the use of pesticides is unnecessary and could be stopped without seriously damaging our food production and that the use of pesticides has often exacerbated the problem. I consider that this is nonsense. Worldwide it is calculated that nearly 40 per cent of all food produced is lost to pests of some sort. While it is true that some pest problems have been exacerbated by the unwise use of certain chemicals, it is not true that this damage is a new problem. Thus in the Bible there are many references to the devastating effects of insects, and of weeds. To quote a few: in Psalm 105, we read'He spake and the locusts came, caterpillars, and that without number. And did eat all the herbs in their land and the fruit of their ground.' In the book of Job, it says, 'Let thistle grow instead of wheat and cockle instead of barley' (another version puts 'noisesome weeds' instead of cockle). There are many other examples. Several of the plagues of Egypt described in Exodus were caused by insects, and the importance of weeds is stressed in several of the New Testament parables. The israelites would certainly not have agreed that they faced no problems from pests in their agriculture.

Similar stories can be found throughout the world's literature. Today in many tropical countries we see whole crops being destroyed when pesticides are not available. In Britain such total destruction is seldom experienced, but pesticides have greatly increased yields, and have made farming more flexible so that farmers have a greater choice of crops to grow at any time, so they are better able to meet the needs of the customer as well as the possibility of obtaining better returns for their efforts.

It is unfortunate that people often talk of 'pesticides' as if they were all very similar chemicals with similar properties, including toxicity. In fact many of the most widely used herbicides are almost non-poisonous to man and other vertebrates, while some insecticides can cause serious damage if ingested in quite small amounts. Thc situation is well illustrated by the way the European Community has proposed draft regulations for the levels of pesticides which may be allowed in drinking water throughout the community. These regulations suggest that no pesticide should be permitted at a level above one microgram per litre of water, and that the total level of all pesticides should not exceed five micrograms per litre. One microgram per litre is equivalent to one part in a thousand million (or one part in an American billion or in a European milliard). There has been a considerable outcry from some environmentalist groups in Britain, as some samples of our water have contained pesticides in amounts above the levels permitted by the draft directives. They have castigated the British water authorities, and have suggested that consumers are in danger of being poisoned.

In fact no pesticide has been found in any drinking water at a level which his anything like that liable to cause any harm. The most commonly found pesticides (which have only been found in a tiny minority of cases) are the herbicides atrazine and simazine, persistent herbicides used on a variety of crops. These, and other pesticides, have been extensively tested by Fish and Wildlife Service of the United States Department of the Interior. They found that they obtained no mortality at all when four species of birds were exposed to high doses of these herbicides, including water containing 5,000 parts per million. This harmless concentration is 5,000,000 times the maximum level permitted by the EC draft regulation. For most foods, a rule of thumb is to ban levels of toxic substances more than one fiftieth of that which causes damage. It is certainly time that the EEC should think again, and produce recommendations relating substances to their actual toxic properties.

The trouble is that most people are unable to distinguish between the data relating to high and low levels of pesticide contamination. If a chemical analysis shows that some pesticide can be detected at all, they assume that this is in danger, whereas if the level is very low it may be a reassurance. I was recently asked by the Royal Society of Chemistry to address one of their conferences on the subject: 'The analytical chemist is the cause of most pollution'.This title was not of course meant to be taken literally, but was worded so as to draw attention to what is a very real problem. Twenty years ago, when there was concern about the insecticide DDT getting into foodstuffs, in the USA there was a regulation allowing only 'zero tolerance' of this chemical in any sample of food offered for sale. This was quite sensible, for the analytical methods then in use could not detect DDT at levels as low as one part per million, a level not so far below that which could be toxic. However, today our analytical techniques are as much as a million times more accurate, and these can detect infinitesimally small amounts of DDT almost everywhere. at levels very much lower than those which might cause damage.

Some people may take a different attitude, and say that we cannot be absolutely sure that even the lowest concentration of a poison cannot do some harm. What they should realise is that almost every foodstuff, no matter how carefully produced, naturally contains far higher levels of potentially toxic substances, for instance of metals such as lead and mercury, than it does of insecticides or similar chemicals, In vegetables including potatoes poisons like solanin occur at levels which could harm someone eating ten kilograms or more in a day; to obtain the same harmful effect from pesticide residues would usually necessitate consuming an impossible ration ten or more times as great.

In fact in certain cases low levels of pesticides in animals may actually be beneficial. There is a quite common phenomenon known as 'hormesis' which describes the way in which substances harmful at high concentrations may have the opposite effect at low. This is true of many poisons, some of which are used as drugs in small quantities. In many experiments low levels of DDT have been shown to have beneficial effect on metabolism. As the dose increases we first reach the threshold value where, in these cases, no effect can be detected. Only above the threshold can any damage be detected. The general conclusion is that only relatively high levels of most pesticides do any damage to man and other vertebrates. In this chapter we are therefore concerned with the ways in which waste pesticides can occur at levels where their effects can be demonstrated. These effects, as will be shown, may sometimes arise even when particular substances are properly and correctly used. However, the most serious environmental effect usually only arises from some accident, when high levels of toxic substances are accidentally introduced into the environment.

Herbicides

As will be seen from Table 18.1 in England and Wales some eight thousand tonnes of the 'active ingredient' of herbicides is used every year. This is nearly eight times as much as the quantity of all other pesticides—insecticides, fungicides etc—which are used. Although, as indicated-above, some herbicides such as paraquat produce little waste material to enter the environment, other chemicals are not so efficiently material to enter the environment, other chemicals are not so efficiently applied. Nevertheless herbicides have comparatively little harmful effect on the countryside or on its wildlife, except of course they do kill many wild plants—weeds—which is the reason for their application.

Before the 1939-45 war, most of the weedkillers available were highly toxic substance, like sodium arsenite or copper salts. These left persistent residues in the soil, and eliminate earthworms and soil fungi. Fortunately they were only used over comparatively small areas, and so they did little harm to wildlife, though they could kill those organisms with which they came in contact. The most commonly used 'total weedkiller' was sodium chlorate, which was applied to paths and other areas where it was intended to eliminate any plants. This chemical was effective for many months. It was washed out by rain, and got into watercourses where, unless considerably diluted, it continued to kill the plants. It had the disadvantage that under dry conditions it readily exploded, and could cause serious fires when used around buildings.

With such a poor collection of weedkillers at their disposal, farmers relief mainly on cultivation and hand hoeing where practicable. Improved seed cleaning eliminated the infestation of cereal fields by such weeds as the corn cockle. *Aprostemma githago*, something often deplored by conservationists who appreciated the beauty of this species, even though it was poisonous. However, a range of small weeds was generally tolerated in most crops, when little harm was done and some benefit was obtained as they improved the conditions for partridges and other birds.

In 1942, an entirely new type of herbicide, the phenoxyacetic acids, or 'hormone weedkillers' started to be used, and are the main methods of controlling weeds in cereal crops today. Probably the substance most widely used is MCPA(4- chloro-2-methyl- phenoxyacetic acid) which is not at all toxic to insects and vertebrates. This is sometimes as disadvantage, as insect pests such as the fruit fly,

Oscinella frit, appear to survive unharmed when infested oats are sprayed with MCPA to eliminate broadleaved weeds. Tests by the US Fish and Wildlife Service found that concentrations of MCPA of 5,000 parts per million had no harmful effect on qual, pheasant or mallard, and tests on other vertebrates gave similar results.

Under favourable conditions for spraying, that is when there is little wind, most of the MCPA falls on the crop, its weeds or on bare soil. It is not very persistent, and in soil is broken down by bacteria. However, spraying does take place when the wind is blowing, and so quite a proportion is 'wasted' and lands outside the crop. There are many complaints when this happens, although the damage is seldom serious. Herbaceous plants and even shrubs may show the typical distortions these hormone weedkillers produce, and some plant deaths may take place, but the spray becomes more dilute the further it strays from its target, and damage is generally slight and transitory.

Total weedkillers like paraquat and diquat are very poisonous, and have caused deaths when they have been accidentally drunk by children when they have been stored in soft drinks bottles (generally when their parents have 'won' small amounts from their farmer employers to use on their garden). When applied to plants or to the soil, these chemicals are normally absorbed rapidly and so they do not cause any damage except if spraying is careless and the wrong area is sprayed. Recently glyphosate ('Roundup') has been increasingly used. This is particularly lethal to difficult weeds like couch grass and bindweed. It has the advantage that it is a systemic herbicide, that is if part of a plant is wetted with it, the chemical is translocated to other parts of the plant including underground rhyzomes. Glyphosate is rather expensive, so little is wasted and it has few environmental effects. It is also of very low toxicity to vertebrates.

The long-lasting herbicides simazine and atrazine have been mentioned above. They are widely used on soils on paths and railway tracks where it is hoped to prevent any plants from growing for several months after each application. The chemicals are for the most part bound to the soil, and there is no serious waste problem caused by their being leached out. Some losses do occur, and they have been detected in drinking water but at such low levels as to be quite harmless.

Herbicides are useful mainly because they make the practices of ordinary farming much easier. Weed control which otherwise would take many hours of labour can be quickly achieved in a short time by spraying the appropriate compound. As a rule the side effects of using herbicides are not serious. Environmentally the main complaint is that they make possible substantial environmental changes which might otherwise be difficult to achieve. There are some effects from the materials which are wasted when they do not reach their targets. More serious damage has been seen in ditches and watercourses where containers not quite empty have been dumped. There have been accidents where serious spillages into similar places have occurred, but generally, when properly used, the harmful effects have been small. Modern herbicides, though used in great quantities, are mostly non-toxic as compared with the substances used even twenty years ago.

Fungicides

Fungal diseases of crops can be very serious. The infamous potato blight reduced the population of Ireland by nearly 50 per cent in the middle of the nineteenth century. This was caused by the Fungus *Phytophthora infestans.* This parasite is still common, but it can be controlled by several modern fungicides, so serious crops losses are generally avoided. Cereals suffer from seed- and soil-bolund fungal diseases, from mildews and rusts, all of which can greatly reduce yields. Fruit trees are subject to scab,mildew and canker, all caused by fungi. There is therefore a real need for efficient fungicides.

In the nineteenth century Bordeaux mixture, prepared by mixing copper sulphate, lime and water, was used to protect vines in Europe from fungal attack. It and other preparations based on the fungicidal properties of copper were used on potatoes and other crops. When copper was used year after year as in apple orchards, it remained in the soil in concentrations which eliminate most of the soil fauna including earthworms. Sulphur in various forms, as dusts and washes, was widely used. This seems to have had few detrimental effects excepts to the target fungi.

Most cereal crops are treated with seed dressings to prevent such diseases as bunt, which is caused by the fungus *Tilletia caries*, which turns the grain into a revolting black ball of spores. This has been almost completely eliminated by mercurial seed dressings. As the mercury in the seed dressing is left in the soil, it qualifies as a waste product, and might be expected to have some harmful side effect. However, the amount used is so small as to be harmless—only about one milligram is added to every square metre of the field in which the seed dressing is used. This is a small fraction of the amount of mercury present naturally in most soils.

These seed dressings have killed birds which have dug up the seed and eaten it. Various organomercurial salts, with different vertebrate toxicity, have been used. In Britain, phenylmercury compounds, of low toxicity, have normally been used. In Sweden, where much more toxic methylmercury compounds were employed, bird deaths have occurred. Predatory hawks have also been affected when they ate seed eating birds which had fed on treated grain. However, the situation has been complicated in Scandinavia, where large amounts of mercury have been used as 'slimicides' by the timber industry. This has adversely affected life in fresh water.

Most cereals are regularly treated with systemic fungicides. Unfortunately some of the disease-causing fungi have developed resistance to the fungicides and farmers have had to ring the changes to find a now chemical which is still effective.

Although fungicides are so widely used, they have comparatively little effect on the environment outside the target area. This is because they are generally used in small doses, and also the modern organic compounds like Benomyl have a very low toxicity to both vertebrate and invertebrate animals, and they have little or no effect on growing plants.

Insecticides

Although, as Table 13.1 shows, we in Britain used comparatively small amounts of insecticides, public concern about these substances is greatest. In fact for many people 'pesticides' is a synonym for 'insecticides'. Britain is fortunate in that it is rare for any crop to be totally destroyed by insect pests, something which commonly occurs in the tropics. If we had no insecticides farming could go on much as at present. There would be serious losses in cereals, beans, vegetables and fruits, and the cosmetic quality of fruits and vegetables would be reduced (as it is with some of the 'organic' produce I see on sale in Ca nbridge) but our population would not starve.

Before 1939 our supply of effective insecticides was limited. Some were very poisonous, including hydrocyanic gas used as a fumigant, and copper arsenite ('Paris green') for application against caterpillars and Table 18.1 Tonnes of active ingredients of various pesticides used on cereals and on other arable crops in England and Wales. Note that the areas treated refer to treatments. Some were sprayed more than once. Almost all the 3,600,000 ha of cereals were sown with decreased seed

(3,358,000 ha), an operation which can only be done once! Some had no herbicide treatment, some had more than one spray, hence the total of 4,408,000 beetles. Sodium fluoride was used against ants invading houses, and nicotine to kill aphids in glasshouses. The safest preparations were made from natural products, rotenone from *Derris elliptica,* and pyrethrum from the flowers of *Chrysanthemum cinerariaefolium.* These vegetable products had the advantage that they were comparatively

	Hectares treated	*Tonnes of active ingredient*
Cereals		
Organochlorine compounds	1,000	6
Organophosphorus compounds	294,000	107
Other insecticides	272,000	43
Seed treatments	3,358,000	48
Fungicides	978,000	588,
Herbicides	4,408,000	8,026
Other pesticides	188,000	263
Other Arable Crops		
Insecticides		
Organochlorine compounds	40,000	35
Organophosphorus compounds	274,000	99
Other insecticides	180,000	416
Seed treatments	430,000	2
Fungicides	616,000	882
Herbicides	918,000	6,131

non-poisonous to mammals and birds, though derris was used as a fish poison, which did not render the catch dangerous to man. These vegetable insecticides break down quickly, especially if exposed to light, so they never became long-term environmental pollutants. In view of the concern at the persistence of new insecticides in the post-war period, it is somewhat ironical that research was going on to try to prolong the activity of pyrethrum. The disadvantage of the vegetable insecticides, which became very real during the 1939-45 war when there was a risk that insectborne diseases like typhus and malaria might cause more casualties than enemy action, was that they could only be produced in limited amounts, and that to increase the crop would have taken years to organise.

During the 1939-45 war two groups of new insecticides came into use. The first was the organochlorines, or chlorinated hydrocarbons, of which DDT is the most familiar. The second group, the organophosphorus compounds, was discovered in work with allied compounds like mustard gas which were being tested for use in chemical warfare. Parathion (a deadly poison to man) was the first member of this group to be widely used. The great virtue of these synthetic chemicals was that industry could produce them in unlimited quantities.

DDT was synthesised in the laboratory in 1874, but its insecticidal properties were only discovered by Dr Paul Mller of the Swiss Geigy company in 1939; a patent application was made in Switzerland on 7 March 1940. It was first widely used to control the Colorado beetle, which threatened to destroy the Swiss potato crop, in 1941. Samples became available to the British and American authorities, and the chemical was manufactured in considerable quantities in those countries. It was widely used during the latter days of the war, when its use saved many millions of lives. In 1943 when the Allies occupied Naples they found that a typhus epidemic, the disease being carried by the body louse *Pediculus humanus*, had begun, and was causing serious civilian casualties. Had the conditions which existed in the war of 1914-18 been present, the epidemic would have spread with horrendous casualties among both civilians and the military. As it was over two and a half million individuals were rapidly treated by having dust containing DDT liberally blown into their underclothing. This was particularly effective in the least hygienic members of the population who seldom changed their garments, but everyone appeared to be protected. The cases decreased and stopped entirely, and the military appeared to be completely protected. In other theatres of the war, DDT was extensively used to control malaria-carrying mosquitoes.

The most notable feature of the insecticide DDT was its lack of toxicity to man. No serious side effects were noticed in Naples even among those most heavily dosed. In fact DDT has been found to have an acute toxicity similar to aspirin. No deaths from its use have ever been reported, except where it has been eaten in large quantities having been mistaken for flour. At this date it seemed to be the perfect insecticide for use against a wide range of medical and agricultural pests. However, there were those who sounded a word of caution. The earliest published warning which I have discovered was one which I gave at a meeting of the Royal Society of Tropical Medicine in London on 15 February 1945. I said:

"DDT was probably the greatest advance in insect control which had ever been made. But DDT was clearly no panacea, which could be broadcast indiscriminately to kill all noxious pests. A great increase in field research was necessary wherever DDT was used. Fortunately mosquitoes and muscid flies seemed particularly susceptible to this substance, but all arthropods were affected to a lesser or greater extent. Much work should be done on its effects, in the field, on all manner of apparently unimportant insects and other forms of life, to ensure that there was not a serious upset of the 'balance of nature' with subsequent disastrous effects."

I went on to point out that my own experiments had shown that it was less effective against Arachnida, including the scabies mite, *Sarcoptes scabief.*

By the mid-1950s there was more widespread concern about the ecological effects of DDT and the other chlorinate hydrocarbons. In Britain large numbers of seed-eating birds were found dead in the fields in spring, and there were disturbing reports of a fall in numbers of breeding peregrines and sparrowhawks, possible damage to golden eagles and buzzards, and deaths of foxes and badgers. Eventually this was found to be caused by seed corn dressed with aldrin or dieldrin to protect the young seedlings from attack by the larva of the wheat bulbfly, *Leptophylemyia coarctata.*What is ironical is that the damage was done as a result of trying to reduce the risk of widespread pollution resulting from a considerable waste of the insecticides which would have occurred had they been broadcast among the crop. It was decided to stick a minute amount of the insecticide onto each grain, so as to concentrate it at just the spot where the pest attacked. This reduced the total amount of dieldrin or aldrin applied by a factor of ten, but unfortunately the reduced amount of poison was in just the situation where it was ingested by seed-eating birds, who then passed it on to their predators.

There is a comparatively happy ending to this particular story. On the recommendation of the

Advisory Committee on Pesticides, it was agreed that seed dressed with aldrin and dieldrin would no longer be used in spring-sown corn. This was a voluntary but highly effective arrangement, which virtually stopped the deaths of seed-eating birds in subsequent years. Incidentally this effective action in Britain was taken in June 1961, sometime before the publication of Rachel Carson's *Silent Spring*, and it shows that scientists were well aware of the dangers of pesticides long before this book appeared.

In Britain massive deaths from pesticides, such as those which occurred from these seed dressings, have otherwise been uncommon. I know of no cases where DDT has acted as an acute poison, but it has clearly had important ecological effects, and these have resulted in it, and the chlorinated hydrocarbons, being virtually withdrawn from use. Although the acute poisons ceased to be widely used in 1962, DDT was still generally available, notwithstanding increasing ecological evidence that some species of wildlife were having their populations seriously reduced. At first this was difficult to understand. It was found that predatory birds were indeed picking up DDT, but levels in their tissues were generally well below those known to be harmful. This was largely because the DDT was, as we thought, 'detoxified' by being transformed by metabolism to DDE, a compound of much lower toxicity. However, eventually we found that DDE, particularly in birds, has other properties. It is responsible for causing the birds to lay eggs with thin shells which are easily broken in the nest and which reduce breeding success.

DDT was also found to get into freshwater. It was very insoluble, and levels were too low to affect anumals drinking the liquid. However fish, which 'breathe' very large amounts of water to obtain their oxygen, also took the insecticide into their bodies via their gills, and concentrated it by a factor of as much as ten thousandfold. Its use, despite its other values, has therefore been greatly reduced throughout the world, and it has been banned in most developed countries. Its use has continued in countries which cannot afford more costly and less effective chemicals, and where insect-borne diseases are causing many deaths, though even here the search for substitutes continues.

DDT is responsible for making an unimportant pest more serious. As mentioned in my statement to the Royal Society for Tropical Medicine, Arachnida are less seriously affected than insects. Codling moth and other pests of fruit trees have been effectively treated with DDT, but in many cases this has resulted in a great increase in the level of Red Spider Mite, *Tetranychus telarius*, which has proved more difficult to control.

Because of its persistence, many people expressed the fear that even if DDT ceased to be used, its levels in the environment would continue to rise and give serious results for many years. Thus in *Limits To Growth* we read that DDT levels in fish will continue to rise for twenty or so years even after its use stops. Fortunately this has not occurred. Since the 1970s DDT levels in birds and in the environment have decreased considerably. Today the predatory birds usually lay eggs with normal shells, and populations of such species as the Peregrine are back to pre-1939 levels—with the result that racing pigeon owners are once more demanding that their number should be controlled!

The first organophosphorus insecticides, parathion and TEPP, were acute poisons which killed many operatives. Unfortunately when DDT was prematurely banned in some tropical countries, parathion was reintroduced and many humans died. These substances are quickly broken down, so do not act as long lasting pollutants. Fortunately less toxic products such as malathion are now widely used, with little damage to operatives or the environment.

The general trend in recent years is for the use of less poisonous pesticides, and for substances of only limited persistence. The public is much less at risk than it has been in the past. Even before 1939 there were cases of poisoning with arsenic which has been used on apple trees. It is somewhat ironical that there is probably more public concern about the danger of almost non-existent pesticide residues in our food than there was when this was a real problem.

19

Waste Heat

All energy ends up as waste heat. Our own metabolism, which keeps our bodies warm, also produces heat which escapes into the surroundings. In temperate or cold countries this waste heat produces a local warm micro-climate which supports its own fauna and flora. Thus the body louse lives in human clothing and feeds by sucking blood. But it is equally dependent on a warmer environment, for this insect could not normally survive or breed at the sort of outside temperatures usually found in Britain. In fact, people who live in very cold countries, with temperatures well below zero, produce body heat and support parasitic insects and other forms of life on their skin or among their garments which would not otherwise survive.

Metabolic heat, however, is only a fraction of what we liberate by burning fuels and by exploiting the various sources of energy on which civilisation depends. There have been worries that the greatly increased amount of heat might be enough to affect the whole globe, with unpredictable results. This widespread effect is unlikely, for man's present emissions of waste heat, though very substantial, amount only to about 0.01 per cent of the heat received from the sun. Even if the standard of living everywhere was raised to that of the current level enjoyed in the USA, where energy is spent and wasted prodigally, the total heat given out would be less than 0.1 per cent of the sun's contribution.

Nevertheless, the heat we produce can have important local effects. Our cities are significantly warmer than the surrounding country side; the mean annual figure for this warming in London is about 1°C (Fig.19.1). In calm, cold weather in winter, when heating of buildings is at a maximum, the warming is greater and may even rise to about 11°C.

In New York the heat emitted is said to exceed that received from the sun by a factor of 7. The temperature rise there may be greater in summer than in winter, caused by the excessive use of air conditioning, which cools the insides of the buildings, but which transfers this heat, plus what is released by the inefficient use of electricity, to the outside. One other cause of high temperatures, particularly at night, in cities is that the buildings absorb heat from the sun during the day, and this is radiated out during the night.

Warm city centres may support populations of plants and animals which are intolerant of cold. Thus frost is less common in the city, so that species which would be killed elsewhere survive at least the earlier part of the winter in urban gardens. These surviving plants may provide food for animals. The massive numbers of Starlings which roost in Central London, and which make such a mess of buildings, are probably the result of the warmer city climate. Feral Pigeons throng our cities largely because of the food available, much of its provided by generous visitors in defiance of requests by the authorities not to feed the birds. The warmth and the extra light (partly a form of waste energy) enables

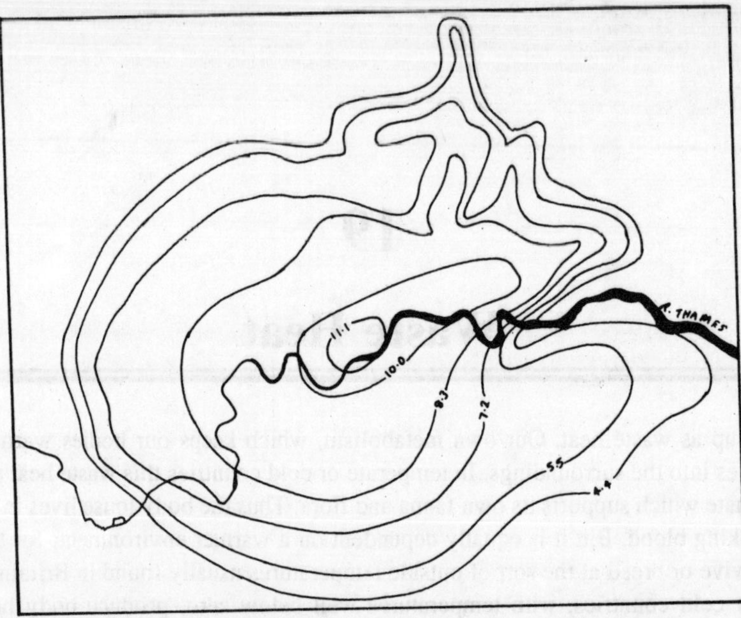

Figure 19.1. Minimum temperature distribution in London on 14 May 1959.

Pigeons and Sparrows to extend their breeding season; also they need less food to keep warm, and so more survive than would under colder conditions.

The waste heat so far considered enters the environment more of less accidentally. The greatest local effects arise when industrial plant, particularly electricity generating installations, are designed so as to use water to cool their machinery. Here the effect is mainly restricted to the water into which any heated effluent is discharged, but the effects on aquatic forms of life may be very considerable. Small power stations have drawn cooling water from streams, passed it through their machinery, and released the warmed water to the stream. They have also used stream water in the operation of their boilers. However, large power stations such as those operated in Britain by the Central Electricity Generating Board need such quantities of water that they would dry up most available natural sources; thus most cooling water is recirculated time and time again, and only comparatively small amounts are returned to affect the river, lake or the sea from which it was originally derived. In these cases the greatest part of the heat loss is by evaporation in cooling towers, but the not inconsiderable residue goes to raise the temperature of the receiving water.

One important effect of raising the temperature of water is that this also reduces its capacity to dissolve oxygen. Thus a litre of water, at 5°C, in free contact with the atmosphere, can absorb about 9 ml of oxygen weighing 13 mgs. As the temperature rises, the oxygen content falls, so that at 20°C it is only about two-thirds of the level at 5°C. As the rate of methodism of coldblooded animals, over the range of temperatures at which they normally live, may double with a rise of temperature of 10°C, one would expect an oxygen shortage. It should be noted that air is much richer than water as a source of oxygen. A litre of air contains 210 ml of oxygen, weighing 300 mgs, i.e. over 20 times as much as is found in the same volume of well-oxygenated water. So a fish has to pass a great deal of water over its gills to breathe, and the lower the oxygen content the greatest the amount of water. As the gills absorb toxic substances as well as oxygen, this may explain why these are particularly dangerous in polluted and thus often partially deoxygenated water.

The question of deoxygenation by sewage and other pollutants is dealt. Cases of deoxygenation by heat occur, but are relatively uncommon. In fact, warm water leaving an electric power station may contain more oxygen than was found in the water, at a lower temperature, when it was abstracted from the river.

At Drakelow on the River Trent, for example, some 850,000 m^3 of effluent from the cooling towers was returned to the river every 24 hours in summer, and 650,000 m^3 in winter. Less water was required in winter as it started off colder, so that it could absorb more heat. The average temperature of the intake in summer was 18.2°C and that of the outlet was 27.4°C. In winter, the corresponding figures were 66°C and 20.4°C. Rises of this magnitude obviously had a substantial local effect, and temperature rise, to 22.4°C in summer, and to 12.4°C in winter, were measured 1.6 km downstream; so a large area of the river, and a large volume of its water, was affected.

It is perhaps surprising that this warming of the effluent did not cause any deoxygenation. In fact, the oxygen level in this river was substantially improved. The Trent is a moderately polluted river, with the oxygen level falling as low as 17 per cent of saturation, and seldom rising above 50 per cent. The water, as it passes through the cooling towers, becomes saturated, or even, to some extent, super-saturated. As it became warmer, the amount of oxygen taken up was limited, but nevertheless at Drakelow between 4 million tonnes of oxygen in summer, and 2 million tonnes in winter, are added to the river every day. Of course, had the river been more highly oxygenated from the start, a different result would have been experienced. But as a rule, heating from cooling towers seldom produces any problem from deoxygenation; as will be seen, high temperatures and toxic biocides are more likely to be harmful.

As mentioned above, waste heat mainly affects the environment by warming water used to cool industrial plant, especially electric generators. Aquatic organisms, other than warm-blooded mammals and birds, have a similar pattern of reactions to temperature. They have a cold death point, where exposure is lethal. Above that is a chill coma temperature, where they are immobilised, but generally survive this state, perhaps for very long periods. At the upper end of the scale is a heat coma temperature, and above that a thermal death point. As a rule, the heat coma temperature is only a degree or two below the thermal death point. The thermal death point may not be sharply defined—organisms may survive for short periods at temperatures which, with longer exposures, are lethal. In between the chill coma temperature and the organisms' upper limiting temperature there is usually a narrower zone where optimum conditions prevail.

The position of these various points on the temperature scale may vary greatly for different organisms. As a general rule, those from warm environments are more liable to damage, and to immobilisation, at temperatures where organisms from cold climates are able to live normally. Thus warm effluents, which may be unsuitable for many native species in Britain, are sometimes invaded by aliens from hotter countries.

In natural freshwaters in Britain, the assemblage of plants and animals in a cold, upland stream differs from that of a warm river at a lower level in the south of England. Many animals found in cold waters cannot survive in warmer water which provide optimum conditions for other species. Thus the flatworm (*Planaria alpina*), which is restricted to cool springs, dies when subjected to temperatures above 28°C, where are both civilians and the military. As it was over two and a half million individuals were rapidly treated by having dust containing DDT liberally blown into their underclothing. This was particularly effective in the least hygienic members of the population who seldom changed their garments, but everyone appeared to be protected. The cases decreased and stopped entirely, and the military appeared to be completely protected. In other theatrees of the war, DDT was extensively used to control malaria-carrying mosquitoes.

The most notable feature of the insecticide DDT was its lack of toxicity to man. No serious side effects were noticed in Naples even among those most heavily dosed. In fact DDT has been found to have an acute toxicity similar to aspirin. No deaths from its use have ever been reported, except where it has been eaten in large quantities having been mistaken for flour. At this date it seemed to be the perfect insecticide for use against a wide range of medical and agricultural pests. However, there were those who sounded a word of caution. The earliest published warning which I have discovered was one which I gave at a meeting of the Royal Society of Tropical Medicine in London on 15 February 1945. I said:

"DDT was probably the greatest advance in insect control which had ever been made. Bit DDT was clearly no panacea, which could be broadcast indiscriminately to kill all noxious pests. A great increase in field research was necessary wherever DDT was used. Fortunately mosquitoes and muscid flies seemed particularly susceptible to this substance, but all arthropods were affected to a lesser or greater extent. Much work should be done on its effects, in the field, on all manner of apparently unimportant insects and other forms of life, to ensure that there was not a serious upset of the 'balance of nature' with subsequent disastrous effects."

I went on to point out that my own experiments had shown that it was less effective against Arachnida, including the scabies mite, *Sarcoptes scabiei.*

By the mid-1950s there was more widespread concern about the ecological effects of DDT and the other chlorinated hydrocarbons. In Britain large numbers of seed-eating birds were found dead in the field in spring, and there were disturbing reports of a fall in numbers of breeding peregrines and sparrowhawks, possible damage to golden eagles and buzzards, and deaths of foxes and badgers. Eventually this was found to be caused by seed corn dressed with aldrin or dieldrin to protect the young seedlings from attack by the larva of the wheat bulbfly, *Leptophylemyia coarctata.* What is ironical is that the damage was done as a result of trying to reduce the risk of widespread pollution resulting from tion. These biocides have proved very effective in reduction asion of mussels in cooling systems that use sea water, and temperature has probably increased their toxic effect.

When an electrict generating station is shut down in winter, the temperature falls substantially. In the USA this has been found to kill fish. Apparently, species which would otherwise migrate downstream in autumn, to avoid the cold, have remained in the favourable conditions of the thermal plume, and have then been subjected to this sudden fall in temperature.

When there is a local rise in temperature to a level which might be harmful, killing sedentary organisms, fish and rapidly moving invertebrates may escape to cooler waters and thus avoid damage. Heat, by increasing the rate of metabolism, will make the fish move more swiftly and thus they are more likely to escape danger.

In Britain, where waters have rarely been heated to dangerous levels, there have been three main effects. First, there has been a change in the composition of the population. Secondly, there has been an extension of the breeding season of organisms which are inactive in cold conditions. Thirdly, there have been given in two volumes on the effects of thermal changes arising from electricity generation by Dr. T.E. Langford (1983, 1990) and only a few characteristic cases will be quoted here.

Where warm waters cause a change in the numbers of different species, as compared with unheated reaches, the variations have generally been less than might be expected. Stonefly larvae (*Plecoptera*), which are particularly susceptible to pollution as well as to raised temperatures, are often eliminated, and the general effect has been a reduction in diversity.

Warming water is apparently beneficial in that it encourages breeding and also extends the period when growth takes place. This has been exploited by those concerned with Oyster and Mussel culture. Many invertebrates evidently continue to breed in autumn in rivers such as the Severn below the Ironbridge power station, when the same species above the power station are more or less dormant.

Important invasions of exotic species in heated British rivers have occasionally been reported. In reaches of the Trent, the exotic Mediterranean snail *Physa acuta* has tended to replace the less heat-tolerant indigenous *P. fontinalis,* and is quite widespread, having been found in more than half the power station effluents studied. The Asian worm *Branchiura sowerbyi* occurs in Britain, but is less common than in America. Considerable numbers of immigrant warm-water marine species have been found in heated waters in the Swansea docks, and the populations appear to be reinforced by specimens brought in by shipping. The exotic barnacle *Balanus amphitrit* and the wood-boring isopod *Limnoria tripunctata* have replaced indigenous species. Damage done by *Limnoria* and also by the shipworm *Teredo navalis* has been exacerbated by the warm conditions which have allowed these animals to extend the period of their activity into the winter. There are few reports of exotic species in rivers, doubtless invasions seldom occur in inland waters.

The comparative unimportance of heat as a pollutant in Britain, and even in America (where many more incidents have been observed), may come as something of a surprise. The position is well summed up by T.E. Langford in the concluding section of his recent book; 'It seems clear that the prediction of the effects of thermal discharges from the new breed of power stations need not be subject to the irresponsible extrapolation and exaggeration which was found in the 1960s and early 1970s. The research and surveys have not borne out the dire predictions of disaster which came from academic and political ambition.'

It is perhaps surprising, in these days when there is so much talk (if so little action) about 'recycling', that more use is not made of all the waste heat generation in our power stations. The trouble is that such waste is so-called 'low grade heat', where temperature are not very high, so that its use is limited. Most of our power stations, too, are some distance from centres of population, where the heat could be utilised. In other countries there are successful schemes where combined use is made of heat and power. For these, the whole installation must be properly designed at the outset, and, as already suggested, the power station must be in an urban situation. In fact, one of the first such schemes was operated successfully in London, but was discontinued when the power station was closed and electricity brought from a large generator outside the city. This was justified because the new generator proved much more efficient than the old one, only using half the fuel to generate the same amount of electricity, more than compensating for the saving made in the previous scheme.

The one area in which surplus waste heat has been used successfully is in the heating of glasshouses—there is plenty of room for these in the rural sites where the power stations are now built. Since the heat would otherwise probably be totally wasted, it does not matter that this heating method is rather inefficient. It is likely that this use will increase, and that tropical crops will become a feature of the Trent valley where the main concentration of power stations is found. The other successful use of waste heat is in fish farms, and in raising shellfish, where growth is accelerated by the raised temperature.This is generally uneconomical where the energy has to be paid for at current rates.

20

Incinerators and Resource Recovery Systems

INTRODUCTION

Definitions and Scope of Chapter

The term *resource recovery* is used to describe the extraction of usable materials or energy from waste. The recovery process may involve conversion of the waste to energy, chemical transformation into a new substance, re-use in the same application after processing, or recycling into another product. *Incineration* is a process used to decompose a chemically complex waste material into its constituent components by the application of heat. The heat generated by the process can be converted to other useful forms of energy and those elemental components of value can be collected for recycling. An incineration system can recover either energy, materials or both.

The term waste is applied to a vast array of items and substances that have no apparent application, ranging from household refuse to by-products of chemical production to asphalt from old roads. What is considered waste is subjective. Today, natural gas is considered a valuable fuel source; previously, this off-gas generated in the drilling of oil wells was considered waste. Most anything can be recycled, recovered, or reused. The question is whether it is economic to do so for a particular waste stream at a given time and place. As natural supplies of materials and inexpensive energy are depleted, the incentives for material reclamation and energy recovery will increase. In this chapter, the terms waste is applied to the materials input into an incinerator for recovery purposes, although as such, the waste becomes a raw material for the process.

The quantity, composition and source of the waste will dictate which recovery options are economically feasible. There are many chemical and physical processes that can be used to recover various types of materials from certain waste streams; a brief survey of noncombustion recovery processes is provided in Table 20.1. Incineration as a recovery option is applicable to a large number of waste streams. Energy can be recovered by burning high-energy content organic wastes. Valuable inorganic materials can be recovered by separating a large-volume organic component from a small-volume inorganic component. The organic component is burnt, and the inorganic metal or mineral is reclaimed from the combustion residue.

The chapter attempts to provide sufficient information on the combustion process, incinerator technologies and pertinent regulatory and economic issues to determine whether resource recovery via incineration is appropriate for a particular application. This chapter focuses on systems that are applicable for industrial and commercial scale applications, rather than municipal treatment facilities. Typical industrial incinerators range from small (3 million Btu/hour units) to large (100 million Btu/hour systems). with feed rates ranging from 100 pounds per hour to 50 tons per day(tpd). Municipal system feed rates may vary from 100 tpd for a small unit to 3,000 tpd for large facility.

Industrial, commercial, and municipal incinerators opreate on the same priniciples and incorporate many of the same design features; however, the large scale of the municipal projects, their predominantly heterogenous waste streams and their public, rather than private, service functions, do require that. additional and unique engineering and operational issues be addressed. Industrial energy recovery systems generate energy that can usually be used at the same plant or a nearby building; municipal systems generate energy requiring distribution to remote customers. Industrial material recovery systems are often designed for a homogeneous waste stream and the recovery of a specfic waste component from the residue. Where material recovery is practiced at municipal incinerators, it usually occurs prior to incineration; recoverable materials such as paper, glass, and aluminum are removed from the waste stream prior to its being burned. The primary function of a municipal incinerator is the efficient volume reduction and sanitary disposal of solid waste, with resource recovery used to offset operational costs. In some industrial and commercial systems, waste disposal is also the main objective. However, the goal of most industrial and commercial material-recovery systems is to tecover valuable substances

Table 20.1

Potential Recovery Processes

Transformation Category	*Process*	*Appllication Examples*
Sorption	Absotption Adsorption	Solvent recovery Metal recovery
Molecular separation	Reverse osmosis Ion exchange Ultrafiltation	Desalination Metal; recovery Mineral recovery
Phase transition	Distillation Evaporation Condensation Refrigeration	Solvent recovery Sulfuric acid recovery Solvent recovery Gasoline vapor recovery
Chemical modification	Cementation Precipitation Catalytic hydrogenation Reduction	Copper recovery Metal recovery Transformation of organic compounds Metal recovery
Physical dispersion and separation	Flotation Liquid/liquid exraction	Oil recovery Solvent recovery
Other	Surpercritical fluid extraction Electrolytic demulsification Liquid surfactant membranes	Organic component recovery Oily wastewater treatment Organic acids and amines recovery

Survey of processes for removal amd possible recovery of components from industrial waste streams.

profitably—at costs lower than the mining and processing costs for the raw materials. Whereas the wastes fed into an industrial or commercial incinerator are usually generated or purchased by the operator, the municipal, incinerator is paid for the disposal of the wastes by the generator.

Incentives for Developing Resource-Recovery Systems

Two factors have contributed, especially in the last two decades, to the growing interest in the recovery of resources from materials that previously would have been discarded: the diminishing supply of inexpensive raw materials, and growing environmental concerns as reflected by regulatory policies.

The incentives to reclaim nonrenewable resources will inevitably increase as decreasing natural supplies and the increasing political constraints restrict access to many of the earth's remaining reserves. Most supplies that could easily be extracted have already been consumed. Imports account for a significant portion of the supply of some militarily and economically strategic materials. Recycling metals, even in small quantities, is becoming an increasingly cost-effective alternative to buying new raw materials obtained from low-grade ores or remote areas. Political interest in decreasing dependence on imported energy soures has lead to policies encouraging the development of alternate sources of energy, including waste-to-energy plants. However, it should be noted even if all the municipal solid waste generated in the United States were to be converted to energy, only amout 1% of domestic energy needs would be met.

Environmental concerns, and the regulations resulting from them, are an increasing influence on resource-recovery systems with conflicting affects in incineration project economics. On the one hand, air pollution regulations have required the addition of expensive air pollution control sustems to incinerators. On the other hand, increasing concern about the long-term environmental effects of landfills has resulted in stricter regulatory controls over landfill construction and operation, reducing the cost competitiveness of landfills as a dispossal option. For example, narionally in 1985, the average tipping fee at municipal resource recovery facilities was about twice thatat sanitary landfill. However, regionally, due to the growing scarcity of landfill capacity, especially in the Northeast, there was much less discrepancy in the fees charged for the two disposal modes. A recent U.S. EPA survey concluded that available hazardous waste land disposal capacity, estimated at about 31 million tons, will be filled by the mid-1990s if it continues to be used at the current rate of 3 million tons annually.

Incineration is becoming a major alternative to land disposal for many waste types. As the costs for the disposal of both hazardous and nonhazardous industrial wastes increase, companies are looking at resource recovery in cases where it may not be economically profitable by itself, but can serve to offset the disposal costs associated with their wastes.

Historical Perspective

Perhistoric man inhabited one location, such as a cave, for centuries without becoming buried in his own refuse. Even after the development of larger communities, most waste materials were separated and either sold or given away for re-use. Data show that the refuse yards to which all the solid wastes from the city of Edinburgh, Scotland, were brought remained the same size for more than a century. It appears that with creative handling of the problems, profitable uses were found for most wastes throughout history until the start of the industrial revolution.

Since the industrial revolution began, except for temporary increases in recycling activity during the two World War, the recycling rate has steadily decreased. The proportion of domestic wastes which have been recycled declined to below 1% by mass. The actual numbers depend on what wastes and recycling activities are included. The decline in recycling was triggered by the concentration of populations and the specialization of industry. As people moved from the farms to the cities and the operations of agriculture and industries grew in scale, the large quantities of wastes generated could no longer be beneficially absorbed locally. In addition to the quantity and centralization of waste generation, modern manufacturing processes created wastes of new chemical composition and often dangerous characteristics. The industrial revolution brought with it a rise in the standard of living that mirrored the decline in recycling rates. In the U.S., recycling could not compete economically with inexpensive lanfill space and abundant raw materials.

The first incinerator, defined as a *facility devoted to the burning of refuse*, was built in England in 1874. By 1914, England had 200 incinerators, over 30% of which produced electrcity from associated steam plants. The first U.S. incinerators were built in 1885 for the government . In 1905, a refuse-fired steam plant powered the lights on the Williamsburg Bridge in New York City. Subsequently, many incinertors for waste disposal and/or energy and resource recovery were built and operated in the U.S. until the combination of air pollution regulation and inexpensive landfill space in the 1960s and 1970s made them obsolete. In many areas, small incinerators for apartment houses and commercial applications were banned because of problems associated with reducing stack emissions.

In Europe, a number of factors combined to encourage the disposal of wastes and the recovery of energy via incineration after World War II. In Denmark, for example, 60% of the municipal solid waste is currently converted to energy. Increasing populations required most of the available land for housing and farming; waste disposal in landfills soon became impractical. Simultaneously, the higher cost of fossil fuels in Europe encouraged energy recovery. Also, utilities in Europe were generally state owned, so the interests of the energy and waste disposal industries could be coordinated. In the U.S., where a mix of private and public ownership exists, it is much more difficult to guarantee a steady supply of waste to, and energy from, incinerator facilities. Only as energy costs rose in the 1970s and supplies became unreliable did generation of energy from incineration become economically attractive in the U.S. In 1982 there were over 500 refuse-fired energy systems in Europe, while only about 30 were operating in the U.S. By 1987, more than 300 systems were operating, under construction, or in the planning stages.

After the sudden price increases of the late 1970s, the relatively low energy and raw material prices of the early 1980s, together with the high costs associated with pollution control, resulted in unfavorable resource-recovery project economics. Currently, on both the municipal and industrial scales, resource recovery is often a break-even proposition at best, if seen strictly in terms of the costs of energy and materials. However, the economics improve when energy and material recovery are seen as a means of offsetting disposal costs while simultaneously providing a buffer against expected future rises in energy and raw material prices.

INCINERATION FUNDAMENTALS: A PROCESS OVERVIEW

Combustion Fundamentals

Incineration separates combustible materials from noncombustible residue. In the physical sense, as indicated in Figure 20.1, burning is the joining of carbon atoms from the waste with oxygen atoms from the air, forming carbon dioxide amd releasing energy. Heat is required to initiate the reaction and is generated by it. Combustion, defined as the rapid union of a substance with oxygen accompanied by the evolution of light and heat, transforms the carbon, hydrogen, and oxygen present in the solid waste stream into gases which can be safely returned to the environment: carbon dioxide and water.

Any other components of the waste stream are discharged from the process as either gases (flue) or a solid residue (ash). The composition of the exhaust gases and ash depends on the inorganic components of the waste stream. Certain elements in the waste will result in the generation of toxic compounds which must be removed, before the exhaust gases are released into the atmosphere, or treated, before the ash is landfilled. The actual combustion products from a particular waste stream in a given type of incinerator will be determined by numerous chemical and physical processes, some of which are not well understood. However, the following assumptions and generalizations are used when designing incineration systems:

1. All of the carbon the waste and fuel is converted to carbon dioxide (CO_2). In actual systems, compelete combustion may result in the generation of carbon monoxide (CO) and minute quantities of a wide range of organic compounds.

2. Incineration of chlorinated hydrocarbons generates hydrogen chloride (HCI), assuming sufficient quantities of hydrogen are present, and small amounts of chlorine (Cl_2). A significant percentage of the HCI (20-40%) can be absorbed by alkaline particulates within the flue gases.

3. Incineration of organic fluorides generates hydrogen fluroide (HF).

4. Incineration of organic bromines generates bromine (Br_2) and small amounts of hydrogen bromide (HBr).

5. Incineration of organic iodide compounds generates iodine (I_2).

6. Incineration of organophosphorus compounds generates phosphorus pentoxide (P_2O_5), which is considered quite hazardous.

7. Most of the sulfur in the waste and fuel is converted to sulfur dioxide (SO_2) and about 2%-4% sulfur trioxide (SO_3). The ratio of SO_2/SO_3 depends on the chemistry of sulfur in the fuel, time sequence of temperature, composition of the flue gases, and on the presence of catalytic ash material. Higher proportions of SO_3 result from the burning of organic sulfonates, some heavy metal sulfates, and discarded automobile batteries, which contain free sulfuric acid. Some sulfur may remain in the ash, as inorganic sulfates or through gas-solid reactions of sulfur oxides.

8. Most of the nitrogen in the waste, fuel, or combustion air is exhausted as nitrogen (N_2); however, comparatively small, but potentially significant quantities of nitrogen oxides (NO_x), mostly NO_2 and some NO, are generated by thermal fixation of nitrogen in the combustion air or organic nitrogen compounds. At flame temperatures above 800°C, burning of organic nitrogen compounds generates nitric oxide; at flame temperature above 1800°C, nitrogen fixation in the air becomes important. At temperatures below 600°C, oxidation to NO_2 occurs.

9. All hydrogen not accounted for above is exhausted as water vapor (H_2O). Hydrogen (H_2) is a possible product of imcomplete combustion.

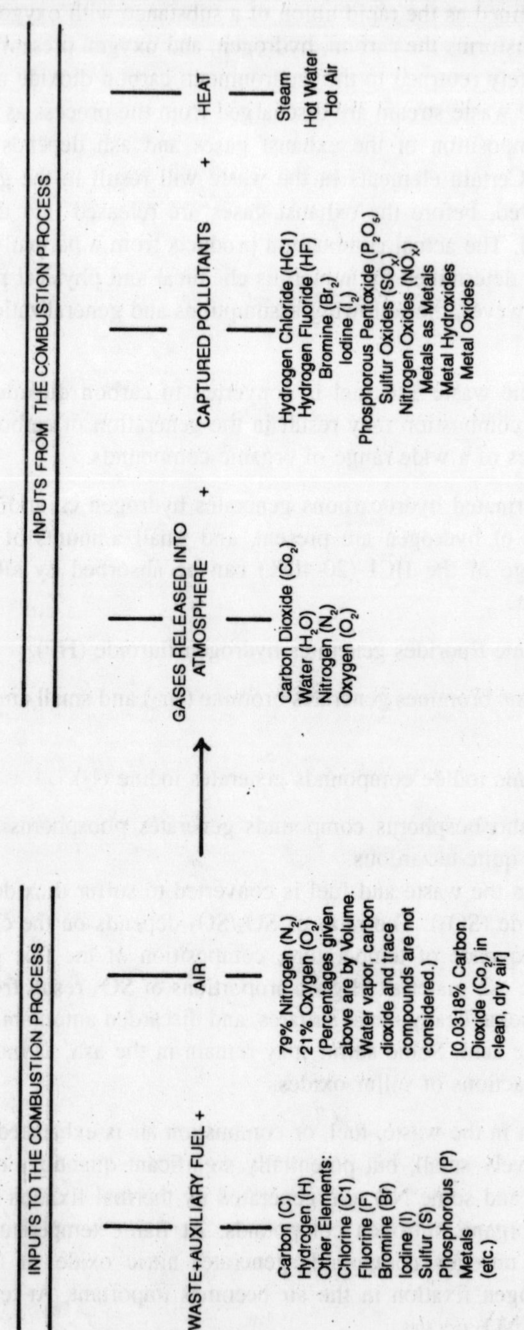

Figure 20.1. Major categories of reactants and products encountered in waste incineration processes.

Table 20.2
Analysis of Waste Characteristics Influencing Incinerator System Design

GENERAL INOFORMATION

Source of Waste Stream
Homogeneous vs. Heterogeneous
Change in Composition with Time
Change in Disposal Rates with Time
(average, minimum, maximum, seasonal variation)
Change in all Significant Parameters with Time
(btu values, etc.)
Container Characteristics
(Shape, size, weight)
Storage Conditions
(temperature, humidity, compactibility)
Physical State of Waste at Ambient Temperature
Change in Physical State of Waste with Temperature
Handling Precautions
(radioactive, infections, toxic, explosive, hazardous according to EPA definition)
Nature of Toxic Hazard
(ingestion, inhalation, dermal contact)
Potential for Mixing of Waste Streams to Increase Process Efficiency

GENERAL PHYSICAL PARAMETERS

Total Waste Quantity
Physical State
(solid, liquid, gas, mixture)
Characteristics of Incinerator Feed
(quantity, shape, size, weight, volume, density, stratification, surface area)
Compositional Weight Fraction
(% metal, paper, glass, textiles, food waste, plastics, rubber, leather, wood, etc.)
Process Weight Fraction
(% combustible, noncombustible, recoverable, etc.)
General Description
(age, odor, colour, pH)

ADDITIONAL PHYSICAL PARAMETERS FOR SOLIDS

Percent Soluble
Percent Suspendable
Hardness
Melting Point
Heat Conduction
Particle Characteristics
(size, distribution, shape, aggregation
Surface Area
Sorption Characteristics
Porosity
Density

ADDITIONAL PHYSICAL PARAMETERS FOR LIQUIDS

Temperature/Viscosity Relationship
Turbidity
Boiling Point
Vapor Pressure
Percent Total Solids
Percent Soluble Solids
Percent Suspended Solids
Percent Aqueous
Stratification
Dissolved Oxygen
Gel Formation

ADDITIONAL PHYSICAL PARAMETERS FOR GASES

Temperature/Pressure Quality
Particulate Content
Liquid Content
Density

COMBUSTION CHARACTERISTICS

Heat Content (btu value)
Specific Heat
Specific Gravity
Stoichiometric Oxygen Requirement
Flame Temperature
Combustion Products
Pyrolysis Products
Combustible Product Content
Mositure Content
Volatile Content
Fixed Carbon Content
Ash Content
Ash Characteritics
(fusible, nonfusible, metallic)
Minmum Time/Temperature Requirement for Complete Waste Decomposition

CHEMICAL PARAMETERS

Percent Carbon
Percent Hydrogen
Percent Oxygen
Percent Sulfur
Percent Nitrogen
Percent Chlorine
Percent Fluorine
Percent Bromine
Percent Iodine
Percent Heavy Metals
(mercury, lead, cadmium, copper, nickel, etc.)
Percent Precious Metals
(gold, silver, platinum, etc.)
Percent Toxics
(chromium, arsenic, selenium, beryllium, asbestos, etc.)
Percent Special Compounds
(benzene, PCBs, dioxins, furans, etc.)

10. All alkali metals are converted to hydroxides (e.g., NaOH, KOH).

11. All non-alkali metals are converted to oxides (e.g., CuO, Fe_2O_3).

12. All oxygen provided as excess air not accounted for above is exhausted as oxygen (O_2).

Waste Analysis and Characteristics

Waste streams are usually categorized by their source: municipal (household), commercial (stores, hospitals) or industrial (manufacturing plants, refineries). Waste streams can vary significantly in composition. Solid wastes can consist of non- renewable resources such as metals or renewable resources such as wood products. Waste generated directly by a manufacturing process can be homogeneous in nature. The refuse generated by society, including various discarded manufactured products, is very heterogeneous in form and composition.

Proper analysis of the waste stream is critical to designing the best system for an application. In addition to determining the capacity required of the system, three major parameters must be identified: physical form, heating value, and chemical composition. Table 20.2 lists some of the data required from a comprehensive waste analysis. For heterogeneous waste streams, such as from municipal and commercial sources, characterizing the feed accurately can be quite difficult; for wastes generated by specific industrial processes, this information is often already available. If necessary, commercial laboratories equipped with the, appropriate analytical equipment, including gas chromatographs, infrared scanners, and mass spectrometers, can perform the required analysis. Standards and techniques for the proper sampling and analysis of materials have been promulgated and documented by various trade organizations and government agencies, such as the American Society for Testing and Materials (ASTM) and the U.S. Environmental Protection Agency (EPA).

Physical Characteristics

Incinerators can process wastes in various forms: solids, liquids, sludges, gases, or combinations thereof. However, not all incinerator designs can accept all forms of waste; some types can handle large bulky containers, others require preprocessing to reduce the size of the waste feed. Some incinerator designs are more susceptible to damage or operating problems from ash fusion; they would not be appropriate for waste streams with a fusible ash content. The nature of the ash content also indicates the potential for slag formation and particulate emissions. Not all incinerators are designed for high temperature operation (above 800°C); certain waste streams, especially those containing hazardous wastes, could not be handled.

Heating Value

The heating value of a waste, usually expressed as Btu/Ib, is the quantity of heat released when the waste is burned. The energy content of the waste must be quantified to determine the auxiliary fuel requirements and to establish processing rates in relation to equipment size. The heating value also indicates whether special equipment or operating conditions are required. For example, a heat recuperator may be needed to improve process economy when burning endothermic materials or a cooling system may be required to maintain the mechanical integrity of the system when burning exothermic wastes. Fuel is required to preheat the equipment and initiate the combustion process. Subsequently, additional auxiliary fuel may be required if the waste's combustion energy is insufficient to produce the design operating temperature. The higher the heating value of the waste, the less fuel is required for operation. Depending on the design of the equipment and the composition of the waste, combustion is self-sustaining at values above 3,000 to 5,000 Btu/lb. Table 20.3 lists some typical heating values for selected waste streams.

Chemical Characteristics

A complete chemical analysis identifying the percentages of carbon, hydrogen, oxygen, nitrogen, sulfur, halogens, metals, etc., is required to calculate stoichiometric (theoretical) combustion air requirements, predict combustion gas flow, and identify the nature and quantity of pollutants that will be generated. A system for processing wastes with a consistent composition would require one type of pollution-control system tailored to the disposal of the particular pollutants generated. A system for an application with potentially heterogeneous waste streams may require a more flexible pollution abatement system, capable of handling a variety of pollutant types and loadings.

Incineration System Fundamentals

System Design Objectives

There are three main reasons for burning wastes: material recovery, energy generation, or waste disposal. The main goal in each case is quite different. Consequently, although the fundamental principles are the same, each unit is equipped with additional or modified equipment for its particular application.

When disposal is the main purpose, the process is expected to provide maximum volume reduction and/or material detoxification. Through incineration, it is often possible to reduce the mass, volume, and weight of the waste by more than 95%, substantially reducing subsequent transportation and landfill disposal costs. Significant reductions in waste quantity are possible only for waste streams with a high carbon content; the carbon is converted to CO_2 and released into the atmosphere. The remaining residue is usually less toxic and easier to handle than the original waste stream. Detoxification of hazardous wastes is accomplished by the thermal decomposition of the complex toxic substances into simpler compounds or elements that are either inert or can be managed more easily than the original hazardous materials. When incinerating wastes with a predominantly inorganic content, changes in physical and chemical characteristics, rather than volume reduction, are the main objectives.

The cost of the additional equipment, associated operational expenses, and lack of a nearby market for the recovered products, often discourages resource recovery in systems dedicated to disposing of wastes as inexpensively as possible.

Although material-and energy-recovery systems are often lumped into one category, they do impose unique requirements on a system's design and operation. Material-recovery systems may require significant amounts of auxiliary fuel and their cost effectiveness depends on the price of the recovered material versus the price of the required fuel. The recovery of energy in material recovery systems is usually limited to reducing auxiliary fuel requirements.

Energy can be recovered from waste-disposal and material-recovery systems regardless of whether the waste stream is exothermic or endothermic. For an energy-recovery system to be profitable, however, the waste stream must be exothermic. requiring auxiliary fuel only to preheat the equipment and initiate

Table 20.3
Energy Content of Selected Waste Streams

Material	*Btu/lb as Fired**
Elements and Compounds	
Carbon(C)	14,093
Sulfur(S)	3,983
Hydrogen (H_2)	61,100
Carbon monoxide (CO)	4,347
Ammonia (NH_3)	9,668
Hydrogen sulfide (H_2S)	7,100
Methane (CH_4)	23,879
Propane (C_3H_8)	21,661
Isobutane (C_4H_{10})	21,257
Ethylene (C_2H_4)	21,644
Benzene (C_6H_6)	18,210
Acetylene (C_2H_2)	21,500
Napthalene ($C_{10}H_8$)	17,298
Methyl alcohol (CH_3OH)	10,259
Ethyl alcohol (C_2H_5OH)	13,161
Acetic acid ($C_2H_4O_2$)	6,280
Consumer Products	
Polyethylene	20,000
Polyurethane (foamed)	13,000
Polystyrene	17,700
Newspapers	7,975
Magazines	5,250
Coated milk cartons	11,330
Meat scraps	7,623
Coffee grounds	10,000
Latex	10,000
Nylon	13,620
Rubber (synthetics)	14,610
Shoe leather	7,240
Asphalt	17,000
Wood (average)	9,000
Rags (linen of cotton)	7,200
Rags (silk or wool)	8,400—8,900
Textile waste (non-synthetic)	8,000
Textile waste (synthetic)	15,000
Linoleum scrap	11,000
Typical Waste Categories	
Trash(type 0 waste)	8,500
Rubbish (type 1 waste)	6,500
Refuse (type 2 waste)	4,300
Garbage (type 3 waste)	2,500
Pathological (type 4 waste)	1,000

** Gross or higher heating value.*

combustion.

Major Subsystem Functions

The block diagram in Figure 20.2 identifies the major equipment subsystems involved in resource-recovery incineration systems. The basic functions of a modern resource- recovery incineration system can be categorized as follows: feed processing, waste incineration, residue handling, air pollution abatement, process monitoring and control, and energy recovery.

Wastes are charged into the incinerator by an appropriate feed system. Within the heart of the system, the incinerator, the materials are thermally decomposed into their elementary components. Various incinerator designs can be used in energy- or resource-recovery applications. Modern incinerators usually consist of an ignition chamber, combustion chamber, and associated peripheral equipment. Wastes are ignited and solids decomposed into gases in the ignition chamber. The combustion chamber is really the first stage of pollution control, completing the oxidation of any remaining combustible components in the flue gases leaving the ignition chamber.

Products generated by the combustion process are either exhausted into the atmosphere through the stack, collected as residue from the ignition chamber, or removed from the flue gases by the air pollution control (APC) system and collected as solid or liquid APC system residues. Residue handling systems collect the residue for disposal or further treatment. Materials, usually metals, are recovered mostly from the ignition chamber, although the APC residue may also contain recoverable substances. The pollution-abatement equipment must minimize the entry into the atmosphere of anything other than carbon dioxide, water, oxygen and nitrogen. In energy recovery systems, energy recovery equipment transforms the heat contained within the combustion gases into other usable forms. A control system supervises the entire operation and monitors the state of the equipment and efficiency of the process. As Figure 20.2 illustrates, the inputs to the combustion process are the waste, auxiliary fuel, combustion air, and possible, water and neutralizing chemicals. The outputs from the process are clean flue gases, solid and liquid residues that may contain concentrated, recoverable materials, and energy.

Candidates for Resource Recovery Via Incineration

Any compound containing carbon, hydrogen, and oxygen is a potential incineration candidate. Optimum candidates for material recovery are wastes requiring separation of an organic component from one major inorganic component of value, such as the recovery of gold from gold-bearing ion exchange resins. Any waste stream with a high carbon and low moisture content would be a good candidate for energy recovery, e.g., organic waste, plastics, and hydrocarbon liquids such as solvents and waste oils.

Characteristics such as a low heating value, high water content, and significant percentages of inorganic salts, sulfur, or halogens indicate that the wastes may be poor incineration candidates. Wastes with a high moisture content require significant quantities of auxiliary fuel or special heat recovery equipment. Inorganic, alkaline salts become airborne and collect on furnace surfaces, creating a slag that may interfere with the proper operation of some incinerators. Chlorides and sulfides result in the generation of acid gases that must be cleaned before being exhausted into the atmosphere; also surfaces within the system require protection from acid attack. The technology for overcoming these problems

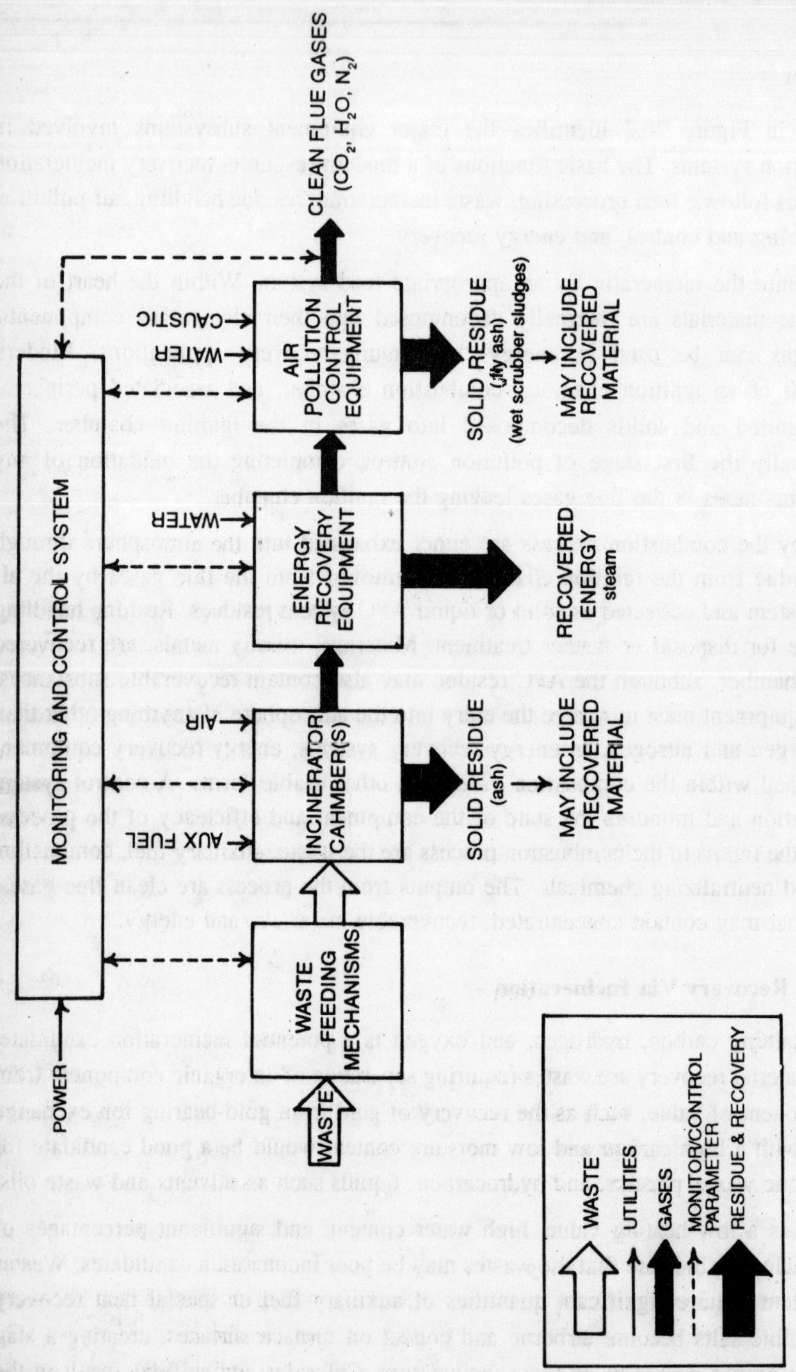

Figure 20.2. Block diagram view of resource recovery incineration process.

exists; however, the costs of auxiliary fuel and additional equipment may be too high for certain applications.

Energy Recovery

Burning is associated with the generation of energy in the form of heat contained within the high-temperature gases. The energy can be used internally within the system to minimize operational costs or it can be converted into other forms for use elsewhere.

Within the system, combustion air and boiler feedwater can be preheated by the flue gases. Some systems make use of the hot flue gases to dry incoming high moisture content waste streams. Energy recovery for use within the process when burning endothermic wastes reduces auxiliary fuel costs. This operating cost reduction can offset the cost of the additional equipment required and improve the overall economics of the project.

When planning to use or market energy recovered from waste incineration, a number of factors must be considered. The quantity, quality, reliability, and availability of this energy source may not compare favourably with traditional energy supplies. This is especially true if future changes in waste stream composition and consistency cannot be accurately predicted. Consequently, provisions for back-up or alternate energy sources may be required. In order to overcome these disadvantages, recovered energy may have to be marketed at prices below those for conventional energy sources.

The advantages of waste-to-energy facilities, however, usually outweigh the disadvantages and there appears to be an increasing trend toward commercial and industrial establishments developing on-site refuse to energy facilities. One case study projected a payback in less than 20 months on a waste-burning boiler investment, and a manufacturer of bedding products reduced fuel oil consumption 50% and avoided disposal costs of $110,000/year by burning approximately 18 tons of fiber scrap weekly.

A number of different approaches can be used to recover energy for use in other applications at the same facility, for sale to another customer, or for sale to the local utility. The destination of the recovered energy will influence what equipment should be used and into what form of energy the heat should be converted. The heat contained within the flue gases can be used to generate steam, hot water, or hot air.

Steam is the most frequently used media for heat recovery. Steam generated by incineration can be used for the same types of applications as steam produced by conventional fuel sources: heating, driving steam turbines to produce electricity, air compression, etc. Specifications for steam output, such as the temperature and pressure, depend on the final application and must be identified before designing the energy recovery system.

In addition to steam and electricity, energy can be recovered from waste in other forms. The waste can be used in the production of a solid fuel for subsequent use in incinerators, conventional boilers, furnaces, etc. Refuse-derived fuel (RDF) is composed of the combustible fraction of solid waste streams, processed into cubes, pellets, powder or fluff. Pyrolysis, the thermal degradation of wastes in a low-oxygen or oxygen-free environment, generates a number of products that can later be incinerated to recover energy: hydrogen, methane, carbon monoxide, methanol, acetic acid, organic liquids and oils, and a solid char. Since the use of RDF applies primarily to municipal systems and practical applications of pyrolysis have had limited success, these forms of energy recovery are not covered in this chapter.

Material Recovery

As discussed in previous sections, incineration concentrates the inorganic components of a waste stream into the solid residue collected either from the combustion processes or the solid or liquid residues collected by the pollution-control system. Practically any component of this residue can be recovered by separation and purification through some mechanical or chemical process. Whether it is economical to do so depends on several factors: the scale and cost of both the incineration and residue processing operations, the current market value of the recovered material(s), and the cost of alternative recovery technologies for the particular waste stream.

In the past, because of their relatively high value, metals have been the materials most frequently recovered by incineration. Past industrial applications have included the recovery of gold and silver from printed circuit boards and resins, the burning off of insulation from copper wires and lead cables, the processing of X-ray film to recover silver, and the recovery of aluminum and zinc from automobile scrap. However, incineration is not limited to the recovery of precious metals. For example, carbon black is produced from the pyrolysis of tires and chlorine can be regenerated from salts in APC residues. Sulfur, when recovered by APC equipment as sulfur, sulfuric acid or sulfur dioxide, can be marketed. Current applications, estimated reserves and reserve bases (which include currently sub economic resources), supply/demand issues, and recovery trends are also surveyed. In these tables, where the supply of scrap is discussed, the term refers to old, post-consumption scrap; clean scrap generated by manufacturing processes and reused or sold is not included. Since some uses are dissipative, all of the raw materials consumed are not eventually available for recycling.

A number of international, political, and economic factors have combined to keep prices of many raw materials, including metals and fuels, down during the early 1980s after experiencing sudden increases in the late 1970s. With some exceptions, the lack of markets, combined with the simultaneous increase in recovery costs due to APC regulations, has discouraged resource recovery during this period. Improved mining techniques have made mining of lower grade ores more attractive. Advances in surveying technologies lead to new sources. Development of new processes and the use of substitute materials in place of the rare or expensive elements also reduced the demand for some materials Projected shortages of some materials, such as gold and zinc, predicted in the early 1970s have failed to materialize. In 1970, known reserves of gold were estimated at 353 million troy ounces, sufficient for 11 years at the 1970 rate of consumption. Known reserves of zinc were put at 123 million tons, or 23 years' worth of supply. However, by 1985, known reserves of gold were placed at 1,280 million troy ounces and zinc at 170 million tons, sufficient at current rates of consumption well past the year 2000 . Large areas of the world, especially potential ocean sources of metals and minerals, have not yet been explored. Modern remote sensing techniques such as multispectral satellite and aircraft scanners may lead to many unexpected sources of raw material.

However, several factors are beginning to combine to make recovery and recycling, even of abundant materials, attractive. As the supplies of easily accessible high grade ores become exhausted, raw material prices will reflect the higher costs of processing lower grade ores or transporting materials from remote mining sites. Many mining and refining operations are very energy-intensive; for example, aluminum is one of the most abundant elements, but the high energy requirements involved in processing the ore encourages recovery of discarded aluminum. Advances in incineration technologies, improvements in APC equipment, and new residue-processing techniques should continue to improve the cost-competitiveness of incineration as a recovery technology. Finally, even if large reserves of raw materials still remain undiscovered, their supply is finite.

In recovering one or more components from a waste stream, incineration is usually the first stage. Some subsequent processing of the recovered components will often be required. In addition to the component(s) being recovered, the residue will contain any other noncombustible elements of the waste stream and a small quantity of products of incomplete combustion. When printed circuit boards or other electronic scrap is burned, a mixture of metals is obtained. The plastics and resins are burned off; the fiberglass and bits of metal, such as pieces of wire and component casings, remain. The mixture may contain recoverable quantities of gold, silver, copper and lead. After the insulation is burned off of copper wires, bundles of clean copper wires remain in the incineration chamber. If lead-bearing cables are incinerated, the molten metal is drained into forms. When X- ray or photographic films and papers are burned, the fly ash and bottom ash contain silver metal or silver oxide which can be separated from the ash by melting and fluxing.

The end use of the recovered material will dictate what additional processing is required. Quality standards for a large number of categories of recovered materials have been established by various trade associations, professional organizations, and government agencies, including the American Society for Testing and Materials (ASTM), the Institute of Scrap Iron and Steel, National Association of Secondary Material Industries, etc. These standards specify characteristics such as type and quantity of impurities that can be accepted by specific end uses and the physical form required of the recovered product. Analytical techniques for verifying compliance with the standards have also been promulgated. The additional processing required to use or market the recovered products will be reflected in the project's operating costs. Identifying the final product specifications and post-combustion processing alternatives is a critical element in the overall economic feasibility analysis of a resource recovery incineration facility.

The method of separation of the component (s) to be recovered from the combustion residue will be determined by the chemical and physical characteristics of the residue, the quantities involved, and the best available technology for the type of processing required. The theory of many separation techniques is well established; due to the potential competitive advantages involved, the practical application of those techniques on an industrial scale often involves proprietary information, trade secrets, or patents. Frequently, the processes may include one or more of the following: magnetic separation, grinding, multistage screening, melting and fluxing, electrolysis (ion exchange), and evaporation and distillation.

Although the U.S.Bureau of Mines developed a continuous process for separating the mixture of metals, glass, slag, mineral-containing ash, and unburned combustibles from municipal incinerator residues, practical applications have been limited by economic constraints. Processing of residue from heterogeneous waste streams, such as from municipal incinerators, results in relatively low recoveries and poor quality materials. Ferrous scrap is sometimes recovered from municipal incinerator residues, but few U.S. facilities recover other products .For municipal mass burn systems, preprocessing of the waste stream to remove metals, glass, and other noncombustibles prior to incineration is preferable, increasing combustion efficiency and decreasing the quantity and toxicity of residue needing to be landfilled.

REGULATORY ISSUES AFFECTING RESOURCE-RECOVERY INCINERATION SYSTEMS

Regulatory requirements have a significant impact on the design and operation of modern resource-recovery incineration systems. The regulations fall into three general categories:

1. Pollution control regulations that restrict the entry into the environment of pollutants contained within either the solid, liquid, or gaseous residues generated by the system. Usually, the regulations restrict pollutant emissions to below some measurable standard.
2. Health and safety regulations limiting worker's exposure to dangerous compounds.
3. Regulations promoting energy and resource recovery through various economic incentives such as tax benefits or simplified access to markets for recovered energy or resources.

Most of these regulations have an administrative component, requiring the reporting of information relevant to the system's design, operation, and compliance with the regulatory limits or standards. Together, these regulations specify the minimum performance required of systems and procedures involved in pollution abatement, materials handling, monitoring and control, and the final treatment and disposal of solid and liquid residues.

There are a number of federal regulations that apply to all incinerators. Other regulations apply to a subset of resource-recovery incinerator systems: those of a particular size or type; such as large municipal resource-recovery systems or small cogeneration facilities. There are additional, and often more restrictive, state and local regulations. The individual regulations are too numerous to be presented here; also, the regulations are constantly being modified as our knowledge of the environmental and health effects of various pollutants increases and new technologies for handling them are developed.

An overview of major environmental issues and federal regulations effecting the design and operation of industrial resource-recovery incinerators is presented below. The regulations assume the development of a new facility, not the retrofitting of existing facilities. Table 20.4 and 20.5 list some pollutants associated with waste incineration systems, the normal ambient concentrations of these substances, several federal regulatory limits, and, as an example of additional state limits, the ambient air concentrations allowable in Connecticut.

Air Emission Regulations

Gaseous emissions from incinerators, depending on the chemical composition of the waste stream, auxiliary fuel, and air, can include various pollutants: particulates, unburned hydrocarbons, carbon monoxide, sulfur oxides, nitrogen oxides, halogen acids, and various metal salts and oxides. Some level of pollution abatement is required when burning most wastes to reduce the particulates and inorganic acids generated to below levels specified by the Clean Air Act, Resource Conservation and Recovery Act, Toxic Substances Control Act and applicable state and local regulations. In some states, such as Connecticut, the Maximum Allowable Stack Concentrations (MASC) for a source are calculated based on specific ambient air standards, stack height, flow rate, and distance to the closest property line. Other states, such as Illinois and California, require that the best available control technology be applied

Table 20.4

Typical Regulatory Limits and Ambient Concentrations of Selected Nonmetallic Pollutant

Pollutant	*Symbol*	*Ambient Concentrations*	*Some Federal Standards and Emission Limis*	*Connecticut Hlv*
Particulates	N/A	21.655 ug/m^3 (AC) 22.680ug/m^3 (AF)	OSHA: 0.0066 g/ft^3, 15 mg/m^3 (dust) 5 um dia., 0.0022 g/ft^3, 5 mg/m^3 NAAQS:75ug/m^3 (annual geometric mean) NSPS:0.08gr/dscf*12%CO_2	0.08 g/dacf 12%CO_2 (2-hr avg) 0.015g/dscf 12% CO_2 (M)
Organic Compounds		0-1 ug/m^3 for non-CH_4 hydrocarbons in rural	RCRA: 99.99% destruction of principal organic hazardous constituent.	(42 or 54% C1 Polychlorinted biphenyis) 0.01 ug/m^3
Carbon Monoxide	CO	.00001% volume, .1 ppm 0.1-0.2 ppm (N.Hemisphere, 0.04-0.06 ppm (S.Hemisphere,	OSHA: 55 mg/m^3, 50 ppm NAAQS: 10 mg/m^3, 9 ppm (8-hr avg)	$\frac{CO}{CO_2}$ ratio of 0.002 (M) 10 mg/m^3, 9 ppm
Sulfur Oxides	SO_x	0.1 ppb (A) (SO_2) 0.00000002% volume, 0.002 ppm (A)	OSHA (SO_2) 13 ug/m^3, 5 ppm NAAQS: (SO_2) 80 ug/m^3, 0.03 ppm (annual arithmetic mean)	(SO_2) 0.32 Ib/MBTU (M)
Nitrogen Oxides	NO_x	(NO_2) 0.0000001% volume, 0.001 ppm (A)	OSHA (NO) 30mg/m^3, 25 ppm (NO_2) 9 mg/m^3, 5 ppm* NAAQS: (NO_2) 100 ug/m^3, 0.05 ppm (annual arithmetic mean)	(NU) 600 ug/m^3, 0.5 ppmv (NO2) 700 ppm (NOx) 0.6 Ib/MBTU (<)
Hydrogen Chloride	HCI	N/A	OSHA : 7 mg/m3, 5 ppm* NOWS: (chloride) 250 mg/1	90% reduction (H)
Hydrogen Fluoride	HF	N/A	NPDES: (flouride) 1.4 mg/1	50 ug/m^3, 0.06 ppmv
Sulfuric Acid	H_2SO_4	N/A	OSHA: 1 mg/m^3	20 ug/m^3

Chlorine	Cl_2	20-900 ppm, 100 ppm avg(5) 0.444 ug/m^3 (AC) 0.155 ug/m^3 (AF)	N/A	60 ug/m^3, 0.02 ppmv
Bromine	Br_2	0.019 ug/m^3 (AC) 0.077 ug/m^3 (AF)	N/A	14 ug/m^3, 2 ppbv

Legend

HLV	Hazard Limitjng Values (B-hour average, unless otherwise specified)
MASC	Maximum Allowable Stack Concentrations
OSHA	Occupational Safety & Health Administration (8-hour time weighted average)
NAAQS	National Ambient Air Quality Standards
NSPS	National Source Performance Standards
NDWS	National Drinking Water Standards
RCRA	Resource Conservation Recovery Act EP Toixicty Test Concentrations
NPDES	National Pollutant Discharge Elimination System (typical limit)
N/A	Not applicable or not available
(M)	Municipal refuse to energy plant limits (not 8-hours HLV)
(S)	Common trace chemical ranges in natural soils
(A)	Atmospheric trace gas concentrations
(AF)	National 1980 mean urban air concentrations of fine particulate (<2.5 (um)
(AC)	National 1980 mean urban air concentrations of coarse particulate (2.5-15um)
*	Ceiling (ceilling concentrations cannot be exceeded at any time)

Calculation Of Conecticut Masc

If the discharge point is 20 meters or less measured vertically from the ground elevation at the discharge point, refer to (A). If the discharge point is more than 20 meters, refer to (B).

(A) $MASC = \frac{1}{F}[0.885H\ (D + 1.08F^{0..64})\ 1.56]$

(B) $MASC = \frac{1}{F}[0.885\ (D + 1.08F0.64)^{1.56}\ \exp\ (10.33(E-20)^2(D + 1.08F^{0.64})^{-1.56})]$

F = average flow rate (cubic meters per second) from discharge point.

D = 10 meters or distance from discharge point to the closest property line, whichever is greater.

H = hazard limiting value (micrograms per cubic meter) or (parts per million).

E + heigth (meters) of the discharge point.

Table 20.5

Typical Regulatory Limits and Ambient Concentrations of Selected Metallic polutants

Pollutant	*Symbol*	*Ambient Concentrations*	*Some Federal Standards and Emission Limits*	*Connecticut Hlv*
Aluminum	Al	1.797ug/m^3 (AC) 0.353 ug.m^3 (AF)	N/A	(metal/oxide)200ug/m^3 (soluble salts) 40 ug/m^3
CadmiumCd		.01-.7 ppm, .06 ppm avg (S) 0.006 ug/m^3 (AC) 0.007 ug/m^3 (AF)	NPDESL 0.05 mg/1 OSHA: (Cd, Cd0) 0.1 mg/m^3 NDWS: 0.01 mg/1 RCRA: 1 mg/1	(dusts) 0.5 ug/m^3 (Cd0 fumes) 0.1 ug/m^3
Chromium	Cr	1-1000 ppm, 100 ppm avg (S) 0.008 ug/m^3 (AC) 0.006 ug/m^3 (AF)	RCRA: 5 mg/1 NDWS: 1 mg/1	(metal) 2.5 ug/m^3 (chromic acid) 0.25 ug/m^3
Copper	Cu	2-100 ppm, 30 ppm avg (S) 0.019 ug/m^3 (AC) 0.026 ug/m^3 (AF)	OSHA 0.1 mg/m^3 NOWSL 1mg/1 NPDES: 0.02 mg/1	(dusts) 20 ug/m^3 (fumes) 2 ug/m^3
Lead	Pb	2-200 ppm 10 ppm avg (S) 0.083 ug/m^3 (AC) 0.314 ug/m^3 (AF)	RCRA: 5 mg/1 OSHA 0.05 mg/m^3 NAAQS: 1.5 ug/m^3 (quarterly arithmetic mean) NOWS: 0.05 mg/1 (50 ppb) NPDES: 0.1 mg/1	(dust, fumes) 3 ug/m^3
Mercury Hg		.01-3 ppm, .0^3 ppm avg (S) 0.003 ug/m^3 (AC) 0.003 ug/m^3 (AF)	RCRA: 0.2 mg/1 OSHA: 0.1 mg/m^3 NPDES: 0.005 mg/1 NPWS: 0.002 mg/1	(alkyl compounds) 0.2 ug/m^3 (vapor) 1 ug/m^3
Nickel	Ni	5-500 ppm, 40 ppm avg (S) 0.004 ug/m^3 (AC) 0.007 ug/m^3 (AF)	OSHA: (Ni) 1 mg/m^3 (Ni(CO)$_4$] 0.001 ppm, 0.007 mg/m^3 NPDES: 1 mg/1	(oxides) 0.3 ug/m^3 (metal) 5 ug/m^3 [Ni(C)$_4$] 1.75 ug/m^3, 0.25 ppbv

Silver	Ag	.01-5 ppm, .05 ppm avg (S)	NPDES 0.005 MG/1 NDWS: 0.05 mg/1 RCRA: 5 mg/1	0.2 ug/m^3
Tin	Sn	0.006 ug/m^3 (AC) 0.005 ug/m^3 (AF)	N/A	(organic compound) 2 ug/m^3 (metal) 40 mg/m^3
Zinc	Zn	10-300 ppm, 50 ppm avg (S) 0.038 ug/m^3 (AC) 0.067 ug/m^3 (AF)	OSHA: (ZnO) 5 mg/m^3 NDWS: 5 mg/1	(ZnO fumes) 100 ug/m^3

Legend

HLV Hazard Limiting Values (8-hour average, unless otherwise specified)

MASC Maximum Allowable Stack Concentrations

OSHA Occupational Safety & Health Administration (8-hour time weighted average)

NAAQS National Ambient Air Quality Standards

NSPS National Source Performance Standards

NDWS National Drinking Water Standards

RCRA Resource Conservation Recovery Act EP Toxity Test Maximum Concentrations

NPDES National Pollutant Discharge Elimination System (typical limit)

N/A Not applicable or not available

(M) Municipal refuse to energy plant limits (not 8-hour HLV)

(S) Common trace chemical ranges in natural soils

(A) Atmospheric trace gas concentrations

(AF) National 1980 mean urban air concentrations of (fine particulate (<2.5 um) course

(AC) National 1980 1980 mean urban air concentrations of coarse particulate (2.5-15um)

* Ceiling (ceiling concentrations cannot be exceeded at any time)

Calculation of Connecticut Masc

If the discharge point is 20 meters or less measured vertically from the ground elevation at the discharge point, refer to (A). If the discharge point is more than 20 meters, refer to (8).

(A) $MASC = \frac{1}{F}[0.885\ H\ (D + 1.08F^{0.64})\ 1.54]$

(B) $MASC = \frac{1}{F}[0.885H\ (D+1.08F^{0.64}exp(10.33(E-20)2(D+1.08F^{0.64})-1.56)]$

F = average flow rate (cubic meters per second) from dis charge point.

D = 10 meters or distance from discharge point to the closest property, line, whichever is greater.

H = hazard limiting value (micrograms per cubic meter) or (parts per million).ug/m^2

(BACT), and limits and operating conditions are specified during the permitting process for each installation. Sources of air pollution other than stack emissions must also be considered, including dusts generated by material handling operations.

Effects of Pollutants

Each type of pollutant or combination of pollutants creates its own type of damage to plants, animal and materials. Primary pollutants are dangerous in the form emitted; secondary pollutants are those formed through reactions with emitted substances, even far from the source.

Incomplete combustion results in the presence of carbon monoxide (CO) in the flue gases. Carbon monoxide has the ability to pass through the lungs directly into the blood stream; there it destroys the ability of the red blood cells to carry oxygen. In addition to CO, incomplete combustion can also result in the generation of various organic compounds. Most are low-molecular-weight hydrocarbons and partially oxygenated compounds (aldehydes, organic acids). Many of the hundreds of hydrocarbons which form air pollutants are possibly carcinogenic and might be at least partially responsible for increased lung cancer rates. Some of the organic compounds are hazardous by themselves; others form complex and harmful compounds in various reactions with nitrogen and sulfur oxides. Trace quantities of polynuclear aromatic hydrocarbons (PNHs), polychlorinated byphenyls (PCBs), polychlorinated dibenzo-p-dioxins (PCDDs), and polychlorinated dibenzofurans (PCDFs) may be found significant because of their suspected impact on health, even at very low concentrations. As more information is obtained regarding formation mechanisms and environmental effects, new regulations governing various paraffins, olefins, aromatics, furans, and dioxins can be expected.

Because they are suspected carcinogens, the trace quantities of PCDDs and PCDFs sometimes found in incinerator residues have been the foundation of much opposition to the incineration of wastes containing chlorinated plastics. However, there is no evidence that plastics serve as precursors for formation of chlorinated dioxins and furans in incinerator residues. They appear to be unburned residuals of compounds originally present in the waste stream. Ensuring complete combustion by proper incinerator design and operation is the best way to avoid discharge of carbon monoixde and potentially hazardous organics into the atmosphere.

Particulate emission include particles of mineral and metallic matter, unburned combustible compounds, soot, heavy tar, and noncombustible aerosols such as sulfuric acid (H_2SO_4) or hydrochloric acid (HCl). The environmental effects of particulates depend on their sizes, which can range from below one micrometer (μm) to over 100 micrometers :

1. Less than 0.1 μm: the particulates induce nucleation of water droplete causing weather modifications.
2. Between 0.4—0.8 μm: the diameters are approximately equal to wavelength of light, leading to greatest restriction of visibility by scattering and absorbing light.
3. Between 1-5 μm: maximum deposition in the lungs upon inhalation.
4. Between 3-15 μm: deposited in the upper respiratory system.
5. Between 10-100 μm: result in dust and dirt.

Particulate that is inhaled can damage the respiratory system and may be toxic. Although metallic particulate discharges into the atmosphere account for less than 1% of the total particulate discharge, metals can have severe toxic effects. Heavy metals lead to direct biochemical reactions. The inhalation of metals has been found to have a greater effect on the human body than receiving the same metals by ingestion. Lead, inhaled or ingested, accumulates in body and can cause headaches, anemia, birth defects, permanent brain and central nervous system damage, and can be fatal. Cadmium, associated with the incineration of cadmium-containing products such as tires and some plastics, is toxic to humans; it interferes with the natural processes of zinc and copper metabolism and can cause cardiovascular diseases and hypertension . Mercury compounds are highly toxic, resulting in severe mental and physical symptoms; poisoning results in blindness; progressive weakening of the muscles, numbness, paralysis, coma, and death . Although exposure to small concentrations of nickel through food and water does not appear to cause illness in humans, nickel carbonyl is extremely toxic.

Particulates can act as a catalyst in the conversion SO_2 to SO_3. About 90% of the sulfur dioxide and 10% of the nitrogen oxides found in the atmosphere are the result of human activity, mostly related to combustion sources. More than 95% of SO_x is SO_2, which slowy oxidizes to SO_3 . SO_2 is an eye irritant, causes and aggravates respiratory diseases such as emphysema and bronchitis, and is linked to hung cancer. SO_3 is toxic to humans, especially to older people and those with respiratory problems. SO_3reacts with water vapor to produce sulfuric acid (H_2SO_4), and acid that damages lung tissues. Nitric oxide (NO) is not considered harmful in itself but it converts to NO_2. Nitrogen dioxide also acts as an eye irritant and can damage the respiratory system . Both NO_2 and the sulfur oxides contribute to smog. Nitric oxide (NO) and sulfur trioxide (SO_3) react in the atmosphere forming nitric acid (HNO_3) and sulfuric acid (H_2SO_4), respectively, contributing to the acid rain phenomenon. Acid rain damages fabrics and metals, as well as the environment. Calcium carbonate ($CaCO_3$), a major ingredient in marble, mortar, and concrete, reacts readily with H_2SO_4 to produce water- soluble calcium sulfate ($CaSO_4$). The calcium sulfate is washed away, slowly dissolving ancient monuments and newer buildings.

Ozone (O_3), generated by photochemical reactions between various nitrogen oxides and oxygen in the atmosphere, aggravates lung problems such as asthma and lowers resistance to infections. Although nitrogen oxides do not usually reach the stratosphere, if they do, they react with ozone to form oxygen ($NO+O_3 \rightarrow NO_2+O_2$), destroying the ozone layer that helps shield us from ultraviolet radiation.

Normal air contains various gaseous and particulate components; plants evolved in an environment with a set-range of component concentrations. Even small deviations from background concentrations can exceed the boundaries of adaptation and tolerance in plants. Plants are especially susceptible to damage from SO_2 (in concentrations as low as 0.5 parts per million (ppm), NO_2, HF, and secondary pollutants, such as ozone. Nitrogen oxides rarely occur in concentrations that are directly harmful to plants but they serve as precursors to ozone and other pollutants. Nickel has been reported to be toxic to some plants in concentrations as low as 40 ppm .

Clean Air Act and the National Ambient Air Quality Standards

Until the early 1970s, facilities could be built almost anywhere without considering their effect on the ambient air. Today, consideration of environmental effects is a standard step in the siting and design of new facilities. The Clean Air Act (CAA) of 1970 and subsequent amendments provided the foundation for the air pollution control regulations currently in place, giving the EPA the responsibility and authority to establish a nationwide program of air pollution abatement and air quality enhancement. In general, the CAA provides for a combination of state and federal stationary source emission standards designed to achieve and maintain federally prescribed ambient air quality. This air quality is defined by the National Ambient Air Quality Standards (NAAQS) for certain pollutants that were of immediate concern with respect to the public's health, called the criteria pollutants: CO, SO_2, NO_2, ozone, total suspended particulates, and lead. Regulations governing permitting, testing, and record- keeping were also enacted. The limits effect all potential sources; the monitoring and permitting requirements also affect state and local governments, manufacturers of testing equipment, fuel suppliers, and the general public. For a resource-recovery project, in addition to costs associated with the air pollution control equipment, the legal and administrative costs associated with compliance can be quite substantial.

Both primary and secondary limits are set for each of the criteria pollutants; the primary limit is based on health effects, and the secondary limits are based on the detrimental effects of the pollutants on vegetation, visibility, and architectural surfaces. A mathematical model is used to determine the effects of emissions from a facility. Incorporating such factors as stack height, prevailing winds, and topography, the model predicts the ground level concentrations for each pollutant. These predictions are compared with the NAAQS to determine if additional controls are required.

The responsibility for attaining the NAAQS rests with the states, under federal supervision. Since air quality is not consistent across the nation, each state's State Implementation Plan (SIP)—its approach to achieving and maintaining the NAAQS while still allowing for industrial and population growth—is different. State regulations cover new and existing sources of air pollution not subject to other specific regulations.

To minimize the creation of new large sources of pollution, the New Source Performance Standards (NSPS) were established. NSPS apply to new or modified sources of air pollution specifically designated by the regulation. Several dozen major source categories are currently covered by these regulations; most of the industrial incinerators of the scale being discussed here are curently not covered, but new categories may be added. These regulations are standards, not just emission limits; therefore, they may require process modifications or specific emission reduction techniques. NSPS require the best system of continuous emission reduction that is available, considering economic and energy implications for the overall project. In cases where the NSPS limits a noncriteria pollutant, each state is required to develop regulations limiting emissions of that pollutant from all existing sources.

Prevention of Significant Deterioration (PSD) and Nonattainment (NA) regulations address the preconstruction review of major new or modified emission sources. With respect to a given pollutant, a region either has an ambient air quality better than the applicable NAAQS and/or is a nonattainment area. The goal of the PSD and NA regulations is to preserve the clean air of the attainment areas and to improve air quality in the nonattainment areas. A major new source of pollutants may locate in a clean area only after showing that its emissions will not violate ambient air standards; it may locate in an nonattainment area only after showing that its emissions will be more than offset by emission reduction measures taken at other facilities. Lowest Achievable Emission Rate (LAER) control technology, without regard for economic factors, is required for sources wishing to locate in an NA area and having potential regulated pollutant emissions over 50 tons per year. In general, only facilities that have the potential, determined after the control device and including all fugitive emissions, to emit over 250 tons per year of any regulated pollutants are subject to PSD or NA review requirements.

For the scale of incinerators being discussed here, federal PSD and NA regulations would not normally apply but similar state local reviews and restrictions, with lower limits, may exist for these smaller sources. In addition to the criteria pollutants, PSD and NA regulations also cover volatile organic compounds (VOC), beryllium, mercury, vinyl chloride, fluorides, sulfuric acid mist, hydrogen sulfide, and total reduced sulfur.

National Emission Standards for Hazardous Pollutants

Hazardous pollutants not listed as criteria pollutants are subject to the National Emission Standards for Hazardous Pollutants (NESHAP). Hazardous pollutants are those identified as resulting in an increase in mortality or an increase in serious irreversible or incapacitating reversible illnesses. NESHAP regulations apply to both new or existing emission sources of any designated pollutant and to a list of specifically designated sources. Pollutants for which NESHAP standards have been established include mercury, cadmium, and polycyclic organic matter (POM). These standards include not only specific emission limits, but design and operational specifications as well. Because of the severe health implications from uncontrolled emissions, NESHAPs do not require consideration of economic effects and include extensive monitoring and reporting requirements .

Liquid Residue Disposal Regulations

There are a number of potential sources of aqueous residues in incineration systems. Water is used:

1. In wet scrubbers to wash particulates and acid gases from the flue gases.
2. To cool and/or reduce dust during waste feeding and ash collection.
3. As a cooling medium in non-contact water-cooled screw feeders, ash conveyors, flue gas coolers, etc.
4. For heat recovery in boilers.

Much of the water introduced into the system will be discharged into the atmosphere as (clean) steam. Depending on the application and the local availability of water, water can be recirculated within the system. For example, the water used during the operation of wet scrubbers can be recirculated through an appropriate water conditioning system, generating small amounts of sludge that can be processed through the incinerator. However, in some cases, aqueous residue from the system will need to be continuously discharged.

Water is pure only in the vapor phase; gases, including O_2 and CO_2, dissolve in the water as soon as droplets form. Mineral acids are formed as SO_2 and NO_x dissolve in water. Contact with soil and minerals results in naturally occuring minerals dissolving in groundwater and surface waters. Chemical impurities commonly found in water in significant quantities include calcium, magnesium, sodium, potassium, bicarbonate, chloride, sulfate, and nitrate. Trace amounts of other ions such as lead, copper, arsenic, manganese, and a wide spectrum of organic compounds are also common. Although contact of water with minerals, soils, rocks, atmospheric pollutants, and decomposition of organic matter in the environment account for much of the dissolved impurities in water, surface waters are also used for disposal of most of the world's liquids wastes.

Federal laws concerned with protecting the nation's water supplies include the Safe Drinking Water Act and the Clean Water Act. The Federal Water Pollution Control Act Amendment of 1972 provides for a National Pollutant Discharge Elimination System (NPDES) to regulate the discharge of contaminants into the waters of the U.S. The Safe Drinking Water Act provides for an Underground

Injection Control (UIC) program to regulate the underground injection of contaminants. NPDES and UIC permits issued by the federal EPA or an approved state agency are required for discharge into all waterways, municipal sewers, and deep wells. Allowable discharge limits vary depending on the industry, the best available economically achievable control technology, and where the effluents, are discharged. Pollutants are classified as either conventional, toxic, or nonconventional. Conventional pollutants include biological oxygen demand (BOD), total suspended solids (TSS), and pH. Toxic pollutants are defined as those listed as priority pollutants, and include metals, asbestos, total cyanides, pesticides, total phenols, purgeable compounds, compounds extracted under acidic conditions, and neutral extractable compounds. Nonconventional pollutants can be substances (nitrogen, ammonia, oils) or physical conditions (thermal characteristics). Along with limiting the maximum quantity of certain pollutants that can be discharged, the minimum permissible concentration of dissolved oxygen and other desirable matter in the waters is specified.

In some cases, the water from incineration systems may be discharged directly into the minicipal sewer system. This will be determined by the characteristics of the water being discharged, the capabilities of the municipal sewage treatment facilities and local regulations. Process waters, such as boiler blowdown, scrubber residues, and ash quench water, may contain contaminant concentrations too high to be discharged into sewage systems without treatment. Since process waters are often recirculated, the quantities requiring disposal are usually small. The volumes of non-process waters, such as once-through cooling system water, can be significant, but the only difference between the incoming and outgoing water may be temperature. If the cooling water is to be discharged directly into the environment, the effect of the thermal plume on aquatic organisms will need to be evaluated. Detrimental effects can be avoided by designing the discharge ports to maximize dilution.

Solid Residue Disposal Regulations

Eventually, some portion of the residue, even after materials recovery, will need to be disposed of in a landfill. In some applications, the quantity of fly ash and bottom ash generated can be quite substantial and its subsequent disposal is an important component of overall operating costs. Currently, most ash can be disposed of in a sanitary landfill, although new regulations are being considered.

The major problem facing the landfilling of incinerator ash is leaching. Ground water which infiltrates the solid waste can dissolve some of the waste materials. The leachate can percolate through the soil and find its way into the water supply. Since incinerator ash may possess trace quantities of toxic metals, long-term leaching effects are a concern. Disposal of the ash in landfills designed not to produce toxic leachate is preferable, but such landfills are expensive and usually not located nearby. The Solid Waste Disposal Act and numerous state and local regulations govern the disposal of nonhazardous solid residues from incineration facilities.

Ash with hazardous characteristics can be generated by burning nonhazardous wastes, since incineration concentrates the noncombustible components by burning off the combustible elements. Since fly ash usually contains a higher concentration of metals, it may be classified as hazardous while the bottom ash from the same process may be considered nonhazardous. The quantity of fly ash can range from less than 1% to over 15% of the total ash content, depending on the incinerator's design and operating conditions. The quantity of metals that makes the ash hazardous is far below the quantity that is economically recoverable. Disposal in special (secure) hazardous waste landfills is required for ash from commercial or industrial incinerators exhibiting hazardous characteristics. The ash from municipal refuse-to-energy plants is currently not regulated as a hazardous waste, although new regulations and options such as the use of monofills are being explored.

Hazardous ash may need to be stabilized or solidified before land-filling. Stabilization converts the toxic components into stable forms that will resist leaching. Stabilization of heavy metals involves the addition of sulfur-containing chemicals. Various solidification techniques exist: portland cements, asphalt, natural cements, silicates, soil-binding agents, and ion-exchange resins. Two types of solidification processes exist: encapsulation (the solidifying agent physically surrounds the waste the waste particles, as in asphalt) and chemical fixation (a chemical reaction occurs between the waste and a solid matrix, as with ion exchange resins). Promising tests have been conducted on stabilizing municipal incinerator ash by adding hydrated lime or portland cement and fabricating concrete—like blocks suitable for disposal in the sea or artificial reef construction . It appears that the blocks can maintain their structural integrity in a marine environment, that leachate levels are acceptable, and that the economics may be viable. Ash has been used to make concrete blocks and other structural materials used in construction. The feasibility of melting ash to stabilize it is also being investigated. Although treatment of the ash may increase its volume, overall disposal costs are decreased.

Other Regulations

Hazardous Waste Regulations

Estimates vary on the quantity of hazardous wastes generated by industry from 10% to 50% of the total industrial discharge . Industries that generate substantial quantities of hazardous wastes include primary metals, organic chemicals, electroplating, inorganic chemicals, textiles, petroleum refining, rubber, and plastics. The hazardous components are mostly toxic metals (arsenic, chromium, lead, mercury, cadmium), salts, acids, bases, and synthetic organics (halogenated hydrocarbons, pesticides, PCBs, phenols). Many of these waste streams are suitable candidates for resources recovery via incineration.

Certain wastes are defined by the EPA to be hazardous and their handling and disposal are regulated by provisions of the Resource Conservation and Recovery Act (RCRA) of 1976 and the Hazardous and Solid Waste Amendments (HSWA) of 1984. These wastes can be solids, liquids, sludges, or gases, and the EPA has regulatory authority over them "from cradle to grave." The EPA has identified a number of lists, occasionally updated, of hazardous substances:

1. Wastes containing any of the substances on a list of priority pollutants including compounds of cadmium, copper lead, mercury, nickel, silver and zinc.
2. Acutely hazardous wastes which are fatal to humans in low doses.
3. Toxic wastes having carcinogenic, mutagenic, or teratogenic effects.
4. Wastes from designated nonspecific sources.
5. Wastes from designated specific sources.

Altogether, hundreds of elements, compounds, discarded chemical products, and industrial process residues have been identified as hazardous and are regulated by RCRA. (Reference includes these lists]. Substances not on the lists but that exhibit the following characteristics, defined in detail by the regulations, are also considered hazardous:

1. Ignitability: waste that burn easily under standard conditions.
2. Corrosivity: very acidic or basic liquids, including aqueous liquids with a pH value less than 2 or greater than 12.5.
3. Reactivity: unstable substances that undergo violent physical or chemical changes under standard conditions.
4. EP toxicity: wastes from which an extract obtained by a specified procedure contains concentrations of certain contaminants above allowable limits.

Anything, such as clothes or containers, that has become contaminated by a hazardous waste is itself considered hazardous. Ash, or any other residue generated by systems incinerating hazardous wastes, is also considered hazardous. If the residue is nontoxic, it can be delisted by following the appropriate analytical and reporting procedures.

RCRA applies certain criteria to the incineration of hazardous wastes, including:

1. Identification of the principal organic hazardous constituent (POHC).
2. Verification that the operation provides for a Destruction and Removal Efficiency of at least 99.99%, usually shown by a trial burn prior to issuance of an operating permit.
3. Verification of 99% HCl removal efficiency for applications where HCl emissions are in excess of 4lb/hr.
4. Verification that particulate emissions do not exceed 0.08 gr/dscf, corrected to 50% excess air.

RCRA regulations require the continuous monitoring of the following operational parameters, at minimum: combustion temperatures, waste feed rate, combustion gas flow rate, and carbon monoxide in the exhaust gas stream. Additional requirements and/or restrictions may be placed on the permit by the regional EPA office issuing the permit.

Since the incineration systems being discussed in this chapter are primarily smaller scale industrial systems, it should be noted that there is a small generators exclusion from strict compliance with RCRA disposal regulations. The exclusions include generators of less than 100 kiograms (220 pounds) per month of hazardous wastes and wastes containing less than 1 kilogram (2.2 pounds) per month of any substances listed as an acutely hazardous waste . Although small quantity generators are exempt from certain RCRA regulations, the waste generator must still determine if the waste is hazardous and if so, comply with the applicable reporting, transportation, and disposal regulations.

A number of waste categories, including radioactive wastes, have been excluded from classification as hazardous under RCRA. Incinerators have been used for the volume reduction of radioactive wastes; both energy and material recovery are possible in these systems. However, incinerators for handling low-level radioactive wastes require special safeguards, filters, containment of material-handling mechanisms, and additional treatment of all effluents. Different regulations and regulatory agencies are involved than for the incineration of other types of hazardous wastes.

One additional non-RCRA category of hazardous wastes should be mentioned. The incineration of substances containing Polychlorinated Biphenyls (PCBs) is currently covered under the Toxic

Substances Control Act (TSCA). The following criteria must be applied to the incineration of these wastes:

1. Combustion at 1200°C (2192°F) for 2 seconds at 3% excess oxygen in the exhaust or at 1600°C (2912°F) for 1.5 seconds at 2% excess oxygen.
2. The waste feed must be automatically shut off if the temperatures or excess oxygen values fall below those specified.
3. A destruction and removal efficiency (DRE) of 99.9999% must be demonstrated.

TSCA also requires monitoring and recording, at specified intervals, of the feed rate, combustion temperature, CO,CO_2, and O_2. Additional parameters, including NO_x, HCl, total chlorinated organic component, PCBs, and total particulate matter, must be monitored during the trial burns. Additional requirements may be established by the permit.

Health and Safety Regulations

The Occupational Safety and Health Administration (OSHA) and other government agencies regulate worker exposure to potential hazards in the work environment. In incineration facilities, especially those handling hazardous wastes, a number of these regulations apply. Protective clothing and special procedures may be required to prevent exposure to toxic chemicals or infectious agents during waste and residue handling operations. Precautions against explosion and fire hazards, oxygen deficiency in enclosed working environments, and heat stress may be required. Excessive noise may also be a problem at some facilities; since damaged inner ear hair cells cannot be replaced, noise-induced hearing loss is irreversible. If employees are subjected to noise exceeding an 8-hours, time-weighted average sound level of 85 dBA, a hearing conservation programme is required. State and local regulations on noise emissions beyond the boundaries of a facility may dictate the construction of noise suppression structures and the monitoring and reporting of noise levels.

Legislative Incentives for Resource-Recovery Facilities

Federal, state, and local regulations have, at various times, been passed to encourage resource recovery. For example, the Public Utilities Regulatory Policies Act (PURPA) of 1978 encourages power production by requiring utilities to purchase electricity from qualifying small power production facilities at an appropriate rate. Among other requirements, to qualify, power production capacity must be less than 80 megawatts, and no more than 25% of the total energy input of the facility may be from oil, natural gas or coal. PURPA regulations are implemented by the individual state's regulatory authorities.

Whereas fees and taxes are often placed on other means of waste disposal to fund state disposal and research activities, tax credits are often used to promote resource recovery and pollution control activities. However, regulations encouraging the purchasing of recycled materials by government agencies have not been successfully applied, tax incentives are frequently altered, and PURPA regulations, challenged in many cases, are subject to change. The long-term financial benefits to a resource recovery project from these types of regulations should be considered with caution.

Summary of Regulatory Issues

Federal, state, and local regulations currently have a significant impact on the design, operation, and economic viability of resource-recovery facilities, regardless of whether they are small 10 tpd industrial units or large 1,000 tpd municipal systems. Regulations determine the requirements for major process activities, especially air pollution control and residue disposal. Significant costs, both in terms of equipment and personnel, are associated with compliance.

Prior to building and operating a facility a number of permits must be obtained, including, as applicable, permits for construction, operation, air discharges, water discharges, solid waste disposal, hazardous waste transport, etc. The permiting process for incineration systems can be quite involved; permitting processes lasting over a year are not unusual, especially when dealing with hazardous wastes. Permit applications and information on current regulations can usually be obtained from the individual state's Department of Environmental Protection or its equivalent. Technical information regarding the incineration system's operation, provided by the system's manufacturer, is often required for the permit. Public hearings may be required and siting and environmental issues must be carefully addressed. As sampling and analytical techniques have improved, a wide variety of organic and inorganic compounds have been identified in incinerator effluents. Although many of these compounds are generated only in picogram quantities, the fact that they are suspected carcinogens and that information for complete environmental and health risk assessments is not available often results in substantial public opposition to local sitings of any incinerators, including ones used for resource recovery.

Regulations also serve as the driving force in trends that promote resource recovery. Tariffs on imported materials, economic incentives for recovery facilities, and tight restrictions on landfill disposal all contribute to increasing the economic viability of resource-recovery facilities.

INCINERATORS: AN EQUIPMENT OVERVIEW

Design Fundamentals

Temperature, Time, and Turbulence

All incinerators use air and heat to degrade the waste materials into their constituent components. A solid or liquid must be converted into the gaseous phase before burning will occur. Complete burn, or burning with 100% efficiency, occurs when all the carbon in the combustion gases is converted into CO_2. Inefficient combustion result in free carbon, carbon monoxide (CO), or other carbon compounds remaining within the flue gas. The objective of incineration is, of course, to obtain the highest possible combustion efficiency.

Combustion efficiency depends on the three T's of combustion: temperature, time, and turbulence. The temperature in the furnace, the time at that temperature, and the turbulence in the chamber control the rate of conversion.The time and turbulence are usually determined by the incinerator design and can be controlled only within a limited range. During operation, temperature is the variable that offers the most convenient control of the process. At higher temperatures, more complete breakdown of the combustion products occurs.

The temperature within the incinerator is a function of the waste and auxiliary fuel heating values, the furnace design, the quantity of combustion air supplied, and combustion control. The minimum temperature must be higher than the ignition temperature of the materials being processed. The maximum temperature is usually determined by the type of refractory material used. Temperature is controlled by moderating either one or more of the following parameters: the excess air supply, the waste feed, or the auxiliary fuel supply. When burning highly exothermic materials, additional cooling may be needed; for example, water can be added or the furnace walls can be water-cooled.

Turbulence expresses the degree of physical mixing of combustible gases and vapors with air. The increased rate or of collisions between carbon and oxygen resulting from a high degree of turbulence increases burning efficiency. With insufficient turbulence, stratification of the combustion gases may occur and some gases may pass through the incinerator unburned. Techniques used to increase turbulence include forced blowing of combustion air and the construction of baffles within the combustion chamber, redirecting the flow of combustion gases.

Since combustion does not occur instantaneously, sufficient space must be provided to allow the fuel and combustible gases the time required to burn completely. This time, called residence or retention time, is a function of furnace temperature, volume of gases passing through the furnace and chamber size. Retention time, as defined here, concerns the burning reaction between combustible gases and combustion air and is measured in seconds. It should be distinguished from processing time, which concerns the completion of burning of the waste. Depending on the material burned, the processing time required for complete decomposition may range from a fraction of a second for gases to minutes or hours for solids, which must be gasified before burning.

Construction Materials

A refractory material is a substance used to line furnaces because of its ability to endure high temperatures and resist abrasion. There are a variety of refractory and insulation products available for incinerators, both metallic and nonmetallic. The refractory material must be compatible with the waste type and the operating conditions. For example, operating in a reducing, rather than oxidizing, atmosphere will degrade refractories containing iron or silicon . In the selection of a refractory material for a particular application a number of factors must be considered, including:

1. Strength under high (operating) and low (shipping, installation) temperatures.
2. Performance characteristics and structural integrity under the cycle of expansion and contraction resulting from repeated heating and cooling.
3. The refractory's ability to resist penetration by slags or gases.
4. Resistance to erosion and corrosion of the material due to physical contact or chemical reactions with the waste components.
5. Insulation characteristics.

There are two general categories of refractories: firebrick and castable. Standard refractory operating temperatures range through 1600°C (3000°F). Special ones for higher temperature operation are also available. Good insulating refractories have conductivity values of 0.2-3.0 Btu-in./ft^2-hr-F. Relatively poor insulators, such as standard firebrick, have values above 6.0 Btu- in./ft^2-hr-F.

Firebrick is kiln baked to a uniform, controlled consistency, shape and quality. Anchors are normally provided to hold the brick in place. Typically, the refractory brick is composed of silica (SiO_2), alumina (Al_2O_3), and titania (TiO_2) in varying percentages, depending on the manufacturer and its specific performance characteristics.

Castable refractories are supplied dry and must be mixed with water before installation by pouring (being cast into place requires a form); ramming (with plastic castables, form not required); or other means. Castable refractories form a smooth and continuous mass that can be placed in irregular areas such as burner openings. Castable refractories must be either airset or heat-cured to establish the ceramic bond creating the strength of the refractory matrix. During air setting the refractory must not be subjected to mechanical or temperature changes; during heat curing the refractory must be heated at a controlled rate at temperatures sometimes above the incinerator's operating temperature, requiring special procedures. To obtain a good quality refractory, special care must be taken during the casting and curing stages.

The frequency of refractory replacement depends on the operating conditions. The typical period between complete refractory replacement ranges from 5 to 15 years. Good maintenance procedures can extend the life of the refractory substantially. A number of factors contribute to refractory damage. Abrasion can result from the impact of moving solids. The mechanical shock caused by falling waste from entry via the feed mechanism or thermal shock due to turbulent flow near air inlet ports can damage the refractory. Spalling, the flaking away of refractory surfaces, can be caused by thermal stresses or mechanical action. Uneven temperature gradients can degrade the refractory. Steam, trapped by rapid drying of the refractory, does not have an opportunity to escape the surface but expands, causing cracking and mechanical spalling.

At high temperatures, certain materials—usually ash, metals, or glass—deform and physically change to a more amorphous state and begin to flow as a heavy liquid. If the temperature of this material is then reduced, the material will solidify into a hard slag. Molten ash may become airborne and attach to a refractory or metal surface which is cooler than the air stream. This slag can be acidic (as a result of silicon, aluminum, or titanium oxides) or basic (due to oxides of iron, calcium, magnesium, potassium, sodium, or chromium). Compatible refractory materials must be selected to ensure long refractory life. For acidic slags, fireclay, high alumina, or silica firebrick could be used and for basic slags, chrome, magnesite, or forsterite brick.

As with refractory materials, selecting construction materials to be used throughout the unit requires consideration of possible interactions with the waste stream at all stages. Grates and other material handling elements may be susceptible to corrosion, especially if used in the high temperature portions of the system. The acidic components of flue gases pose corrosion problems in wet scrubbers and boilers. At tempertures below 150°C (300°F), steel can experience corrosion caused by sulfuric or hydrochloric acid condensation. At temperatures above 370°C (700°F) a series of reactions takes place between the chlorine, chloride, and sulfide in the exhaust gas and steel. Chlorine reacts with the iron to produce ferrous chloride, which combines with the oxygen in the flue gas to form iron oxide. The iron oxide (rust) leaves the surface of the steel. Other components that may be present, such as alkali salts, promote this corrosion.

Major Incinerator Types

Incinerators come in many different sizes and shapes. However, fundamentally all incinerators are based on the same principles, providing the time, temperature, and turbulence for complete combustion to occur. Major incineration technologies, including those used in the past, those currently popular, and those with future application, are surveyed briefly below.

Open burning and open pit incineration. The oldest technique for disposing of wastes consists of placing the materials on the ground and burning them without any special equipment. Of course, heavy smoke carrying chemical and particulate pollutants results, well in excess of current emission regulations. Open burning has been used to dispose of highly explosive wastes such as TNT. Open pit incinerators were developed for controlled incineration of explosive wastes which would create an explosion hazard or high heat release in an enclosed system. The chamber was built either above or below ground. Air, blown from the open top down into the incinerator chamber, created a high degree of turbulence. With temperatures in excess of 1090°C (2000°F), smoke and particulate discharges were relatively low, but still above current standards.

Single-chamber incinerators. In a single-chamber incinerator, usually a refractory-lined box-like enclosure, waste is loaded onto a hearth (floor) or grate and fired. The waste can be side-loaded through a door or top-loaded via the flue (stack). In the past, top-loaded single chamber incinerators were used for apartment house waste disposal. Single-chamber units were also used in material-recovery applications such as the incineration of insulated copper wires. Considerable smoke was generated,

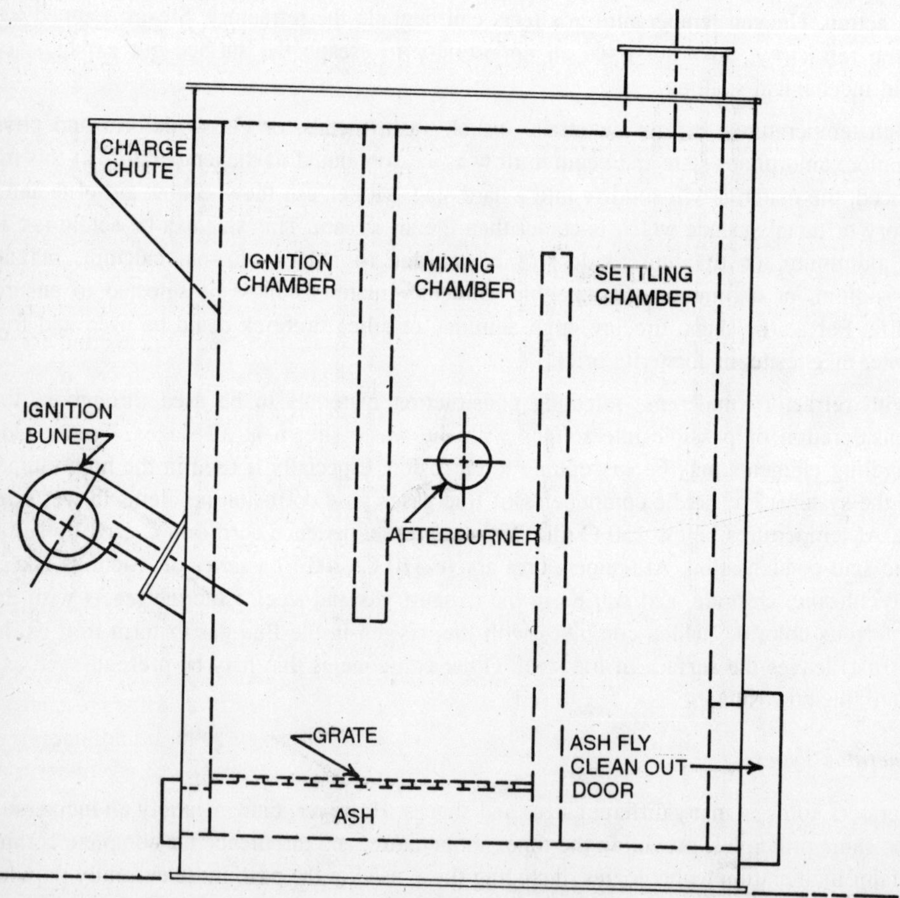

Figure 20.3. This simple box incinerator is illustrative of designs that were popular in the past. Its unsophisticated pollution control methods would not meet most current regulatory requirements.

containing visible and chemical pollutants. Tightening air pollution control regulations terminated the use of most such systems. Single-chamber incinerators will not meet current air pollution control regulations without additional APC equipment.

Multiple-chamber incinerators. Multiple-chamber incinerators are usually rectangular refractory-lined enclosures with several partitions. The primary (ignition) chamber is used for combustion of the solid wastes. The secondary (combustion or afterburner) chamber provides the additional residence time and supplementary fuel required to complete the combustion of any unburned gaseous products and airborne combustible solids not oxidized in the primary chamber. Some designs include mixing or settling chambers. Retort incinerators are compact cubes in shape, with multiple internal baffles positioned to guide the combustion gases through 90-degree turns, forcing the fly ash to drop from the flue gas stream. The abrupt direction changes are in both the vertical and horizontal directions. Wastes are placed either directly on the fixed floor or on some type of fixture (grate) in the primary chamber. Waste feed and ash removal are usually done manually. In in-line incinerators, the flow of combustion gases is straight through the incinerator; abrupt changes in gas flow occur only in the vertical direction. A simple box incinerator is illustrated in Figure 20.3. At temperatures around 760°C (1400°F), multiple chamber incinerators were able to meet particulate emission standards untill recently. Most current standards have been tightened to such a degree that some additional APC devices are required.

Moving tray or moving basket incinerators. Moving trays or baskets, containing the waste materials are passed through a single long refractory-lined process chamber. To obtain a large total capacity from the continuous processing of multiple small batches, the trays or baskets are automatically conveyed, one at a time, through the chamber. A moving basket incinerator, used for copper recovery. Copper wire bundles are placed in the baskets; as the basket moves through the chamber, the insulation is burned off and ash falls underneath the baskets onto the hearth. Temperatures within the process chamber are around 650°C(1200°F). Baskets of clean, copper wire bundles exit at the opposite end of the chamber. An afterburner completes the oxidation of flue gases leaving the chamber. When the insulation is made of polyvinyl chloride, additional APC equipment is required to capture the gaseous hydrochloric acid.

Ashing- and slagging-type rotary kilns. A rotary kiln is a cylindrical refractory-lined shell, mounted at a slight incline from the horizontal plane, revolving around its horizontal axis. Waste enters at the raised end of the kiln and ash is discharged at the opposite end. Control of processing time and mixing of the waste stream are obtained by rotation of the chamber. Processing times are typically seconds for liquids and gases, hours for solids; temperatures range from 800-1600°C (1500-2900°F). There are usually no moving parts within the kiln.The kiln serves as the primary or ignition chamber, and is usually followed by a secondary combustion chamber (afterburner). Rotary kilns are one of the most flexible of incinerator designs. They are applicable to a large variety of solid, liquid, and gaseous wastes and are adaptable for use with a wide variety of feeding mechanisms. There are two varieties of rotary kilns: ashing and slagging. Ashing-type rotary kilns discharge the noncombustibles as ash. Slagging-type kilns are similar, except that the walls are wet with a viscous slag and the noncombustible residue is discharged as a solid. The slag offers some protection to the kiln's inner walls and higher operating temperatures are easier to maintain in a slagging-type operation, but operating problems frequently occur, especially in removing the solid residue.

Liquid injection incinerators. Liquid waste is fed into a refractory lined chamber usually cylindrical in shape, through a burner or an atomizing nozzle which atomizes the waste and mixes it with the air into a suspension. Since burners are susceptible to plugging, particle size is a critical design parameter. Liquids that dry and cake as they pass through the nozzle may cause problems. Temperatures are typically in the range of 650-1600°C (1200-2900°F); residence times range from 0.1 to 2 seconds.

Fluidized bed and circulating bed systems. A fluidized-bed incinerator consists of a refractory-lined vessel containing a bed of inert granular materials such as sand. Gases blown through the bed cause it to act as a fluid. The waste is injected into the fluidized bed, where it is oxidized. The ash is carried out the top with the air stream and subsequently collected. The air pollution control system may be required to handle large particulate loadings, since some bed material may also exit with the flue gases. For liquids and gases, processing times are measured in terms of seconds; longer times are required for solids. Typical operating temperatures range from 450-980°C (840-1800°F) . Fluidized beds are applicable to various solid and liquid wastes; they are not suited for wastes with a fusible ash content and preprocessing is required for bulk wastes. Fluidized beds are characterized by a high heating rate per unit volume and good mixing characteristics. The high degree of turbulence results in low excess air requirement (20-40%) and allows the use of lower temperatures, minimizing nitrogen oxide formation. The bed material can be selected to suppress acid gas formation. The type of waste and the bed material must be compatible to avoid damage. Bed maintenance and operating costs can be high; removing residual materials from the bed can be difficult. A modified type of fluidized bed, the circulating bed, uses high-velocity air to circulate solids through a vertical loop consisting of a combustion chamber, hot cyclone, and a return section. Times range from 2 seconds for gases to 30 minutes for solids; typical operating temperatures range from 760-870°C (1400- 1600°F) .

Multiple hearth. A multiple-hearth furnace consists of a vertical, cylindrical, refractorylined furnace, with several circular hearths attached to the furnace walls, positioned in series one over the other. A rotating shaft is positioned in the center of the furnace, with rabble arms attached to the shaft over each hearth. Waste is charged onto the top hearth. The movement of the arms pushes the material across each hearth to drop holes, transferring the material slowly from the top hearth until it is discharged below the last hearth. The upper hearths serve as a drying zone; burnout to ash occurs on the lower hearths. Flue gases exit through the top of the furnace. Cooling air, passed through the hollow centre shaft and rabble arms, can be recycled into the furnace is preheated combustion air. Typical processing times range from 0.25 to 1.5 hours and temperatures from 760-980°C (1400-1800°F) . Many types of wastes can be handled, although a multiple hearth system is not suited to wastes containing fuisble ash, wastes requiring very high temperatures for destruction, or irregular bulky solids. Large multiple-hearth incinerators have been used primarily for municipal sewer sludge disposal. Fuel efficiency is high and increases with the number of hearths used. Because of the large number of moving parts, maintenance costs can be high. Gas cleaning equipment is required because the flue gases may carry a substantial percentage of the ash content.

Modified industrial furnaces and kilns. A number of furnaces and kilns originally designed for other industrial purposes have been adapted, with varying degrees of success to waste disposal applications. For example, in a modified glass furnace design, waste is fed into a furnace containing molten glass. Electric current is passed through the glass to maintain temperatures in the range of 1200°C (2200°F). Hazardous organic substances are destroyed while minerals and metals are encapsulated in the resulting glass, which is not considered hazardous and can be used, for example, in asphalt. Cement kilns have also been used for the destruction of wastes, especially hazardous wastes. Infrared furnaces, also called electric or radiant heat furnaces, can be thought of as adaptions of the continuous furnaces used in a variety of heat-treating applications. A long rectangular chamber of carbon steel, lined with layers of ceramic fiber, serves as the primary chamber. The material to be processed is conveyed through the furnace on a woven, steel-alloy wire belt. Infrared energy, provided by silicon carbide resistance heating elements, maintains temperatures between 260-1010°C (500-1850°F) for the 10-180 minutes that the waste resides in the furnace . The low level of agitation as the wire mesh carries the waste result in low levels of particulate in the flue gases. Feed preprocessing is usually required because the furnace requires a small input solid particle size. Maintenance costs can be quite high because of the large number of moving parts inside the chamber.

Special application incinerators. Various types of incinerators have been designed for applications with unique requirements. Much of the "customization" is related to the materials' handling portion of the system. One example is the reclamation of metal drums. The drums are passed on a continuous conveyor through a high-temperature primary chamber, driving off volatile compounds. The volatile compounds are destroyed in the higher temperatures of a secondary combustion chamber while the drums exit, on the conveyor, at the other end of the primary chamber.

Pyrolyzers. Pyrolysis is the thermal decomposition of solids or liquids in the absence of oxygen. It is most applicable to wastes with a purely organic content. Typical processing times range from 1/10 of a second to several hours, and operating temperatures range from 480-815°C (900-1500°F). The products of pyrolysis include carbon monoxide, methane, and carbon in the form of a liquid or solid char that can be recovered. Carbon monoxide and methane both possess heating value and can subsequently be burned in oxygen to release heat. Alternatively, the effluent gases can be cooled and the condensate separated into its components: acetic acid, methanol, light fuel oil, etc. True pyrolysis is an endothermic process which requires the external application of heat, consuming substantial quantities of energy. In practice, some air is introduced to provide, via burning, sufficient heat to sustain the process. Therefore, the term is also used in reference to oxygen-starved operation. Pure pyrolytic process (no oxygen gas feed) produce a higher quality burnable gas than oxygen starved operation. Because of the lower quantity of air used, higher thermal efficiency and reduced air emissions are possible with pyrolysis. In practice, it is difficult to maintain the required conditions within the incinerator chamber to obtain complete burnout of the wastes under pyrolytic conditions. Infiltration of oxygen into the chamber via feeder and residue removal systems degrades performance. Also, with mixed waste streams it is difficult to predict the types and quantities of pyrolysis products that will be generated. Ideally, the goal of pyrolysis is to convert the hydrocarbons in the waste into high-Btu value fuel oil, but this has not been achievable in practice. Because of the operational difficulties with mixed waste streams, pyrolysis has not been used frequently in commercial waste disposal applications. However, as an industrial process used for applications such as producing charcoal from wood chips, it has been used for years.

Enery recovery is possible from most of the incinerator types discussed above. Because of their ability to handle diverse waste streams and good operational characteristics, rotary kilns are frequently used for industrial waste-to-energy applications.

The residue generated in most incineration systems can subsequently be processed to recover materials, although the physical characteristcs of the solid residue from certain waste streams may result in operational difficulties for some incinerator designs. Multiple chamber (retort), moving tray, and rotary kiln designs are most prevalent in material-recovery applications.

The advantages and disadvantages of each type of system for a particular waste stream must be

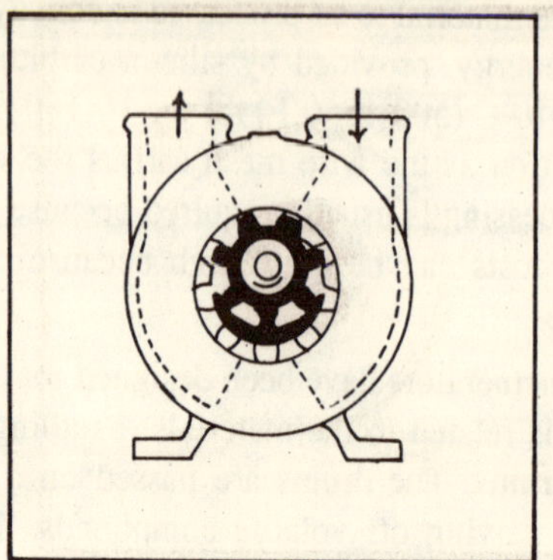

The pump inlet is connected to the storage container and the pump outlet to the burner nozzle (combustible liquids) or to an atomizing nozzle (non-combustible liquids) which protrudes into either the ignition or combustion chamber.

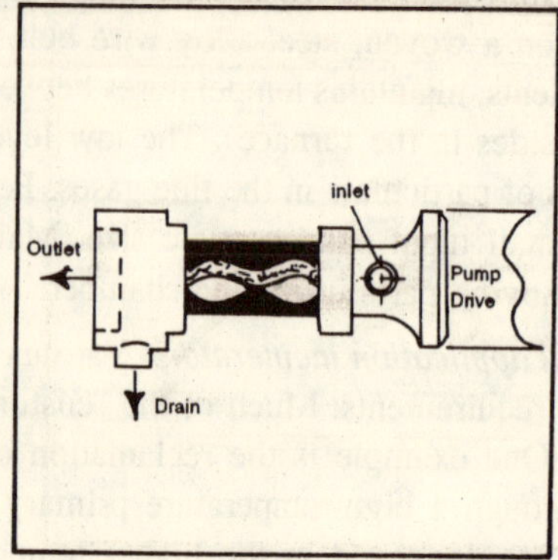

The waste is drawn through the inlet from the storage container. As the rotor turns, waste enters the cavity surrounded by the flexible stator, and exits through the outlet into the feeding nozzle which protrudes into the ignition chamber.

Figure 20.4. Typical pumps used for feeding liquid wastes.

considered on a case-by-case basis. Function and components common to all types of system will be discussed in the following sections. In addition to the basic incinerator (usually consisting of ignition and combustion chamber and the basic peripheral equipment, such as an exhaust fan and stack), modern resource-recovery facilities include systems for:

- Waste handling
- Auxiliary fuel feed
- Air pollution control
- Residue handling
- Energy recovery
- Monitoring and control

Feed Systems

The waste-handling system is sometimes the most overlooked part of an incinerator design; yet it is also the most application- specific aspect and the part of the system most susceptible to breakdown, plugging, and damage if not matched to the waste stream. There are numerous types of feeding systems, each appropriate for a certain type of waste feed. Wastes can be loaded into the chamber manually or automatically, from the top or side, and continuously or in batches. Adequate controls must be provided to prevent feeding if temperatures within the chambers are not within the operating range. Certain waste streams, such as those with infectious or corrosive characteristics, require that special precautions be taken to limit the exposure of operators and equipment to harmful constituents.

Clean, combustible liquid wastes, such as solvents, can be fed into the incinerator by gear pumps. Combustible liquids can be pumped through a burner nozzle; noncombustible liquids are fed through an atomizing nozzle. Progressing cavity pumps can be used for feeding viscous liquids or liquids

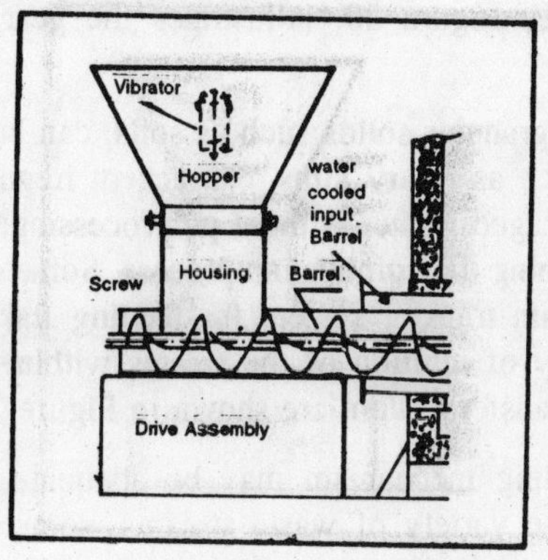

SCREW FEEDER

The waste is deposited into the hopper. A vibrator, mounted on the side wall of the hopper, shakes it and forces the waste into the housing. As the screw rotates, driven by an electric motor, the waste is transported into the ignition chamber.

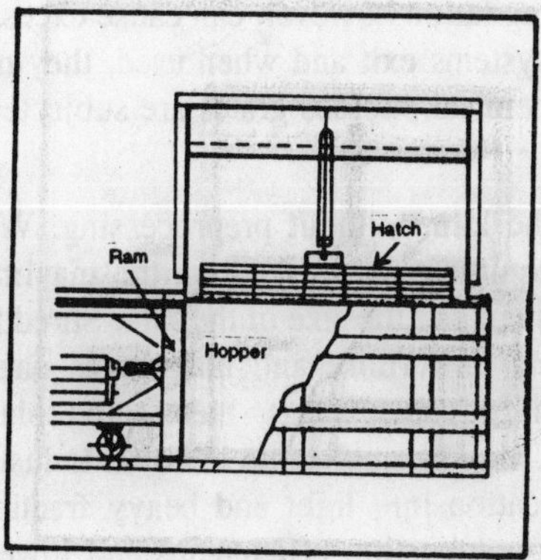

RAM FEEDER

The waste is charged into the hopper. The hatch closes, the ignition chamber's gate (not shown) opens, and the ram thrusts forward forcing the waste into the ignition chamber. The ram returns, the gate closes and the feeder is ready for reloading.

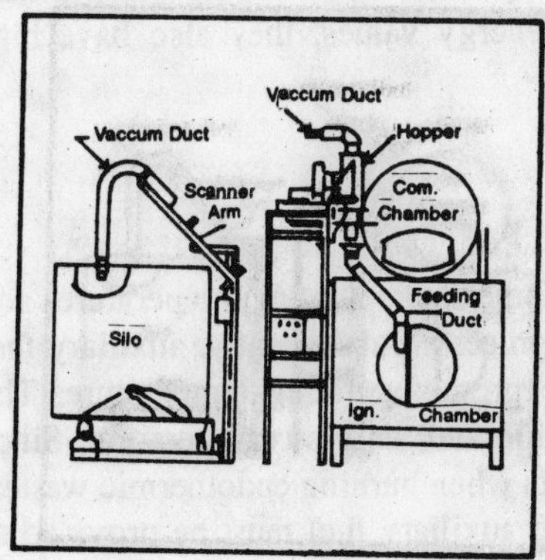

VACUUM TRANSFER

As the silo rotates, the waste is drawn into the vacuum duct and stored within the hopper. As the pressure increases, the slide gate opens. The waste material is blown into the feeding duct and forced into the ignition chamber.

Figure 20.5. Typical feeding mechanisms for solid wastes.

containing a high percentage of solid contaminants such as heavy oils and sludges. In general, a fluid can be pumped if its viscosity is under 10,000 Saybolt Seconds Universal (SSU). It can be atomized and burned in suspension if its viscosity is below 750 SSU. Figure 20.4 illustrates the gear and progressing cavity pumps.

Liquids that cannot be pumped, like some sludges, and granular solids such as soils, can be fed into the incinerator by a screw feeder. Some incinerators, such as rotary kilns, can accept iregularly shaped solids such as automotive scrap, circuit boards, or packaged wastes without preprocessing; ram feeders allow automatic feeding of these wastes without disturbing the combustion process. Some solid wastes may require special transfer systems, such as a vacuum transfer system for feeding flaky or granular wastes, to avoid generation of dust or the possibility of ignition of the wastes within their storage containers. A screw feeder, ram feeder, and vacuum transfer system are shown in Figure 20.5.

For dedicated industrial applications, one type of feeding mechanism may be adequate. For commercial systems that may be required to handle a diverse variety of waste streams, sometimes simultaneously several systems may be needed. For large mass burn incinerators, grate-transport systems are often employed. The grate transports the waste through the incinerator and simultaneously provides turbulance and mixing of the waste the combustion air; too much agitation however, can cause excessive generation of particulate in the flue gases. Many types of grate systems exit and when used, they must be carefully selected and operated. The heavy mechanical and chemical stresses grates are subjected to can cause operational problems.

With rotary kilns, most wastes can be fed directly into the kiln, without preprocessing. When feeding oversized wastes, shredding (see Figure 20.6) may be required prior to feeding; the maximum size of waste components is determined by the type of feeding system and the size of the kiln. Shredding irregularly shaped wastes prior to feeding increase burning rate uniformity and maximizes energy recovery. Sorting of the waste feed, especially the separation of paper, glass, metal, and plastic components, though done more frequently on a municiple scale, may be applicable in some industrial applications. Various techniques are used, including air classification into light and heavy fractions. Removing glass and metal from the waste feed results in higher combustion efficiencies and lowered residue quantities. Excess air requirements and damage from slagging and abrasion can be minimized by removing certain components from the waste stream. Removing chlorine-containing plastics may reduce APC requirements, and although all plastics have high energy values, they also have high recovery and recycling values.

Auxiliary Fuel Feed System

The auxiliary fuel feed system is required to preheat the equipment to operating temperatures and ignite the waste's combustible components, initiating the burning process. Subsequently, auxiliary fuel is supplied, usually automatically, as necessary to maintain the preset operating temperature. The incoming waste stream's characteristics determine fuel consumption and air blower operation. Since excess air carries away heat, to minimize auxiliary fuel requirements when burning endothermic wastes, excess air should be minimized; however, sufficient excess air and auxiliary fuel must be provided to assure complete oxidation of the waste. When wastes with very high (Btu) heating values are fed into the incinerator, excess air may be used to keep the temperature from rising too high above the operating point and damaging the refractory.

The auxiliary fuel feed system consists of burners mounted to the outer shells of both the ignition

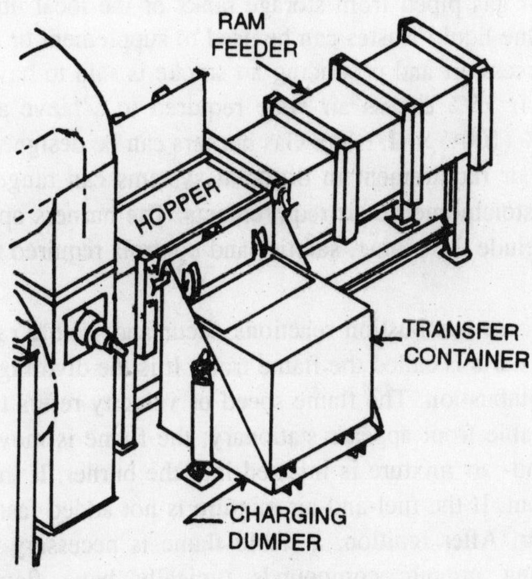

CHARGING DUMPER

To provide additional feeding process automation, solid waste is wheeled or transported by foklift onto the dumper. The dumper is hydraulically tilted upward emptying the contents of the container into a suitable feeding system, such as the ram feeder (shown above).

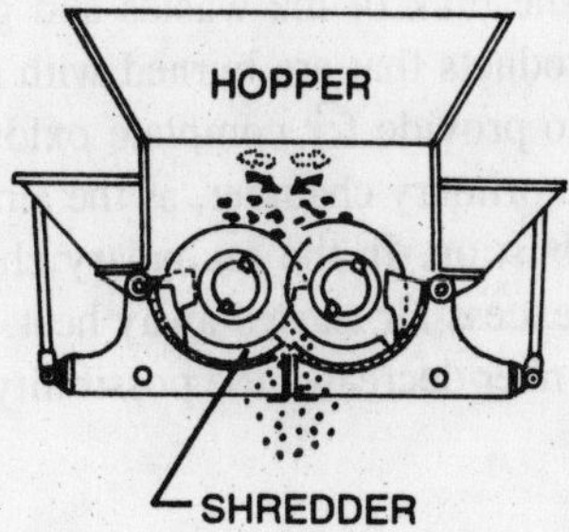

SHREDDER

Oversized, irregularly shaped solid wastes can be shredded prior to feeding into the ignition chamber. Conveyors can be used to bring wastes to the shredder and to automatically transport the shredded wastes to an appropriate feeding system, such as a screw feeder.

Figure 20.6. Auxiliary systems used in waste handling and preparation.

and combustion chambers. Individual or dual fuel type burners can be used. Either natural gas or fuel oil is usually used as the auxiliary fuel. Gear or vane pumps are used with liquid fuels; pressure regulators control the intake of gas piped from storage tanks or the local utility company. Whenever feasible, high (Btu) heating value liquid wastes can be used to supplement or replace the auxiliary fuel.

A burner requiring no excess air and producing no smoke is said to have perfect turbulence, i.e., a turbulence factor of 100%. If 15% excess air were required to achieve a nosmoke condition, the turbulence factor would be 87% (100% × 1/1.15). Gas burners can be designed to produce a turbulence factor close to 100%. Excess air requirement in different systems can range from a few per cent to several hundred per cent over stoichiometric air requirements. The burners operate in conjunction with combustion air blowers and include the piping, safeties and controls required to adjust the fuel/air ratio appropriately.

The flame is the zone where combustion reactions occur and visible radiation is produced. The plane along which combustion starts is called the flame front. It is the dividing line between the fuel/air mixture and the products of combustion. The flame speed or velocity refers to the rate of propagation of the flame. If the flame is stable front appears stationary; the flame is moving toward the burner at the same speed that the fuel-and- air mixture is injected into the burner. If the mixture is injected too fast, the flame may be blown out. If the fuel-and-air mixture is not added fast enough, the burner may flashback, damaging the burner. After ignition, a stable flame is necessary to maintain combustion. Carbon-and hydrogen-containing organic compounds typically have flame temperatures in the 1650-2200°C (3000-4000°F) range at stoichiometric conditions; halogenated hydrocarbons range from 1650°C(3000°F) down to nonflammable. Most materials normally disposed of by incineration have flame temperatures below 1650°C (3000°F). The furnace temperatures of most resource-recovery incinerators are in the range of 650-1200°C (1200-2200°F).

Controlled or starved air operation is usually employed in multiple chamber units. Air injection into the primary chamber is controlled to promote generation of pyrolytic products. After the waste is charged into the primary chamber, only enough air (70- 80% stoichiometric) is provided to allow sufficient burning to occur to decompose the bulk of the wastes and generate heat for pyrolysis. The resulting flue gases contain combustible products that are burned with sufficient excess air (140-200% stoichiometric) in the secondary chamber to provide for complete oxidation. In both chambers, the air flow is controlled by the temperature. In the primary chamber, as the air flow increases, the temperature rises as more oxygen is available for combustion. In the secondary chamber, the temperature falls as more excess air is introduced, because the excess air carries away heat. In metal-recovery applications, the reducing atmosphere of the primary chamber decreases the possibility of metal oxidation and enables recovery of high-quality metal.

Ignition, Process, or Primary Chambers

Various types of ignition chambers, also called process combustion chambers, were discussed earlier. Whether they are rectangular, cylindrical, or irregular in shape, each chamber is a refractory-lined container providing access for feed input, ash collection or discharge, auxiliary fuel burners and combustion air blowers, and an exhaust gas exit. Because of their adaptability to many resource-recovery applications, rotary kilns are discussed in more detail below.

The rotary kiln, or revolving ignition chamber (RIC), illustrated in Figure 20.7, is a cylindrical, refractory lined shell mounted at a slight incline from the horizontal plane. Tracks welded to the outer shell of the chamber guide wheels which rotate the chamber. The speed and direction of rotation is controlled by an appropriate drive assembly mounted to the framework supporting the ignition chamber. The angle to the horizontal is usually less than 3% and the rotation rate is typically in the range of 0.25-1.5 revolutions per minute (rpm). An inspection door is provided for access to the chamber and, if necessary, manual ash removal.

The chamber is typically fabricated from welded steel plates and lined with a high-density, abrasion-resistant castable refractory. High-temperature alloy anchors secure the refractory to the shell. The kiln interior is usually smooth, but if required turbulence can be increased by installing baffles or ridges. Small units are usually factory-built, while large units must be fabricated at the facility. Care must be taken when transporting kilns not to damage the refractory.

The ignition chamber is preheated before the first wastes are charged. Solid wastes are dropped onto the kiln hearth by an appropriate feed system through an opening at the raised end of the kiln. Liquids that will burn in suspension and gaseous wastes can be injected at either end of the kiln. Once the wastes have been fed in, the ignition chamber's burner ignites the material's combustible components. Within the high-temperature air, the materials are decomposed into volatilized gases and ash. The gas is drawn into the combustion chamber and the ash remains in the ignition chamber. As the chamber rotates, the insulating ash layer is broken up and new waste surfaces are exposed to the thermal process. The ash is slowly transferred to the opposite end of the chamber and continually discharged into the ash bin located below the discharge end of the chamber. As indicated by the following equation, by varying the kiln's rotational speed, the burnout of the waste can be controlled:

$$\text{approximate residence time} = \frac{K \times L \times D}{R \times A}$$

where K = Kiln Design Constant

L = Length of chamber (ft)

R = Rotation (rpm)

D = Inside kiln diameter (ft)

A = Inclination angle (inches per foot, per foot of length)

Depending on the materials being processed, RIC temperatures range from 650-980°C (1200-1800°F). The movement of the material over the refractory lining helps retain the required high temperatures. This, together with controlled overfire conditions, allows continuous and complete material burning.

Revolving ignition chambers can be concurrent or countercurrent in design. In a countercurrent design, the burner is located at the opposite end from the feeding system. In a concurrent design, the burner is located on the same side as the feeding system.

Combustion Chamber (Afterburner)

A refractory-lined duct directs the flow of the volatile gases from the ignition chamber to the combustion chamber. Clean, combustible liquid wastes can be fed directly into the combustion chamber. The combustion chamber is usually a cylindrical steel shell, fabricated from rolled steel plates, and lined with a high-temperature, medium-density castable refractory secured to the chamber by alloy steel anchors.

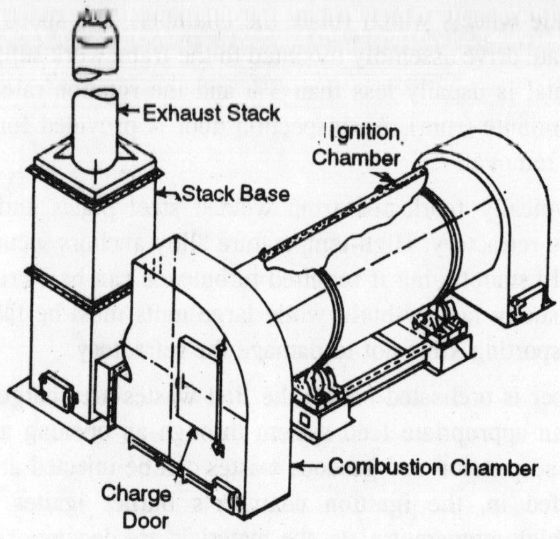

Figure 20.9. Revolving Ignition chamber (rotary kiln).

The combustion chamber (CC) is perhaps the simplest of the major system components, yet it is the most critical with respect to the destruction of toxic hydrocarbons. Sufficient excess air is provided within the combustion chamber to complete oxidation of the smoke's combustible components—any unburned hydrocarbons from the ignition chamber. Consequently, the combustion chamber can be thought of as either the second combustion stage or the first stage of air pollution control. In some cases, the combustion chamber (also called the afterburner or secondary combustion chamber) may be the only form of APC required.

Combustion-chamber temperatures are frequently higher than ignition-chamber temperatures. Depending on the material being processed, temperatures normally range from 870-1315°C (1600-2400°F). A minimum temperature of 815°C (1500°F) is required to eliminate odors. Data indicate that at temperatures of 1200°C (2200°F) and residence times over 2 seconds, all toxic organic compounds will be destroyed. The CC burner is usually positioned so that the incoming gases and particulate pass through the high velocity excess air flame. Auxiliary fuel is proprotioned into the burner as required.

When processing low-energy-content wastes, a heat recuperator can be used to recover and reuse heat energy in the process, reducing auxiliary fuel consumption. A recuperator consists of alloy steel tubes connected to flow-directing manifolds and is located in the path of the hot exhaust gases, such as at the base of the exhaust stack or between the combustion chamber and pollution-abatement equipment. The incoming combustion air is blown through the tubes and heated by exchanging heat with flue gases exiting the combustion chamber. Special burners which operate with hot combustion air and require less fuel to maintain the necessary high flame temperature are used. A properly designed recuperator can reduce fuel consumption by as much as 50%. In some cases, the heat derived from burning even low energy-content wastes can be sufficient to maintain combustion within required temperature limits without supplying auxiliary fuel during operation.

Exhaust Fan and Stack

Clean exhaust gases exiting the system after passing through the air pollution control systems are drawn by an exhaust fan through the exhaust stack. The exhaust fan is mounted between the pollution-abatement equipment and the exhaust stack.

A properly functioning exhaust fan is essential to system operation. Gases being drawn through modern pollution-control systems experience substantial pressure drops. The natural draft, which was used extensively in the past to induce gas flow through the incinerator, is reduced since the temperatures of the exhaust gases are lower after passing through pollution-control devices than they were just passing through an afterburner.

The exhaust fan is expected to operate clean and dry. If the fan becomes fouled by residue contained within the flue gases due to problems with the APC system, the vibrations that are induced may result in mechanical damage to the fan or other system components. The fan must be capable of providing the gas-moving capacity required. Less noise will be generated and fewer operational problems encountered by selecting a fan with a greater capacity than required and operating it at a lower speed to obtain the required flow.

The gases exiting the stack are low in oxygen and high in carbon dioxide and, possibly, moisture. The stack or chimney serves to disperse the vented gases, reducing the local ambient concentration of individual components. A small stack cross section results in unnecessary loading on the exhaust fan. Too large a stack will provide poor dispersion and may cause flue gases to be pulled down by downdraft winds. Exit velocities are typically around 1,000 feet per minute, although greater velocities are sometimes needed to fulfill regulatory requirements.

Gases containing a high moisture content and temperatures near the dew point should be vented through insulated (refractory-lined) stack to avoid corrosion problems caused by condensation on the stack walls. If the gases are dry and vented at temperatures above the dew point, stainless steel stacks can be used. The wind forces which will act upon the stack must be considered; tall stacks may require additional structural support. The stack must be provided with test ports as per EPA monitoring regulations.

Air Pollution Control Equipment

If generation of pollutants is below emission limits, on APC system is required. In most cases, however, the process will generate some type of pollutants based on the composition of the incoming waste stream, as discussed previously. Since federal emission restrictions on smaller units have generally been less restrictive than for large facilities, some multiple chamber controlled air designs may be able to meet particulate standards without further APC devices. However, state and local regulation of smaller sources is increasing.

In designing resource-recovery incineration systems, opportunities to eliminate or minimize the generation of pollutants should be identified. Pollutant entry into the system can sometimes be prevented; proper system design and careful control of operating conditions can minimize generation of many pollutants. For example, using low sulfur fuels minimizes SO_x, generation. Using pure oxygen rather than air minimizes the generation of nitrogen oxide; however, the use of pure oxygen may not be a practical alternative. The selection of the proper operating temperature is important for pollution control as well as for achieving maximum combustion efficiency with the lowest possible auxiliary fuel consumption. Lower temperatures minimize NO_x formation; however, low temperatures contribute to generation of potentially hazardous organic pollutants. Temperatures over 815°C (1500°F) are required to minimize dioxin formation.

Estimating the uncontrolled emission rate (pre-APC) is a critical step in the design process that relies heavily on experience and measurments; accurate theoretical models are not available. In most cases, this emission rate is a complex function dependent on the waste composition, incinerator design, and actual operating conditions. It has not been possible to reliably predict the composition and quantity of the organics discharged from the stack. For a particular system, a comprehensive analysis of the stack discharges under normal operating conditions is required to identify if hazardous organics are being generated.

Carbon monoxide is the major product of incomplete combustion. Its emission rate is related to the temperature and oxygen concentration as seen by the combustion gases passing through the system. In a properly operating system, CO emissions should be close to zero and no subsequent control of CO required.

The total particulate emissions are the sum of the inorganic particulate and the combustible (soot) particulate. Incomplete combustion can result in a wide variety of combustible particulate pollutants, ranging from simple carbon to complex carbon-hydrogen-oxygen-halogen-nitrogen compounds. Carbon, because of its refractive index, the colour of the particles, and the typically small particle size contributes to capacity problems. Formation of soot is favoured by high temperatures combined with low oxygen concentrations. Since soot can lower the efficiency of and damage subsequent APC equipment, the best form of control for soot is efficient combustion at sufficiently high temperatures.

The inorganic particulate is characterized by its noncombustible (refractory) nature; emissions rates cannot be lowered by improving combustion efficiency. The primary source of the inorganic particulate is the ash portion of the fuel and waste. Small quantities of inorganic pollutants can also be introduced by the combustion air, if drawn from a high dust area, and from degradation of refractory or metal surfaces. Assuming that in most of the industrial resource-recovery applications being considered, the ash content be reduced by pre-treatment of the waste stream, design features and operating procedures must attempt to minimize the generation of fly ash. Agitation of the waste bed, misdirected underfire or overfire air, volatilization of metallic salts, high burning rates, large incinerator sizes, and certain waste-feeding mechanisms (stoking, grate design) all contribute to increased fly-ash generation.

Of special interest in the design of metal-recovery incinerators is the fact that the volatilization and recondensation of metallic salts can be a significant source of noncombustible particulate pollutants. The metallic salts, along with metal oxides mechanically released from the bed, constitute a significant percentage of the small (<5 μm particulate . Most APC systems are more effective at removing large particulate; the percentage of metal-bearing particulate exiting the APC system may be higher than the percentage entering. This does not imply that the equipment is failing to remove metals; the total quantity of metals being emitted is substantially reduced.

The type of air pollution control equipment required for a particular application will be dictated by the types of pollutants that are generated (size, density, shape, electrical characteristics, etc.) and the nature of the gas stream (temperature, moisture content, etc.). Basically, the APC system must remove particulates and gaseous, usually acidic, pollutants. Numerous pollutant scrubbing systems exist, but they are all based on several basic physical and chemical principles and represent variations on a few common design themes. Before discussing specific types of APC systems, an overview of the principles involved will be presented.

Particulate and Gaseous-Pollutant Mechanisms

There are several basic chemical and physical principles, one or more of which is involved in the operation of most gaseous- pollutant removal and particulate-collection mechanisms.

Gravity settling. Particulates are carried along in an air stream moving horizontally through a chamber. The heavier particles can settle to the bottom of the chamber as the velocity of the air stream is reduced. Dry mechanical collectors or gravity settling chambers apply this principle. Adding baffles or deflectors in the path of the gas stream increases the collection efficiency. These systems work best on streams with a moderate loading of large particulate (10-100 μm).

Inertial forces, impingement. If the direction of a moving air stream changes abruptly, the heavier particles tend, due to their momentum, to continue in a straight line. Leaving the air stream, they eventually collide with a wall and settle at the bottom of the collection chamber. Various types of cyclone collectors, operating on the principle of centrifugal force, exist. These systems, depending on the design, are efficient at removing particulate in the 5-50 μm range.

Filration. When an air stream passes through a porous material, particulate matter is trapped on one side of the material and clean air passes through the interstices. Fabric filters are efficient at collecting particulate, even at sizes below 0.5 m.

Interception, particle enlarging, absorption. A solid particle is intercepted by a liquid particle when the two are moving relative to one another and are within one radius of each other. Thus, a water droplet agglomerate is formed when dust particles pass through a water spray. The heavier weight of this agglomerate allows it to be more easily separated from the air stream; the water washes the particulates from the air stream. Wet scrubbers simultaneously cool and wash the air stream, and they can also remove some gaseous pollutants. Wet scrubbers also minimize any explosion hazard due to dry dust and air mixtures. Numerous types of wet scrubber designs exist, efficient at removing different ranges of particle types and sizes.

Diffusion. Dust particles migrate to water droplets in wet scrubbers in a way comparable to Brownian movement.

Contraction/expansion. Contraction of a gas stream condenses the moisture within the stream. Turbulence in the contracted area encourages contact between the solid and liquid particles. The agglomerate particles will have the same velocity as the rest of the gas stream. Entering an area of expansion, these large particles will continue in their current direction, while the rest of the gas stream can be redirected.

Electrostatic attraction. Particles can be electrostatically charged. Charged particles, attracted to objects of the opposite charge, can be removed from the air stream. When the two objects make physical contact, the particles are neutralized and fall to the bottom of a collector. Electrostatic precipitators can be very efficient at removing fine particles, approaching 99.9% efficiency.

Chemical reactivity. The chemical reactivity of certain pollutants can be exploited to increase removal efficiencies by injecting appropriate chemicals into the gas stream or adding them to the scrubber water.

Major Types of APC Equipment

The air pollution control system must maintain emissions below those that could cause pollutant ground-level concentrations to exceed established limits during all phases of a system's operation, including start-up, shutdown, and emergencies caused by equipment failure. Very low particulate (to

0.01 g.dscf) emissions and significant acid gas reductions (65-99%) can be achieved using existing technology. None of these systems are inexpensive; each has its own unique combination of capital, operating and maintenance costs that must be evaluated for each particular application.

Air pollution control equipment can be divided into two broad categories, wet and dry scrubbing systems.

Dry scrubbers range from simple, low-maintenance gravity settling chambers that can remove large quantities of heavy particulate to sophisticated fabric filters, or baghouses, which are very efficient at removing submicron particulate. Although some dry scrubber designs address acid-gas removal, most dry systems focus primarily on removing particulate. Two examples of major types of dry scrubbing systems are discussed in more detail below.

Example 1: Fabric filter (baghouse). In the baghouse illustrated in Figure 20.8, precooled gases enter the lower section of the baghouse. The gases are drawn upward through the filter bags. The filter bags are supported by steel wire frames arranged in rows and suspended vertically from a plate. Particles are retained on the outer surfaces of these bags while the clean gases pass through the interstices of the fabric. The cleaned gases pass through into the upper chamber and are exhausted through the outlet. As the dust collects on the side of the fabric, the collector becomes even more efficient. The dust layer blocks the larger of the fabric interstices, which are typically about 100 μm in diameter, helping to capture the finer dust particles. The fabric must be cleaned frequently or eventually no gas would pass through.

In the system shown, the bags are cleaned by momentary surges of high pressure air blown into the bags in the opposite direction of normal gas flow, periodically activated by a remote cyclic timer. The dust blown off falls into the hopper. Since only a fraction of the collector's filter area is cleaned at any one time, the total air flow through the collector is maintained at rated capacities. A manual slide gate, shown at the bottom of the baghouse, provides access for the removal of accumulated dust. Other approaches to cleaning the bags can be used: the cleaning can be continuous or intermittent, as well as periodic. The filters can be cleaned by manual or powered shaking of the bags.

The type of fabric used will determine the conditions under which the filter can oprate. Where the dust is continuously removed and no dust layer forms, for instance, when handling very heavy dust loads, felt fabrics can be used since they are efficient at collecting very fine particles. Cotton and wool are suitable if gas temperatures are below 90°C (200°F). Nylon is abrasion-resistant. Acrylics resist acid and can operate at higher temperatures. Polyesters have good high temperature stability and abrasion resistance. Fiberglass can be used over 260°C (500°F) but has poor abrasion-resistance and cannot be shaken. Teflon and nomex have bood temperature characteristics but are more expensive; teflon also has a low coefficient of friction and may pass particulates through. Although baghouses are very efficient, even with submicron particulate, they have a few drawbacks: most fabrics cannot be used at very high temperature, and the maintenance costs associated with bag replacements can be high.

Example 2: Electrostatic precipitators (ESP). In an electrostatic precipitator, the dirty gas is channeled between two electrodes: a high-voltage discharge electrode and a grounded collecting electrode. The high negative voltage causes electrons to be emitted from the discharge electrode (corona discharge). The electrons then collide with the surrounding gas molecules, ionizing them. The negative gas molecules are attracted to the relatively positive (grounded) electrode, colliding with and transferring their negative charge to the dust or mist particles. Due to electrostatic forces, the negative dust particles then migrate to the grounded electrode, become neutralized and adhere to the plate. The plate must be

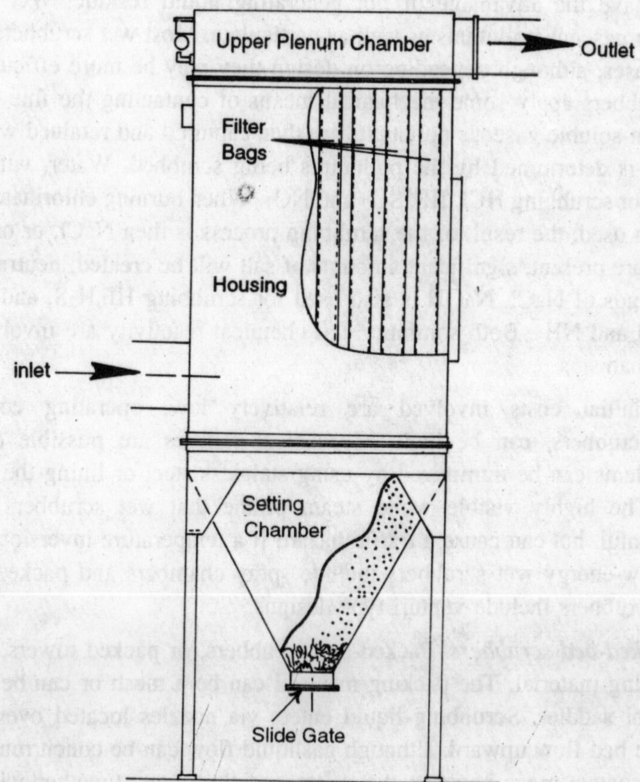

Figure 20.8. Fabric filter (baghouse).

regularly vibrated to prevent build-up of the dust layer which, acting as an insulator, would decrease the neutralization efficiency of the electrode. The particles then collect at the bottom of a hopper. Precipitator designs are usually of either the plate or pipe variety: a series of flat collecting plates with high voltage discharge wires suspended between them or a number of grounded pipes with discharge wires suspended within them. Collection plate area, gas velocity and temperature, the strength of the electric field, and the moisture in the gas stream all effect ESP collection efficiency.

The conductivity of the dust is a limiting factor in the use of precipitators. If the dust has too low of a resistance, it may re-enter the air stream by transferring its charge to the plate; if the resistance is too high, it tends to adhere to the plate and insulate the collecting electrode. The resistivity of the dust can be modified by adding appropriate substances to the air stream. High particulate loads result in collection of dust on the electrode and can lead to spark discharges, causing dust to reenter the air stream. If the precipitator becomes too hot, it could ignite. Electrostatic precipitators can be used for dry particulates, fumes, or mists, and can operate over 540°C(1000°F). Precipitators can be very efficient; high removal efficiencies (98%) have been obtained in removing particulate as small as 0.05 μm from zinc oxide fumes. Dry ESPs cannot remove acid gases and are susceptible to corrosion from HCl. Compared to wet scrubbers, the pressure and temperature drop across ESPs is small and power requirements are low.

Dry scrubbers have the advantage of not generating liquid residue. Wet scrubbers have the advantage of removing gaseous pollutants as well as particulate. Most wet scrubbers try to remove both particulate and acid gases, although depending on design they may be more efficient in removing one or the other. Wet scrubbers apply some mechanical means of contacting the flue gases with a liquid; particulate and solution-soluble gaseous pollutants are then captured and retained within the liquid. The type of solution used is determined by the pollutants being scrubbed. Water, without any additional reagents, can be used for scrubbing HCl, HF, SO_2 and NO_2. When burning chlorinated wastes, a solution of 50% NaOH is often used; the result of the scrubbing process is then NaCl, or ordinary salt. If large quantities of chlorine are present, significant amounts of salt will be created; neutralizing one pound of HCl produces 1.6 pounds of NaCl. NaOH is also used for scrubbing HF,H_2S, and HCN. Other useful reagents include KOH and NH_3. Both solubility and chemical reactivity are involved in wet scrubber acid gas removal mechanisms.

Although the initial costs involved are relatively low, operating costs, especially for high-efficiency wet scrubbers, can be high. Mechanical failures are possible due to erosion and corrosion. These problems can be minimized by using stainless steel or lining the system with rubber or plastic coatings. The highly visible white steam plume that wet scrubbers discharge into the atmosphere is not harmful, but can cause a traffic hazard if a temperature inversion occurs, creating an area of dense fog. Low-energy wet scrubbers include spray chambers and packed towers. High- and medium-energy wet scrubbers include venturi type designs.

Example 3: Packed-bed scrubbers. Packed-bed scrubbers, or packed towers, consist of a bed of randomly placed packing material. The packing material can be a mesh or can be composed of small solids, such as rings or saddles. Scrubbing liquid enters via nozzles located over the bed; the gases usually enter under the bed flow upward, although gasliquid flow can be concurrent, countercurrent, or crossflow. The large contact areas between the water and the gases, together with a relatively long retention time (long gas-flow path) in the packed tower, accomplish absorption of solution-soluble gases and particulate. Packed beds cannot handle heavy particulate loads, but are very efficient at removing submicron particulate and gaseous pollutants.

Example 4: Venturi scrubber. In venturi scrubbers, the dirty gas and scrubbing fluid move at very high velocities, generating turbulence and mixing and therefore attaining good removal efficiencies. With venturi scrubbers there are many ways in which the water and dirty gases can be injected and the sludge and moist gases removed. In some venturi systems, the drop in gas pressure across an orifice accelerates the gas stream. As the scrubbing liquid flows in through the orifice, it is atomized by the passing gas stream. Alternately, the liquid can be sprayed in. The intimate mixing between the liquid and gases results in efficient pollutant capture. If the orifice area is fixed, the pressure drop will vary with the gas flow rate, affecting pollutant-removal efficiency. In other venturi systems, characterized by low power requirements for liquid pumping, the gases pass through an orifice at the surface of a pool of scrubbing liquid. In the (patented) Pumpless Venturi (PV) scrubber shown in Figure 20.9, the gases enter the scrubber through the centrally located intake tube and travel downward. The venturi assembly, located at the bottom of the tube, is partially submerged in the scrubber solution. The gases travel through the venturi and aspirate solution; the scrubbing process takes place here. The pollutants are removed from the gases and released into the solution contained within the reservoir. The solution is induced into rotary motion as the gases encounter the stationary assembly, creating a centrifuge. The

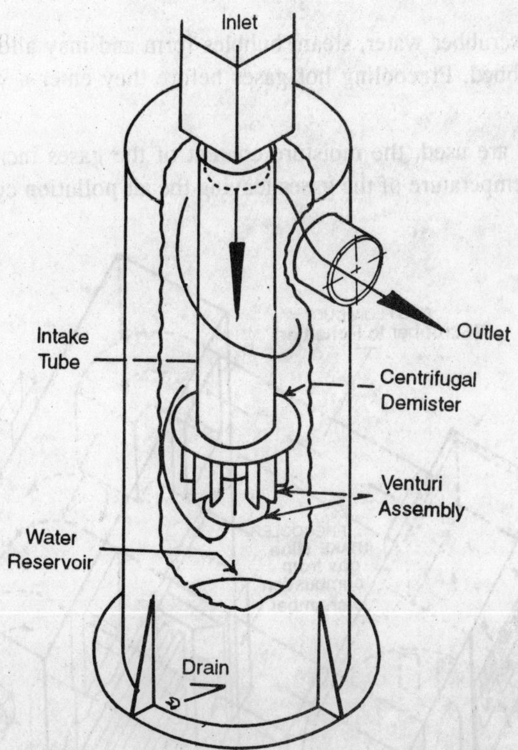

Figure 20.9. Pumpless Venturi (PV) scrubber (patent secured by M&S Engineering & Manufacturing Co., Inc).

solid particles, contained within the solution, settle at the bottom of the reservoir. Periodically, the accumulated sludge and particulate must be drained from the scrubber through its bottom port. The moist gases flow through the centrifugal demister where the solution and gases are separated again. The solution flows back into the reservoir and the gases, reheated by indirect heat exchange with the incoming flue gas traveling through the intake tube, proceed to the next stage of the system. Venturi scrubbers can be better than 99% effective on small-diameter particulate (<0.5 μm); however, the larger water-pressure drops of these systems lead to higher operating costs.

The effectiveness of scrubbing systems is usually directly related to the pressure drop across the scrubber. Higher pressure drops reflect greater turbulence, resulting in more effective scrubbing. Low energy systems (usually less than 12" WC) will remove high percentages of larger particulate, but higher energy systems (<20" WC) are usually required to effectively remove submicron particulate and acid gases. A multi-stage air pollution control system can make use of the advantages of particular types of APC equipment, while minimizing their disadvantages and overall energy requirements. For example, packed bed scrubbers are very effective at removing submicron particulate and acid gases; however, they are susceptible to plugging when exposed to gas streams with heavy particulate loadings. Using a low-energy venturi scrubber to remove the heavy loads of large particulate, as a prefilter to a packed bed scrubber or a baghouse, lower overall capital and operational costs while achieving efficient scrubbing across the full range of pollutants: large particulate, submicron particulate, and acid gases.

Auxiliary Equipment for APC Systems

If not gases contact scrubber water, steam bubbles form and may allow a portion of the gases to escape without being scrubbed. Precooling hot gases before they enter a wet scrubber reduces steam bubble formation.

When wet scrubbers are used, the moisture content of the gases increases. Reheating the gases after scrubbing raises the temperature of the gases leaving the air pollution control system to above their

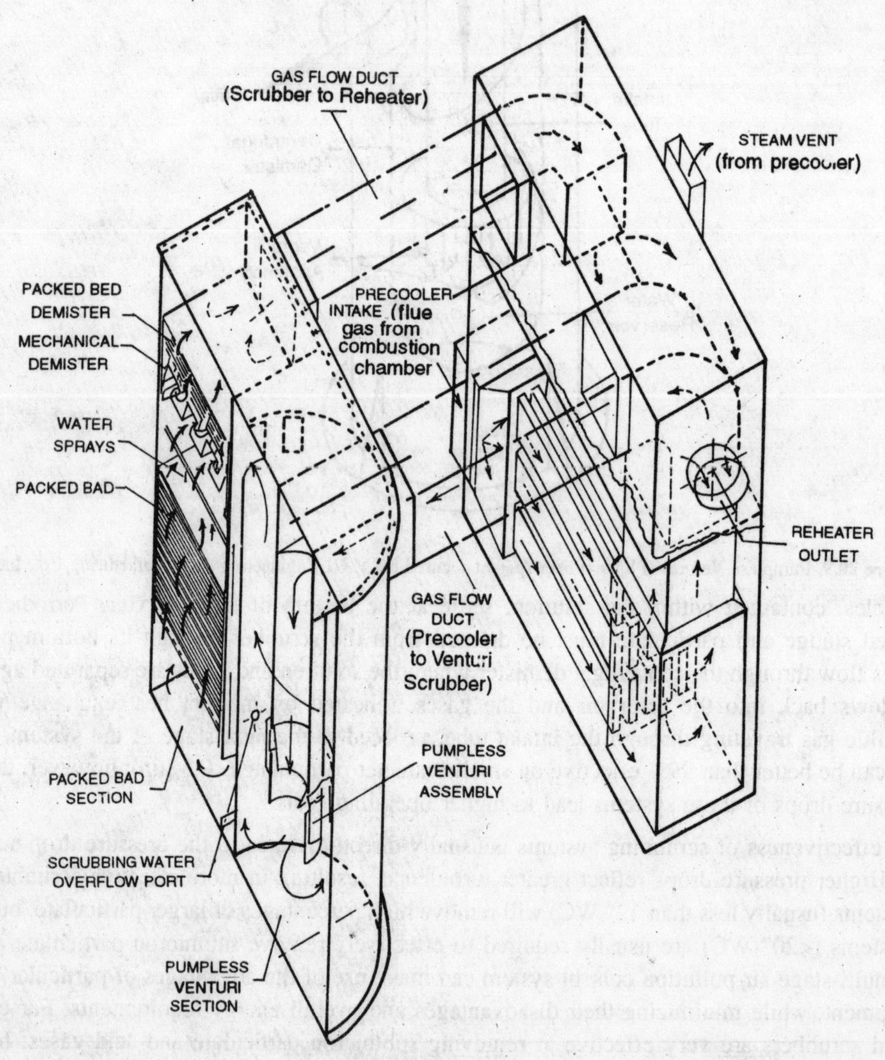

Figure 20.10. A high-efficiency scrubbing system consisting of four stages: a flue gas cooler, a PV scrubber, a packed bed scrubber with demister, and a flue gas reheater.

dew point. This eliminates condensation that may corrode the exhaust fan and stack. Reducing the moisture content also reduces steam plume formation over the stack. Steam plumes are sometimes mistaken for smoke and smoke is an indication of improper incinerator operation.

Flue gas coolers reheaters, and demisters are often used in conjunction with air pollution control equipment, as illustrated in Example 5.

Example 5: High-Efficiency Pumpless Venturi Scrubbing System. The High-Efficiency Pumpless Venturi (PVH) air pollution control system, shown in Figure 20.10, combines a flue gas cooler, reheater, and demisters with a multi-stage scrubbing system consisting of PV and packed-bed scrubbers. A flue gas to water heat exhanger is used to cool the incoming hot gases from the combustion chamber. The gases pass through the PV scrubber (with centrifugal demisting) and then through the packed-bed scrubber. After passing through the second layer of packing which serves as a demister, the gases enter the flue gas to flue gas heat exchangers, where they are reheated before entering the atmosphere. Both the flue gas to water and flue gas to flue gas heat exchangers are noncontact, indirect heat exchangers, with the flue gases passing through stainless steel or carbon steel tubes and the heating or cooling media surrounding the tubes. Fly ash that settles out of the gas stream during its passage through the exchanger is collected in a hopper. The water side of the system must occasionally be cleaned to remove any accumulated sludge. This water is usually once-through water released into the atmosphere as steam, although it can be condensed and recirculated.

When using wet scrubbers, integral water treatment is required if the scrubbers are to continue to effectively absorb pollutants. Water recirculation is recommended to avoid causing a water pollution problem. The water from the scrubbers must be continuously recirculated, filtered, and neutralized. The water-conditioning system consists of some arrangement allowing water to pass through a filter. The contaminants, such as salt generated by neutralizing HCl with NaOH, are collected within the filters and the clean water passes through. Fresh caustic is metered into the water as it is returned to the scrubbing system. Manually or via automatic backflushing, the filters must occasionally be cleaned and the residue collected for treatment or disposal.

Residue-Handling Systems

The design of the residue-handling system will be dictated by the source and physical characteristics of the residue.

In material-recovery systems, the combustion residue includes the material being recovered. Aluminum, mixed with steel, can be recovered from automotive scrap by burning off any conbustible components (oils, fabrics, plastics) and melting off the aluminum. The molten aluminum is drained into forms from the discharge end of the primary (ignition) chamber, while the clean steel scrap remains in the chamber. Zinc is recovered by melting zinc die casting scrap mixed with combustible components and steel scrap. Part of the zinc is collected as metal drained into the forms; the rest of the zinc is captured by baghouses as zinc oxide dust. The clean steel scrap remains in the process chamber. Silver is recovered either as silver metal or silver oxide from the ashes remaining after the incineration of photohgraphic films and papers. After the combustible component of tin-containing sludges has been incinerated, a tin oxide powder remains.

The fine ash continuously deposited in the ignition chamber's ash bin and the fly ash collected in hoppers underneath other components, such as baghouses and flue gas coolers, can be collected manually or removed automatically for transfer into containers for final disposal. Automatic ash collection is preferable in most cases to avoid frequent shutdowns and cooling of the unit for cleaning and ash removal. An auger ash conveyor, suitable for transferring fine ash, is illustrated in Figure 20.11. This conveyor consists of two large screws, one into the ash bin and the other up to the ash container. The screw housing can be water-cooled to reduce the temperature of the ash so that it can be discharged directly into plastic-lined fiber containers. A level controller or weight scale can be provided to automatically inform the operator when the container is full. The entire assembly may need to be covered by a steel shroud to prevent excessive dust generation, especially if the ash contains toxic levels of noncombustible hazardous compounds, such as certain metals. If the conveyor receives ash from several hoppers, gate valves can be sequenced to automatically empty the hoppers into the conveyor. Screen grates should be provided to prevent the entry of large objects, such as steel scrap, into the conveyor.

Another ash conveyor is also illustratd in Figure 20.11. The slat ash conveyor consists of two conveyor chains, parallel to each other, supporting structural steel slats. As the slats move forward, ash is collected from the wet ash bin and transferred to suitable containers; the water drains back into the bin.

Energy-Recovery Systems

Energy can be recovered from the incineration of wastes for conversion into forms that can be used at the local facility or sold for use by a nearby industry or a local utility company. The energy contained within the hot combustion gases can be recovered as hot air, hot water, or steam. Flue gas to air heat exchangers allow space heating by the transfer of energy from the hot flue gases to the room air by convection across the heat-exchanger walls. Commercial boilers often produce steam or hot water for heating in the winter and cooling in the summer. Industrial boilers use steam or hot water primarily for process applications: drying steam in a paper cycle, energy-to-drive presses, or other large pieces of mechanical equipment. A flue gas to air heat exchange is illustrated in Figure 20.11; the rest of this section is concerned with boilers.

Including energy recovery in an incineration system whose main purpose is either the disposal of wastes or material recovery increases equipment and operating costs. Unlike utility boilers, which operate under close control and well-defined load-cycle patterns, commercial and industrial boilers often operate under variable load conditions. The boiler must be able to achieve the required capacity in the shortest amount of time. Also, variations in waste heating value will produce variations in steam generation that must be accommodated by the system. A new set of maintenance and operational concerns are introduced by steam-tube corrosion. These disadvantages must be weighed for the particular application against the benefits of including heat-recovery equipment. In addition to the value of the recovered energy, there are some equipment-related benefits. Heat recovery results in reduced exhaust gas temperatures and volumes. Less excess air is required to control incinerator temperatures in exothermic reaction, smaller chambers can be used because of the reduced gas flow, and smaller air and induced draft fans are required for the smaller gas volumes.

Boiler Equipment

Boilers generate steam or hot water at pressure and/or temperature above that of the atmosphere. The steam or hot water is produced by the transfer of heat from the combustion process, elevating its pressure and temperature. The containment vessel must be designed to safely handle the operating

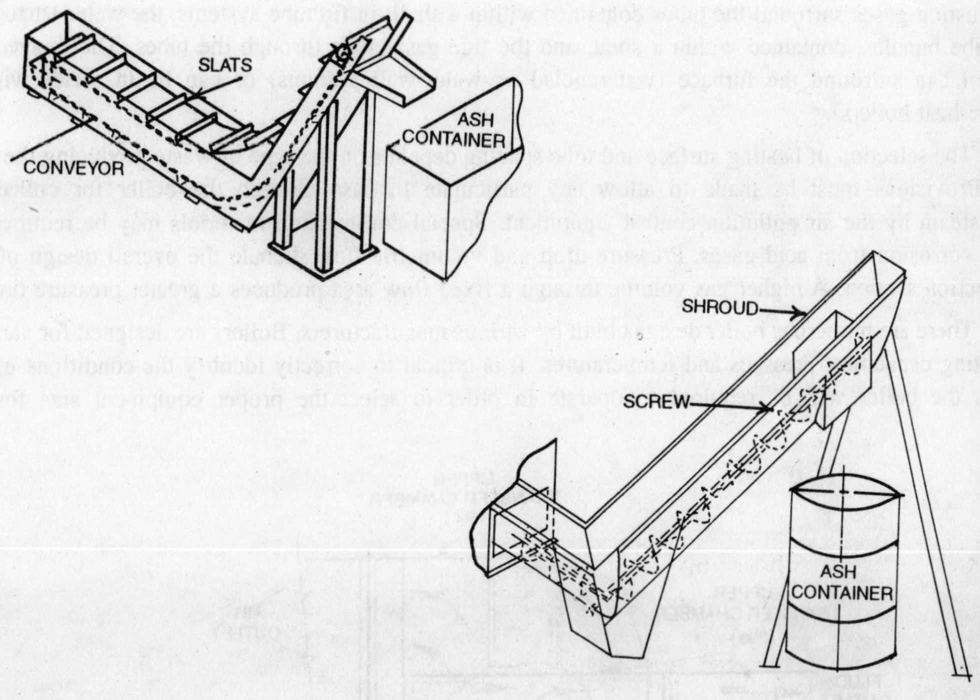

Figure 20.11. Typical ash collection equipment. The slat ash conveyor (top) is suitable for transferring ash containing small particles which require wet collection. The auger ash conveyor (bottom) is suitable for transferring fine ash.

temperatures and pressures. The end-use of the steam or hot water will determine the temperature and pressue that the boiler must be designed to provide. For example, to heat a large facility, a certain capacity at an elevated pressure and superheated temperature may be required to provide the force to transfer the steam to its destination.

The two basic components of a boiler system are the furnace section and the convection section. In the furnace section, the fuel is mixed with air and burned, releasing heat that is carried in the combustion gases. In the convection section, the heat contained in the combustion gases is transferred to the water for the production of steam.

In resource-recovery application, the furnace section would consist of the appropriate waste incinerator, its shape determined by the type of fuel (waste) and the burning and ash characteristics. Furnaces can be divided into either refractory or watercooled categories. The refractory type, with insulation within a casing material, were discussed in detail in the previous sections. In watercooled system, an envelope consisting of tubes placed close together surrounding the furnace, absorbs the heat generated by the combusion process and helps produce steam.

The convection section of the system, often specifically being referred to by the term boiler, consists of closely spaced tubes arranged to allow passing of the combustion gases around the tubes (watertube) or through the tubes (firetube). In watertube systems, the water is inside the tubes and the combustion gases surround the tubes contained within a shell; in firetube systems, the water surrounds the tube bundles, contained within a shell, and the flue gases pass through the tubes. The convection section can surround the furnace (watercooled or water-wall systems) or can be in series with it (waste-heat boilers).

The selection of heating surface and tube spacing depends on the type of waste producing the flue gas. Provisions must be made to allow any particulate to pass through the boiler for collection downsteam by the air pollution control equipment. Special construction materials may be required to resist corrosion from acid gases. Pressure drop and volumetric flow dicbate the overall design of the convection section. A higher gas volume through a fixed flow area produces a greater pressure drop.

There are numerous boiler designs built by various manufacturers. Boilers are designed for various operating capacities, pressues and temperatures. It is critical to correctly identify the conditions under which the boiler will be required to operate in order to select the proper equipment size for an

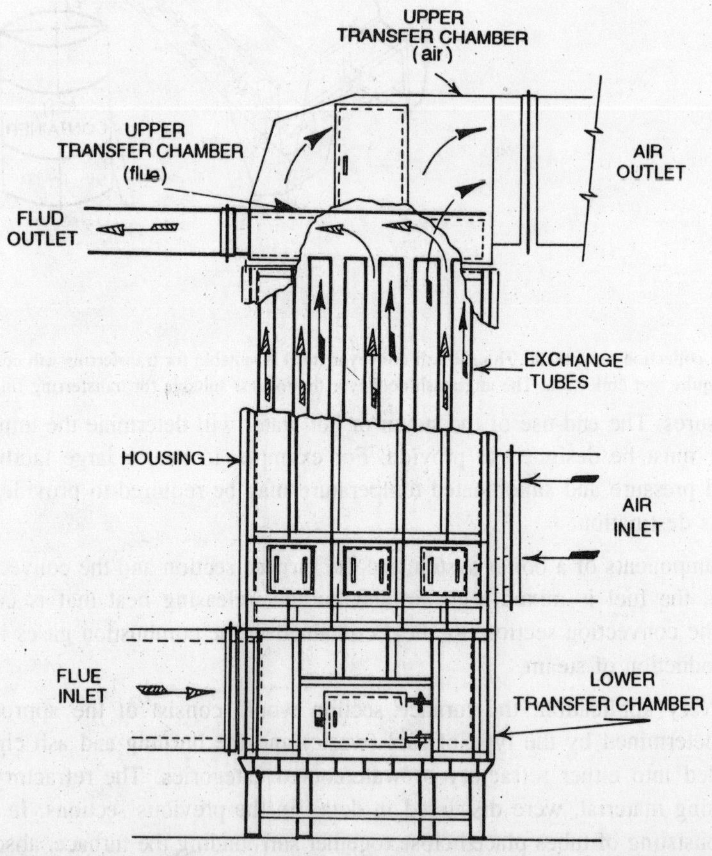

Figure 20.12. Flue gas to air heat exchanger.

application. In addition to minimum, maximum, and normal load ranges, the nature of expected load flunctuations must be specified. Operating conditions for boilers can range from low pressure operation (<15 psig) to 1,400 psig. Steam temperatures can range to 540°C (1000°F) and capacities to 1 million lb of steam per hour. Industrial heat recovery applications with boilers connected to waste incinerators are usually smaller, ranging up to 50,000 lb/hr of steam at pressure up to 250 psig. Example 6 briefly presents the operation of one waste-heat boiler design.

Example 6: Waste heat boiler. The waste heat boiler illustrated in Figure 20.13 consists of three parallel horizontal tubes. encased in a two-inch fiberglass insulating blanket. Hot flue gases enter the boiler through the inlet located in the central tube; the first pass tube consists of one large single tube.

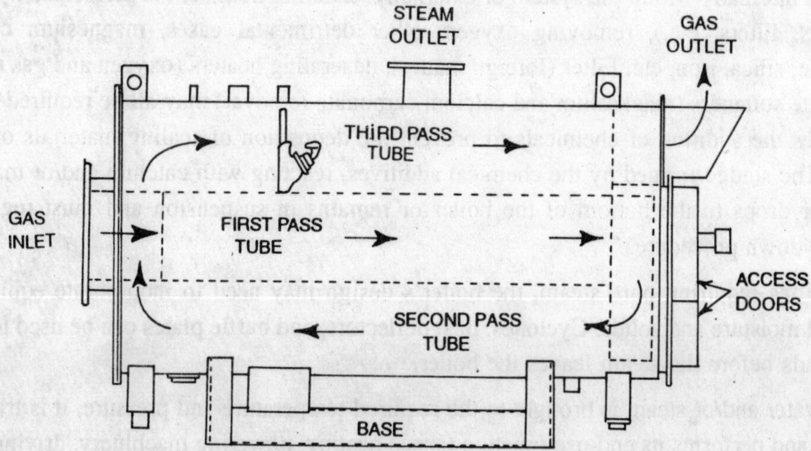

Figure 20.13. Waste Heat Boiler.

The second and third pass tubes consist of multiple tubes surrounded by water. The flow area proportion of all passes assures even gas velocities through the boiler for maximum heat transfer efficiency. As the gases travel through the tubes, heat is transferred to the water and steam is generated. Baffles, located below the steam outlet, promote dry steam generation. The gas outlet may be equipped with an induced draft fan to provide a regulated draft to draw the flue gases through the boiler tubes. The fan outlet would be connected to the stack. Access doors are situated at the rear of the boiler to allow for maintenance inspections. Water is pumped to the boiler by a feed water pump which is controlled by the water level in the boiler. When the boiler is preceded by an afterburner, the hot gases are relatively clean and particulate-related problems, erosion and plugging, are avoided.

Auxiliary Equipment Used with Boilers

Superheaters are sometimes used to redirect the steam back through the flue gases to take on additional heat. Radiant-type units consist of tubes located in the furnace section, receiving heat by direct radiation from the furnace. Convection-type systems are located behind the screen wall of the convection section and receive heat from the passage of gases around the tubes.

Just as air heaters (heat recuperators) can be used to preheat the combustion air via heat transfer from the flue gases, economizers preheat the incoming boiler feedwater via heat transfer from the flue gases exiting the boiler. In both cases, overall energy requirements are decreased, increasing system efficiency.

All natural waters contain some amount of suspended or dissolved inorganic or organic chemicals. Water treatment may thus be required at the input as well as the discharge end of the system. Prior to entering the boiler, the water may be tested to prevent deposition of scale and corrosion from acids, oxygen, and other harmful substances that may be in the water supply. Systems obtaining water from natural sources must take precautions to minimize the impingement of organisms on the water intake. Table 20.6 lists some potential equipment problems resulting from polluted water. Depending on the source of the water, the required treatment may vary seasonally or over time. Boiler feedwater treatment can be performed internally within the system or externally. External treatment is performed by separate equipment (tanks, filters, etc.), removing oxygen, other detrimental gases, magnesium carbonate, calcium carbonate, silica, iron, etc. Filter (foreign matter), deaerating heaters (oxygen and gas removal), and sodium zeolite softeners (magnesium and calcium carbonate removal) may all be required . Internal treatment involves the addition of chemicals to prevent the deposition of scaling materials on the hot boiler surfaces. The sludge formed by the chemical additives, reacting with calcium and/or magnesium carbonates, either drops to the bottom of the boiler or remains in suspension and must regularly be removed by blowdown procedures.

In applications requiring pure steam, the boiler's design may need to incorporate equipment to remove entrained moisture and solids. Cyclones, mist deflectors, and baffle plates can be used to remove moisture and solids before the steam leaves the boiler.

After the water and/or steam is brought to the required temperature and pressure, it is transported to its destination and performs its end-use function (space heating, powering machinery, driving a steam turbine to generate electricity, etc.).When the heat has been taken out of the water, the remaining steam, water, or condensate is either discharged or recycled. As clean steam, it can be discharged directly into the atmosphere. If the passage through the system has introduced chemical contaminants, treatment prior to discharge into the atmosphere (steam), sewers, or bodies of water (hot water) will be required.

Figure 20.14 illustrates an energy-recovery system. A description of this system is given in Example 7.

Monitoring and Control System

The capabilities required of a monitoring and control system in a resource-recovery incineration facility depend on the scale of the operation, the degree of automation wanted, and regulatory requirements. The goal of the control system is to maintain the combustion temperature required for the complete thermal decomposition of the waste being incinerated. This is accomplished by adjusting the auxiliary fuel, combustion air, and waste feed. Systems can use simple direct-control loops to cut off waste feed if critical parameters exceed preset limits or computer-controlled sensors can monitor every aspect of the operation, diagnosing problems, automatically making adjustments, and alerting the operator when routine maintenance activities are required. More comprehensive control systems provide better control over process conditions, conserving energy, and increasing efficiency. However, the increased complexity also brings higher initial costs and increased maintenance and operator-training requirement. When processing wastes containing hazardous components, it may be necessary to use a

Table 20.6

Operational Problems Caused by Polluted Water

Contaminant	*Problem/Consequence*
Alkalinity (bicarbonates, carbonates, various caustics)	Generates scale (e.g., boilers, wet scrubbers)
Hardness (salts, especially calcium and magnesium)	Generate scale (e.g., boilers, wet scrubbers)
Sodium salts (sulfates, chlorides, nitrates and bicarbonates)	Possibility of creating slag which could cause mechanical difficulties.
Silica (SiO_2)	Generates scale
Iron, manganese	Causes stain, objectionable in some cases
Carbon dioxide	Increases alkalinity; enhances corrosive behaviour of dissolved oxygen
Oxygen	Corrosive to iron, steel, galvanized iron and brass
Nitrogen	Retains dissolved O_2, increasing corrosiveness
Hydrogen sulfide	Creates odor, corrosion of iron pipes, fittings, and equipment
Methane	Fire and explosion hazard
Microorganisms and organic pollutants	Forms coating in pipes; cause stain

sophisticated control system, including continuous recording capabilities, to comply with EPA requirements.

To provide for safe and economic operation, certain parameters must be monitored in every system— at minimum, the combustion temperatures in the ignition and combustion chambers and pH of the scrubber water when wet scrubbers are used. The major control system parameters are discussed below.

Temperatures

Temperatures in the ignition and combustion chambers must be monitored and maintained at required levels to ensure complete combustion and to protect equipment from damage due to overheating. Auxiliary fuel must be increased and the waste feed adjusted if the temperatures fall too low. Auxiliary fuel flow and waste feed must automatically terminate if the temperatures rise above the safe operating limits of the chamber. Operating temperature ranges will vary depending on the type of equipment and on the composition of the waste stream. In some material- recovery applications, for example, the temperature may need to be above the melting point of the metal being recovered, as when molten aluminum or zinc is poured into forms, or the temperature may need to be below the melting

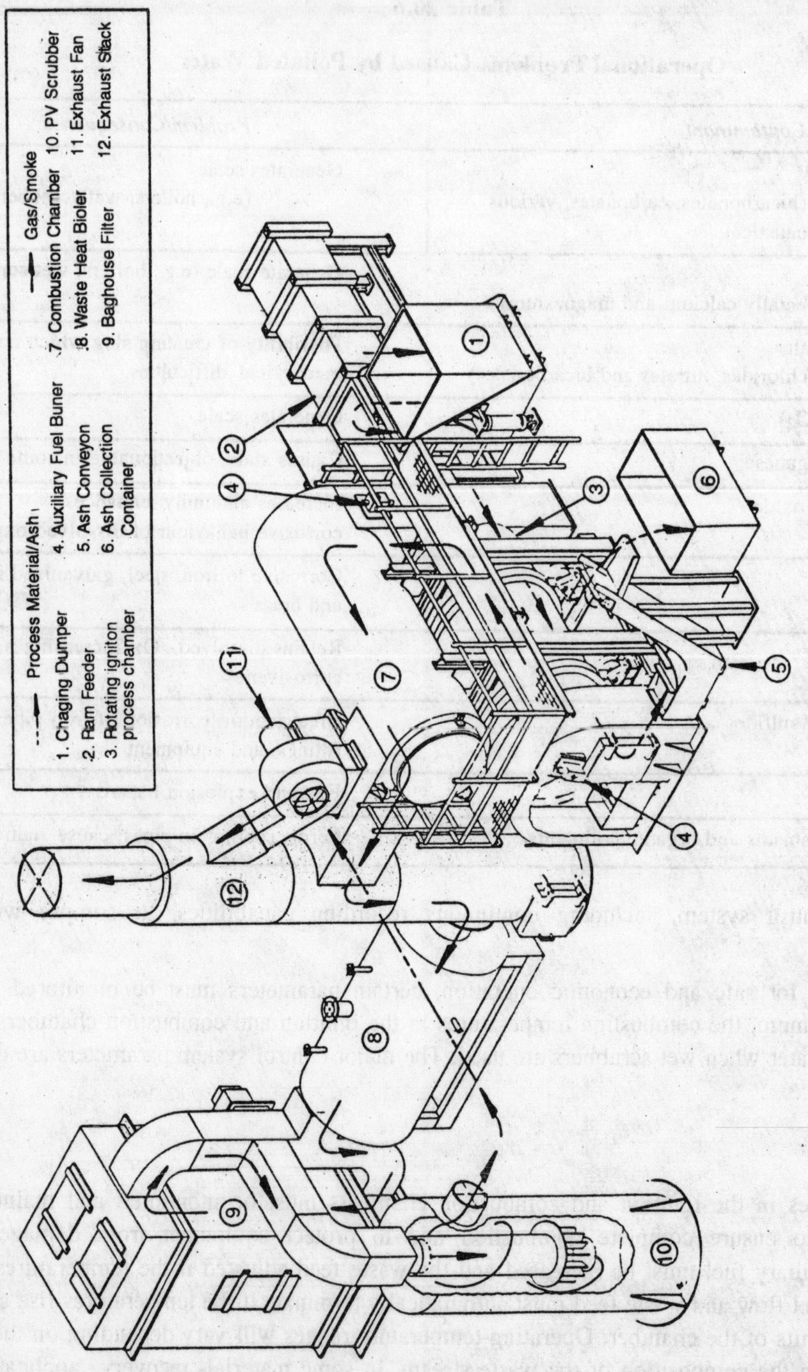

Figure 20.14. An energy recovery system, including a revolving ignition chamber, combustion chamber, a two-stage air pollution control system, and a waste heat boiler.

point of the metal to prevent alloying. When insulation is being burned off copper wires, operating at low temperatures would result in inadequate burning, while operating at high temperatures would result in copper embrittlement. Regulations or permit conditions may also specify the operating temperature ranges required to limit pollutant emissions. Since temperatures throughout the chamber can vary significantly, the location of the thermocouple used to measure the temperature is important.

Temperatures in other parts of the system may also need to be monitored. For example, the temperature of the gases entering the air pollution control system can be monitored to prevent damage to packed-bed packings or baghouse filters from temperatures beyond their design specifications.

Exhaust Gas Concentrations

Calculating the air required for complete combustion based on the quantity of carbon in the combined waste and fuel streams determines the stoichiometric, or theoretical, air requirement. The stoichiometric air value implies that the burning process is 100% efficient and there is instantaneous and perfect mixing of all the components. In reality, to achieve complete combustion, excess air must be provided. The quantity of excess air required depends on the design of the incinerator and the characteristics of the waste; it can range from 5-200%. Burning equipment is rated for operation within a specific range of excess air. Operating below the range specified would result in incomplete combustion of the waste. Operating above the range with too much excess air would reduce the combustion temperature, or increase supplemental fuel requirements.

As both CO_2 and O_2 levels in the flue gas are related to the amount of excess air supplied, either a CO_2 or an O_2 analyzer can be used for monitoring and control. At stoichiometric conditions, oxygen in the flue gas is zero and the percentage of CO_2 can be calculated from the quantity of carbon in the waste and fuel. As the excess air increases, the oxygen component of the flue gas increases, decreasing the CO_2 percentage of the total. However, since a particular CO_2 level can indicate either an excess or a deficency of combustion air, oxygen sensors are preferable.

Carbon monoxide or combustibles analyzers can be used to measure concentrations at very low levels with accuracies better than ± 20 ppm. Since CO and other combustibles are products of incomplete combustion, in a properly operating system, analyzers should read zero. The analytical data can be presented as a fraction of O_2, CO, or CO_2 in the total flue gas by weight or volume or as a fraction of O_2, CO, or CO_2 in dry flue gas by weight or by volume. Stack monitors for many other possible flue gas constitutes, including SO_2 and NO are also available.

Opacity Measurements

Smoke is a suspension of solid or liquid particulate matter within a gaseous discharge. The visibility of smoke is related to the quantity of particles present, rather than the weight of the particulate matter; the colour of the smoke does not necessarily reflect the smoke's density. White or opaque smoke is usually due to insufficient furnace temperatures, not high enough to produce complete combustion. Inorganics in the exit gas may also produce a nonblack smoke discharge. Sulfur compounds will appear yellow; calcium and silicon oxides will appear light brown. Black smoke appears when hydrocarbons are burned in an oxygen- deficient atmosphere and carbon particles remain in the off-gas.

No smoke should be seen from modern incinerators equipped with the necessary pollution controls. When malfunctions occur, smoke or opacity measurements are usually taken visually. For automatic detection of malfunctions, electronic in-line opacity measurements of the exhaust gas stream may be taken.

Gas Pressure and Flow Measurements

Differential pressure gages or gas flow meters are used to measure pressure drops across and gas flows through, the air pollution control system. Problems such as plugged packed beds or spray nozzles can be detected by changes in the pressures experienced throughout the system.

Miscellaneous Monitoring and Control Parameters

In systems where waste feed is automatic, the waste feed needs to be monitored. This measurement can be accomplished in a number of ways, depending on the waste-feed mechanism. If liquid wastes are being pumped into the system, the pressure at the pump can be measured and calibrated to reflect the feed rate. Another approach would be to monitor the change in weight of drums containing liquid wastes being pumped into the system. In addition to feed rate, auxiliary fuel consumption needs to be monitored and regulated. Depending on the system's components, monitoring and control of other parameters may be required: weight or level of residue (ash, molten metal, etc.) collected, level and condition (pH) of the scrubber solution, rotation speed (rotary kilns), and boiler water level and characteristics.

RESOURCE-RECOVERY-SYSTEM DESIGN: ISSUES AND EXAMPLES

Each industrial resource-recovery project is unique and its success and profitability depend on many factors; the actual cost of the equipment used is usually a small percentage of the total expenditure. Revenues from a resource-recovery facility depend on the markets for the recovered resources, which can be very volatile, and the avoidance costs associated witn volume reduction of wastes requiring disposal. Several issues must be addressed prior to beginning equipment selection and analyzing the economic feasibility of the project.

Goals

Is the main purpose of the facility to dispose of waste, with resource recovery being applied to offset operating costs, or is the purpose the long-term profitable recovery of materials and/or energy? Facilities where resource recovery is a supplementary activity would require a system design that allows the material and/or energy recovery equipment to be taken off-line easily during periods of depressed energy or material prices.

Regulations

What are the applicable regulations? State and local regulations can be significantly more stringent than federal regulations. The nature and efficiency of the air pollution control system is primarily regulation-driven. It is important to remember that regulations are subject to (frequent!) change; a system that just meets regulatory standards today, may not be in compliance by the time the facility opens. Although it may be initially more expensive to do so, a more comprehensive APC system, or at least provisions for additional future APC improvements, may be less expensive in the long run.

Waste Analysis

How reliable and complete is the waste analysis? How accurate is the waste quantity estimate over the lifetime of the facility? Since all subsequent cost and revenue estimates are based on the expected throughout, not to mention the selection of the equipment itself, an inaccurate waste-stream analysis will invalidate all subsequent conclusions. Commercial resource-ecovery facilities must consider that the composition of their waste streams may change as new products and technologies are developed. Also, the quantity of supply (waste) may change as industries adopt waste minimization techniques prompted by high disposal and rising raw material costs. Emphasis on waste-source reduction and the development of new recycling technologies for some wastes, such as plastics, that are also excellent incineration candidates, is changing the nature of the waste streams available for incineration.

Table 20.7 brings together the topics discussed in the previous section to provide an overview of resource-recovery system design considerations and economic analysis issues. Different resource-recovery faclity perspectives are : a medium-sized industrial facility and a compact mobile- incineration system. To summarize the many ideas and topics discussed in this chapter, a comprehensive example of a potential industrial resource-recovery application is provided below.

Example 7: Design Example.

Problem description. An industrial plant recovers silver from waste film by washing the film in hot detergent. The resulting sludge, containing silver, must be dried out before the next step in the recovery process. The plants's wastes consist of significant quantities of this plastic film (after washing) and mixed plant trash, mostly paper and wood. During operation, the plant requires 10,000 lb/hr of steam at 100 psi for heating the detergent tanks and sludge driers.

System description. A system, like the one illustrated in Figure 20.14, is used to generate the steam required in the plant by disposing of the wastes generated there. The waste plastic, paper, and general plant refuse is collected into containers at several locations throughout the plant. These containers are brought to the incinerators area by plant personnel using standard fork-lift trucks. At the incinerator, the containers are manually placed onto the changing dumper and the incineration process is manually initiated by means of a start button. The process is automatic and proceeds approximately as follows (the number correspond to the numbers in Figure 20.14):

1. The charging dumper empties the contents of the waste containers into the ram feeder's hopper and the hopper's hatch closes.

2. The ram-feeding cycle is initiated by the limit switch activated by the closure of the hatch. The door to the ignition chamber opens and the ram pushes the charges into the ignition chamber. The ram then retracts back and the chamber door closes.

3. As the waste material slowly moves toward the opposite end of the revolving ignition chamber, new surfaces are uncovered for burning.

4. The auxiliary fuel burners are only required to preheat the chambers and ignite the wastes in this application; the waste stream's high energy content subsequently sustains combustion.

5. By the time it reaches the discharge end of the chamber, the waste material is completely burned into ash and is dropped into the ash conveyor for transfer.

6. The slat ash conveyor transfers the ash to a container, ready for transport to a landfill.

7. The volatilized gas (smoke) produced within the ignition chamber is drawn into the combustion chamber. The combustion chamber is maintained at high temperatures and sufficient excess air is provided to completely oxidize the smoke's combustible components.
8. The exhaust gases carry a significant portion of the energy generated by the system; much of this energy can be reclaimed. Prior to entering the pollution control system, the gases pass through a waste-heat boiler. The steem generated is used directly to provide heating of detergent tanks and sludg driers.
9. The gases exiting the boiler are sufficiently cool to be directed through a baghouse, where the fly ash entrained in the flue gases is captured.
10. Since the system must be capable of burning chlorinated plastics, the gases are also passed through a PV scrubber. The hydrochloric acid (HCl) produced by the incineration of chlorinated organic compounds is captured within the scrubber solution. The moisture-laden gases exiting the venturi scrubber encounter a demisting plate which deflects water droplets back into the scrubber reservoir.
11. The exhaust fan draws the cleaned process gases through the exhaust stack, into the atmosphere.

Calculation of heat recovery. The type and quantity of wastes generated at the plant, as determined by a waste survey and analysis, are as follows:

2,000 lb/hour (waste feed rate)

11,000 Btu/lb

20% moisture

5% ash (noncombustible content)

The equipment is capable of recovering 60% of the heat energy as steam. The feedwater enters the boiler at a temperature of 50°F.

For the calculations given below, the following data (from standard engineering tables) are required:

Boiling Point of Water = 212°F

Specific Heat of Water = 1 Btu/°F/lb

Specific Heat of Steam = 0.5 Btu/°Flb

Latent Heat of Vaporization for Water = 970 Btu/lb

Temperature of Steam at 100 psi = 350°F

1 gallon of water = 8.34 lb

1. % Moisture, by weight, of feed charge:

   0.20 × 2,000 lb/hr = 400 lb/hr

2. % Ash, by weight, of feed charge:

   0.05 × 2,000 lb/hr =100 ib/hr

3. Total Heating Value, Btu per pound of feed

   2,000 ib/hr × 11,000 Btu/ib = 22,000,000 Btu/hr = 2M Btu/hr

4. Total heat available for steam generation:

   22,000,000 Btu/hr × 0.6 = 13,200,000 Btu/hr

5. Heat required to bring the water to the boiling point:

   (212°F–50°F) × 1 Btu/°F/ib = 162 Btu/lb

6. Heat required to convert feed water into steam at 100 psig:

   (350°F–212°F) × 0.5 Btu/°F/lb = 69 Btu/lb

7. Total heat required to generate one lb of steam @ 100 psig:

   970 Btu/ib + 162 Btu/lb + 69 Btu/ib = 1,201 Btu/lb

8. Total steam-generating capacity of the system:

   (13,200,000 Btu/hr)/(1,201 Btu/b) = 10,991 lb/hr

9. Feedwater rate required (for the 10,000 lb/hr of steam required):

   (10,000 ib/hr)/(8.34 Ib/gal) = 1,199 gallons/hr

Conclusions. All of the plant's operational steam requirements can be met by the burning of wastes generated at the plant. If the steam requirements were met by a boiler burning fuel oil at approximately $ 0.01/ib, fuel costs alone would be approximately $ 100/hr or $ 700,000/year, assuming 7,000 hr/yr of operating time. Consequently, equipment costs would be recovered within one year. Of course, this simple analysis does not account for the avoided costs of waste disposal, the equipment costs of alternative steam-generating equipment, or the operating and maintenance costs involved in either this or an alternative system.

Conclusion

Recovery and recycling conserves finite material resources while simultaneously reducing the quantity of waste requiring disposal and of hazardous pollutants discharged into the environment. Incineration is an excellent technique for the recovery of materials and energy from a wide variety of commercial and industrial waste streams. The technology for cost effective, efficient, and environmentally sound resource-recovery incineration facilities is available. New research is expected to continue to decrease the problems and costs associated with pollution control, residue disposal, and the separation of multiple recoverable components from residue streams.

The future of resource recovery incineration systems will be determined by the long-term availability of markets for recovered resources. The general state of the world economy, lower energy prices, the development of alternate energy sources, and improved mining technologies have kept energy and raw material prices from rising as quickly as had previously been expected. However, natural resource supplies are finite and remaining reserves tend to be located in remote or environmentally protected areas. Pollution controls and energy prices affect mining operations as well as recovery facilities. Some elements arc abundant worldwide, but domestic sources are limited; recovery may be advantageous due to transportation costs and strategic considerations. Overall, the incentives for resource recovery are expected to improve and incineration will continue to be an attractive recovery technology applicable to a wide range of industrial and commercial waste streams.

21

Catalytic Incineration of Hazardous Wastes

Introduction

This chapter deals primarily with the oxidation of gaseous criteria pollutants; volatile organic compounds commonly termed VOCs, their liquid precursors, and the intermediate combustion product, carbon monoxide (CO). Dioxins, furans, and polychlorobiphenyls (PCBs) are referenced as applicable reactants for catalytic treatment, but require test runs to finalize designs. The applications discussed are for industrial stationary exhaust gases and not vehicular exhausts, which are mentioned only by reference.

Unless otherwise stated, all catalysts in this chapter are platinum group metals dispersed on a refractory high surface area carrier deposited on a ceramic honeycomb support. Such catalysts provide maximum cost effectiveness resulting from the high activity and durability of platinum and the minimization of the reactor size afforded by honeycombs, hence low pressure drop and maximum equipment cost.

In all design selections, it is assumed that the catalytic incinerator is the preferred process choice vs. thermal incineration whenever prior experience or pilot tests have shown that feed stream constituents do not significanetely affect expected catalyst life. Given this criteria, the driving force for catalytic incineration over thermal incineration is lower temperature operation and the two primary benefits, namely:

- Equipment material selection, cost, and life
- Fuel minimization and energy cost savings

Optimal heat recovery design is presumed for each case.

Catalytic incineration as used in this chapter is a post-treatment, exhaust gas process used to take VOC, CO and other organic emissions to trace ppm concentration levels. Thus, solid waste and bulk liquid hazardous wastes are presumed to have been thermally incinerated or gasified prior to final catalytic treatment.

Scope of Work

This discussion presents an overview of current state-of-the-art catalytic oxidation as used in generally accepted stationary source air pollution control equipment for the chemical and allied process industries. As evidenced by the table of contents, this is a comprehensive treatment of the key aspects of process selection and design.

However, it is outside of the authors' scope to present detailed procedures sufficient to design, build, install, and operate a system. This must be left to the specialized expert or an in-depth full textbook and course.

Sufficient detail is presented to familiarize typical process engineers with catalysis to enable them to raise correct questions of specialists and vendors. Use of the typical data sheets can provide a basis for:

- Preparation of requests for quotations
- Preparation of scoping estimates
- Comparison of alternative designs
- Stand-alone bid evaluations

The sourced references provide detailed back-up for information presented in this condensed form. and also can serve as the basis for self-study in catalyst selection, system design, and new application development. Those introduced to catalysts and catalytic systems for the first time should take heart. Skilled specialists quite often require the time proven methodology of bench tests followed by pilot or slip-stream tests before finalizing design and guaranteeing performance. The usual need to establish catalyst life occurs after determining the most likely deactivation mode. Quite often this can be accurately done from empirically categorized case histories; but in many cases, the slip-stream test at a commercial site provides the final basis for selection and design.

Background

Air pollution control technology hasn't quite made smokestacks obsolete. But equipment manufacturers and users alike have advanced considerably from the days when"smoke control" was the sole design objective. There are a number of alternative methods(or equipment) for pollution control, which generally fall into the following classification:

- Solid adsorbents
- Scrubbers
- Precipitators
- Capture Devices(filters, etc.)
- Thermal Incinerators
- Catalytic Incinerators

The applications for the above control equipment can be roughly categorized as noted in the brief summary below :

Pollutants	**Equipment or Methods**
Particulates	Precipitators, Filters

SO_2, SO_3, H_2S	Liquid Scrubbers Solid Adsorbents
NO_x, CO, Volatile Organics, Ozone, Odors	Catalytic Reactors Thermal Incinerators Non-Catalytic Reactors Liquid Scrubbers Solid Adsorbents

During the last decade, the science of catalysis, particularly utilizing precious metals, has made inroads in air pollution control for both stationary and mobile sources. Automotive catalysts now routinely used on all passenger cars in the U.S are well-documented. In addition, industrial, non-automotive applications use catalysts to significantly decrease the same pollutants, oxides of nitrogen (NO_x), carbon monoxide (CO), and volatile organic compounds (VOCs).

Some milestones in catalyst application developments are noted below.

1964 Ceramic honeycomb supports used in nitric acid plant NO_x abatement. PTx oxidation catalysts used for fork lift trucks, mining equipment*.

1975 Precious metal catalysts are used to significantly reduce automotive carbon monoxide and hydrocarbon emissions on U.S. passenger vehicles.

1976 Widespread use of NO_x and hydrocarbon abatement in Japan for utility and chemical plants, ovens, and furnaces.

1979 Catalysts used to remove NO_x from stationary rich burn engines.

1980 Ozone converters installed on wide-body commercial jet aircraft.

1983 Catalysts used for more complete combustion in home wood stoves.

1985 Oxidation catalysts used to remove CO from cogeneration, gas turbine exhaust.

Various important sources of pollutants that are controllable by using a catalytic reactor are:

Pollutant	**Source or Industry**
Carbon Monoxide	Internal Combustion Engines Gas Turbines Wood Stoves
Ozone	Off-Gases from Water Treatment Plants Cabin Air Commercial Jet Aircraft
Nitrogen Oxides	Engines (Internal Combustion, Gas Turbine) Process Heaters Utility and Industrial Boilers Chemical Plant Process Effluent

*PTX is a tradename of Engelhard Corporation.

Volatile Organic Compounds(VOCs)	Chemical Plants Ovens (Paint, Printing, Coating, Food Processing, Textile) Power Plants Furnaces, Wood Stoves Engines (Internal Combustion, Gas Turbines).

As noted earlier, this chapter will deal specifically with handling VOCs, as well as carbon monoxide, from the sources listed above and many others, by oxidation over catalysts as part of integrated catalytic equipment systems. When off-gases, containing these pollutants are heated to suitable temperatures, combustible components react with oxygen in the exhaust or from added air to form harmless carbon dioxide and water vapor.

For example, when ethylene is heated in the presence of air:

$$C_2H_4 + 3\ O_2 \rightarrow 2\ CO_2 + 2\ H_2O$$

This abatement technique can be effective to the parts per million range, or to below the odor threshold at temperatures considerably lower than thermal incineration.

Catalysts increase the rate of chemical reactions without actually entering the reaction. In a catalytic reactor the catalyst is not consumed, as the fuel is in a thermal incinerator but continues to function, in most cases for years.

Although catalytic reactions can be promoted with other metals, the precious metals especially platinum preferred because of their high degree of chemical activity and durability. Therefore, the real cost of catalysis can be quite low because the same ''charge'' of catalyst lasts many years; ten years is reasonable for many applications.

Catalysts are prepared using various physical support structures. In recent years, the use of honeycomb supports has been preferred for pollution control catalysts because of the substantial advantages it offers over the conventional particulate supports.

For an application in which catalysts are used, there are usually a number of components that constitute the entire system, again depending upon the needs of the user. Typically a system may comprise the following :

- Catalyst
- Reactor
- Preheat burner
- Instrumentation and controls
- Analytical equipment
- Heat exchanger(s)
- Auxiliary equipment (piping, pumps, valves, etc.)

The total system package can be conveniently provided as a compact, skid-mounted unit which is readily implemented in commercial applications. Furthermore, these systems can easily be retrofitted to convert existing thermal incinerators.

Each application of catalytic equipment for air pollution control is designed to meet the needs of the user for his particular process application. In discussing the actual pollution control system, the authors will discuss the methodology for catalyst selection., while the catalytic equipment will be illustrated by describing actual commercial case histories for control of VOCs from plant effluents.

CATALYST PRINCIPLES AND DEFINITIONS

Fundamentals

Catalysis

Analogous to a reactor in a chemical plant, the heart of a catalytic process is the catalyst. By definition. a catalyst enhances the rate of conversion of reactants to desired products without itself being consumed; however, there is considerable interaction between reactants and catalyst, as well as products and catalyst. The net effect, however, is that the catalyst remains unchanged while providing enhanced kinetics to desired end products. Analogously, the catalyst provides an easier alternative path for conversion of reactants to products.

Let's assume the following reaction in which:

$$A + B \begin{cases} \xrightarrow{k_1} C + D \\ \xrightarrow{k_2} E + F \end{cases}$$

A+B reacts to form either C + D or E + F (or a mixture of the four products). The product distribution obtained depends on the energetics for each reaction. One may describe the energetics simplistically, as in Figure 21.1A, in which energy barriers must be overcome for A and B to convert to C + D or E + F. These barriers are called activation energies and play a major role in determining the reaction rate and product distribution. The delta H_2 and delta H_2 represent the energy released in conversion of reactants to products. In general, the path with the lowest activation energy, i.e., E_{A1} (solid line) in Figure 21.1A, will be favored; however, some E + F will be formed.

The presence of the proper catalyst can effectively further decrease the activation energy E_{A1} without altering E_{A2}. In this respect, the catalyst has enhanced the rate of conversion of reactants to C + D without affecting the reaction profile of E + F. Consequently, the product distribution or selectivity will favour C + D. Other catalyst choices can reverse this effect.

It should be noted that the energtetics delta H_1 and delta H_2 which are thermodynamic terms, have not been affected since a catalyst only accelerates reaction rates, Thermodynamic limitations are maintained.

Standard texts should be consulted for details on various models such as Langmuir-Hinshelwood,

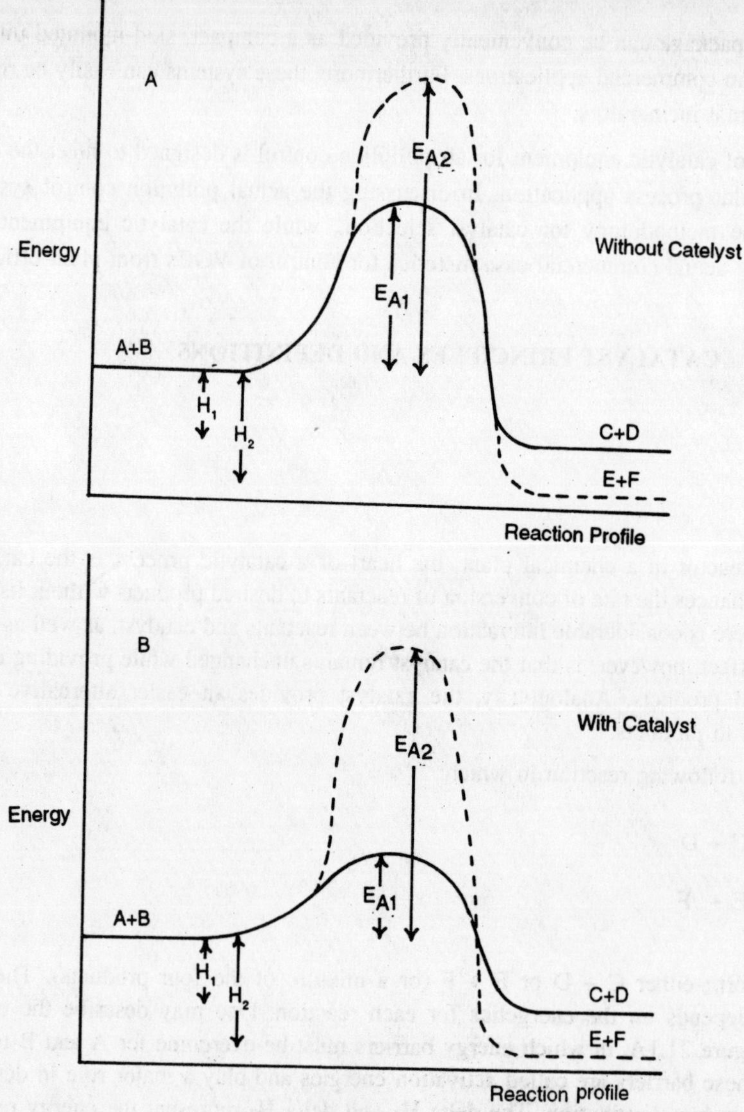

Figure 21.1 Catalytic reaction energetics.

but for convenience a power model is used herein where the rate of consumption of reactants can be described by:

$$\text{Reaction rate} = \frac{-dC_A}{dt} = \frac{-dC_B}{dt} = kC_A{}^{X}C_B^{Y}$$

while the rate of formation of products can be described by:

$$\frac{dC_C}{dt} = \frac{dC_F}{dt} = k_{(1)}C_A{}^{M}C_B^{N}$$

and
$$\frac{dC_E}{dt} = \frac{dC_F}{dt} = k_{(2)}C_A^{G}C_B^{H}$$

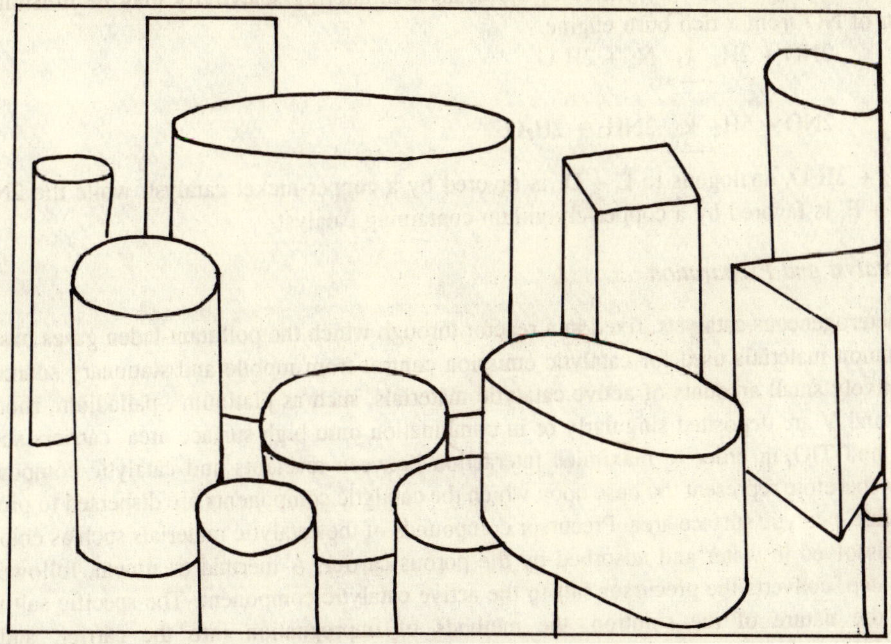

Figure 21.2. Typical ceramic honeycomb geometries.

where dC_z/dt = rate of change of component C_z

C_z^P = concentration of z species at any time raised to a power P

k = individual reaction rate constants

z = components A, B, C, D, E, F

P = reaction order, x, y, M, N, G, H.

The exponents P may vary fractionally since each product distribution pair may require different fundamental mechanistic steps. The rate constants k are described according to the individual activation energies:

$$k = A_0 \exp(-E_A/RT)$$

where A_0 = a constant

R = gas constant

T = °K.

The rate constant k, therefore, increases as the activation energy decreases, compared to a noncatalyzed reaction. In this regard, the catalyst decreases the activation energy by providing a chemical "short cut" for the conversion of A+B to C+D, as illustrated in Figure 21.1B.

This decrease in activation energy allows the catalytic process to operate at a lower temperature, i.e., E_A/T = constant, for the same conversion and rate constant, notwithstanding any compensation effect with A_0.

A realistic example of the influence of the catalyst in altering selevtivity may be illustrated by the reduction of NO from a rich burn engine.

$$2NO + 2H_2 \xrightarrow{k_1} N_2 + 2H_2O$$

$$2NO + 5H_2 \xrightarrow{k_2} 2NH_3 + 2H_2O$$

The $N_2 + 2H_2O$, analogous to C + D, is favored by a copper-nickel catalyst, while the $2NH_3 + 2H_2O$ i.e., E + F, is favored by a copper-chromium containing catalyst.

Nature of Catalyst and Preparation

Solid heterogeneous catalysts, fixed in a reactor through which the pollutant-laden gases pass, are the most common materials used for catalytic emission control from mobile and stationary sources. In practice, relatively small amounts of active catalytic materials, such as platinum , palladium, rhodium, Cu, Co, Mn, and V are deposited singularly or in combination onto high surface area carriers such as Al_2O_3, SiO_2, and TiO_2 in order to maximize interaction between reactants and catalytic components. These carriers therefore represent the base upon which the catalytic components are dispersed to produce highly accessible catalytic surface area. Precursor compounds of the catalytic materials such as chlorides and nitrates dissolved in water and adsorbed by the porous carrier. A thermal treatment, followed by an activation step, converts the precursor salf to the active catalytic component. The specific salt used, the amount, the nature of the solution, the methods of impregnation into the carrier, and the decomposition and activation are all very critical to producing an effective catalyst. Furthermore, the chemical and physical nature of the carrier is also important since it provides the foundation for the active components. All of these parameters are usually maintained in strict confidence by the catalyst manufacturer and thus details are seldom available to the public.

There are two important classes of carriers used in pollution- abatement applications: particulates, porous particles varying in size and shape, and porous powders bonded to the surface of a monolithic to the public honeycomb structure.

Particulates are the most traditional carriers used in petroleum, petrochemical, and chemical industry, and thus are sold in greatest volume. However, honeycombs are becoming more and more commonplace in environmental applications (Figure 21.2). They are particularly valuable for high throughput applications since their more open structure offers little pressure drop for their high surface to volume ratio.

The ceramic or metallic honeycomb is dipped into a slurry of the metal-impregnated carrier and the latter adheres to the walls. The channels are blown free of excess slurry, dried, and calcined. If the carrier is not in-situ catalyzed, the "washcoated" honeycomb must be dipped as a second step into a solution of the catalyst precursor salt, and then dried and calcined. The precise details of all steps remain the proprietary information of the catalyst manufacturer, each of whom has techniques developed for special applications.

Fundamental Process Steps is Catalysis

In a catalytic process, reactants must interact with the active site on the catalyst. The term *active site* refers to that site or sites on the catalyst surface in which chemisorption and reaction occur. At least one molecule involved in the catalytic process must (1) diffuse to the surface of the catalyst particle,

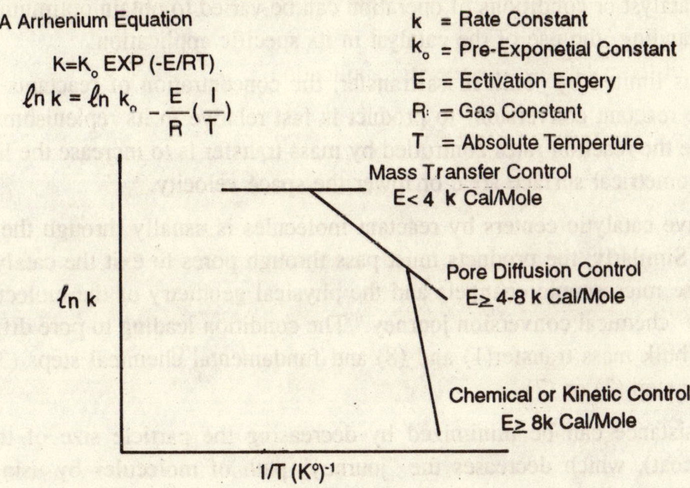

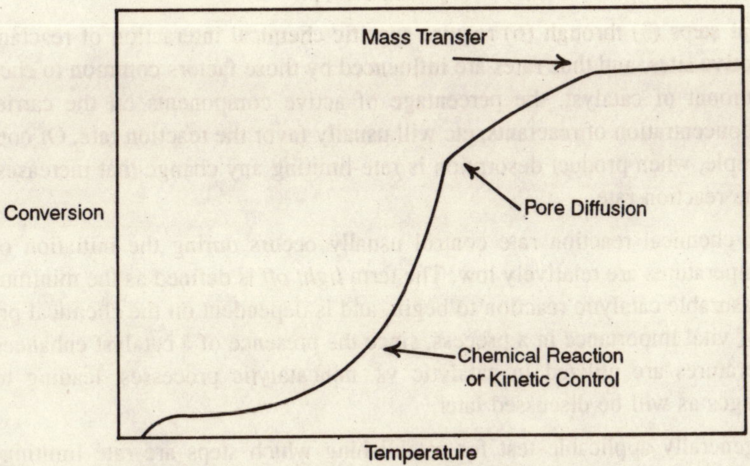

Figure 21.3. Rate limiting steps in a catalyric reaction.

(2) diffuse through the pore network to the active sites, (3) chemisorb, (4) form an activated complex, (5) convert to product while retaining its chemisorbed state, (6) desorb from the active site,(7) diffuse through the pore structure, and finally (8) diffuse into the bulk fluid.

Steps (2) and (7) are controlled by the molecular size of reactants and products and pore structure of the catalyst, while steps (3), (4), (5), and (6) depend on the chemical compatibility of reactant and product molecules (or atoms) with the active sites. The mass transfer steps (1) and (8) depend on reactant/product concentrations, as well as catalyst amount and particle size (i.e., superficial geometric area of catalyst).

One of these steps will control the overall reaction rate. It is essential to understand which is controlling so that the catalyst or conditions of operation can be varied to obtain optimum performance. This is done by understanding the use of the catalyst in its specific application.

When a reaction is limited by bulk mass transfer, the concentration of reactants is zero at the catalyst surface because reactant conversion to product is fast relative to its replenishment. The most practical way to increase the reaction rates controlled by mass transfer is to increase the linear velocity, increase the exposed geometrical surface area, or lower the space velocity.

Access to the active catalytic centers by reactant molecules is usually through the internal pore structure of the carrier. Similarly, the products must pass through pores to exit the catalyst. Naturally, the physical size of these microscopic channels and the physical geometry of the molecules influence the time required for the "chemical conversion journey." The condition leading to pore diffusion control exists when the rate of bulk mass transfer(1) and (8) and fundamental chemical steps (3) through (6) are fast relative to either step (2) or (7).

Pore diffusion resistance can be minimized by decreasing the particle size of the carrier (or thickness of the washcoat), which decreases the "journey" path of molecules by using wider pore carriers or by depositing the active catalytic components nearer the edge of the carrier. By decreasing the carrier particles size of a particulate catalyst, unfortunately the pressure drop will be increased. In the case of honeycomb catalyst, the same effect can be obtained by decreasing the thickness of the washcoat, but this case, there is no effect on pressure drop.

Fundamental steps (3) through (6) require specific chemical interaction of reactants and product molecules with active sites, and thus rates are influenced by those factors common to chemical reaction. Increasing the amount of catalyst, the percentage of active components on the carrier, the process temperature, the concentration of reactants, etc will usually favor the reaction rate, Of course exceptions do exist; for example, when product desorption is rate-limiting any change that increases concentration could decrease the reaction rate.

In practice, chemical reaction rate control usually occurs during the initiation of the catalytic reaction when temperatures are relatively low. The term *light off* is defined as the minimum temperature necessary for measurable catalytic reaction to begin, and is dependent on the chemical properties of the catalyst. This is of vital importance in a process, since the presence of a catalyst enhances kinetics such that lower temperatures are utilized in catalytic vs. noncatalytic processes, leading to considerable economic advantages as will be discussed later.

The most generally applicable test for establishing which steps are rate limiting is use of the Arrhenius plot from which activation energies are determined. An idealized plot is shown in Figure 21.3A and a conversion vs. temperature profile is shown in Figure 21.3B. The calculation of the activation energy, E, from the slope of the Arrhenius plot allows an assessment of the rate-limiting step. From Figure 21.3B it is evident that reactions controlled by chemical kinetics are most sensitive to temperature, while mass transfer controlled reactions are the least.

Summary of Fundamental Process Steps

We have briefly described physical and chemical properties of catalysts that can control the rate of a reaction. It should be evident that considerable knowledge of the application is required to optimize the properties of the catalyst. A feedback network of testing, simulating actual operating conditions and catalyst characterization, is essential for efficient use and manufacture of the catalyst.

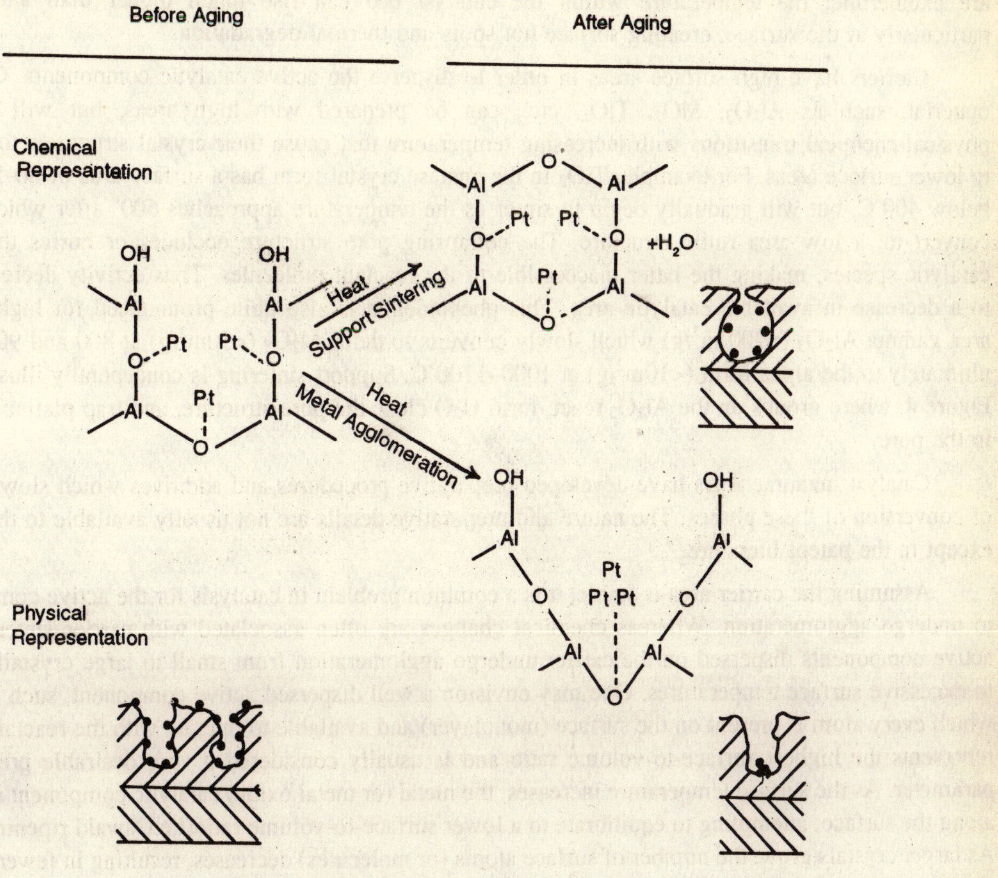

Figure 21.4. Catalyst sintering and agglomeration.

Catalyst Life

We previously defined a *catalyst* as remaining unchanged during the reaction which it is accelerating. In actual application, the catalyst is subjected to a variety of parameters which cause a slow decline in its effectiveness. This decrease in activity is called aging, the extent or rate of which depends on a number of process related conditions. The major conditions leading to catalyst deactivation in pollution-abatement applications, such as overtemperature, deposition of contaminants, etc., will be discussed next.

Thermally Stimulated Deactivation

Essentially all pollution-abatement applications occur at temperatures between 200 and 600°C. Catalyst manufacturers have developed materials which resist thermal degradation at these temperatures. However, conditions can and often do arise that create higher temperatures in the exhaust gas inlet to the catalyst. In addition, upset process conditions can occur, resulting in higher concentration of exhaust gas combustible material, which is then oxidized in the catalyst bed. Since oxidation reactions are exothermic, the temperature within the catalyst bed can rise much higher than anticipated, particularly at the surface, creating surface hot spots and thermal degradation.

Carriers have high surface areas in order to disperse the active catalytic components. Common material, such as Al_2O_3, SiO_2, TiO_2, etc., can be prepared with high areas, but will undergo physical-chemical transitions with increasing temperature that cause their crystal structures to convert to lower surface areas. For example, TiO_2 in the anatase crystal form has a surface area of 80-100 m^2/g below 400°C, but will gradually begin to sinter as the temperature approaches 600° after which it will convert to, a low area rutile structure. The collapsing pore structure occludes or buries the active catalytic species, making the latter inaccessible to the reactant molecules. Thus activity decreases due to a decrease in available catalytic area. This phenomenon is also quite pronounced for high surface area gamma Al_2O_3 (>200 m^2/g) which slowly converts to delta Al_2O_3 (<50m^2/g) at 800 and 900°C and ultimately to the alpha form (<10m^2/g) at 1000–1100°C. Support sintering is conceptually illustrated in Figure 4, where groups on the Al_2O_3 react, form H_2O close the pore structure, and trap platinum atoms in the pore.

Catalyst manufacturers have developed preparative procedures and additives which slow the rate of conversion of these phases. The nature and preparative details are not usually available to the public except in the patent literature.

Assuming the carrier area is stable, it is a common problem in catalysis for the active components to undergo agglomeration. Whereas chemical changes are often associated with carrier sintering, the active components dispersed on the carrier undergo agglomeration from small to large crystallites due to excessive surface temperatures. One may envision a well dispersed active component, such as Pt, in which every atom is present on the surface (monolayer) and available to interact with the reactants. This represents the highest surface-to-volume ratio and is usually considered a very desirable preparative parameter. As the surface temperature increases, the metal (or metal oxide) catalytic component migrates along the surface, attempting to equilibrate to a lower surface-to-volume ratio(i.e.Oswald ripening [14]). As larger crystals grow, the number of surface atoms (or molecules) decreases, resulting in fewer surface sites for reactant interaction. Thus, agglomeration results in a degradation of activity. This phenomenon is illustrated in Figure 21.4.

Again, catalyst manufacturers have recognized the causes and effects of this phenomenon and have formulated their materials to resist these changes to some extent. These formulations and preparative procedures can make a large difference in ultimate catalyst life between changeouts.

Thermally stimulated reaction between the active catalytic components and the carrier can also occur, which lead to deactivation. This phenomenon is most common for nonprecious metal catalytic oxides, such as CuO, CO_2O_3, and NiO, supported on gamma Al_2O_3, which react :

$$CuO + Al_2O_3 \rightarrow CuAl_2O_4$$

to form low-activity aluminates at temperatures in excess of 600°C Precious metals are more inert and thus are more resistant to this mode of deactivation. However, Rh_2O_3 interacts with Al_2O_3 but at much higher temperatures, i.e., >800°C.

Finally, the active catalytic component can decompose to a less active form at higher temperatures. This is quite common for nonprecious metal oxides such as CuO and CO_2O_3:

$$2\,CuO \xrightarrow{700^{\circ}C} Cu_2O + \tfrac{1}{2}\,O_2$$

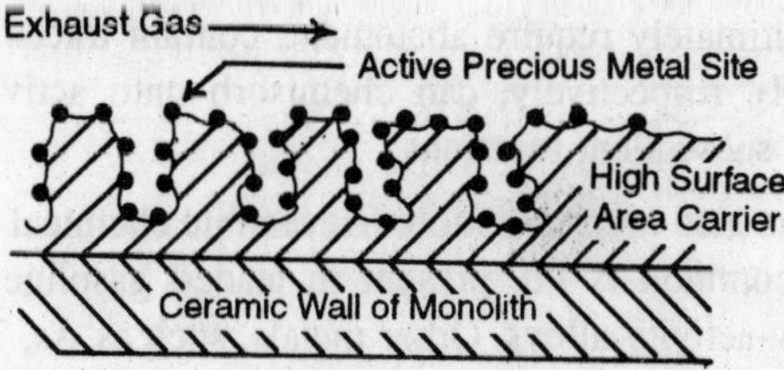

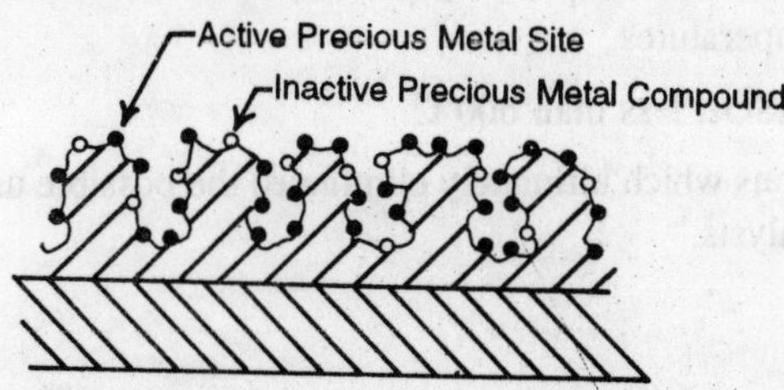

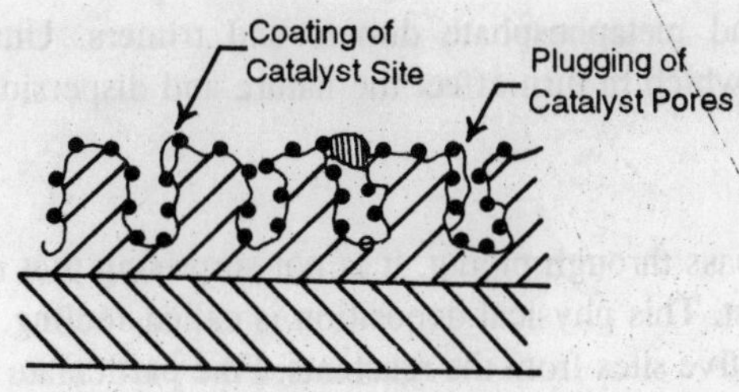

Figure 21.5. Physical representation of aged/fresh catalyst.

$$3\ CO_2O_3 \xrightarrow{800^\circ C} 2CO_3O_4 + \tfrac{1}{2}\ O_2$$

but also occurs for PdO in air:

$$PdO \xrightarrow{800^\circ C} Pd + \tfrac{1}{2}\ O_2$$

Selective Poisoning

Components in the gas stream can strongly adsorb (or react) on the active catalytic species, changing its surface nature and rendering it less active. A selevtive poison discriminantly chemisorbs onto active sites (see Figure 21.5). Chemisorption refers to a strong adsorption, but without necessarily forming a compound. For example, SO_2 or SO_3 will preferentially chemisorb onto Pt at temperatures below about 300°C and inhibit other pollutants from interacting with the Pt. This is a very common phenomenon in pollution, abatement applications, since sulfur oxides are common products of the combustion process with many typical fuels containing sulfur. Furthermore, many solvents used in the chemical industry, which ultimately require abatement, contain traces of sulfur. Other acid and basic gases,such as HCl and NH_3, respectively, can chemisorb onto active sites leading to deactivation, whether directly or through subsequent reactions.

Selective poisoning can also involve a more irreversible chemical reaction between the poison and the active site. The most common is Pb, present in leaded gasoline which reacts specifically with precious metals forming low-activity alloys. Other metals, such as As, Te, Cd, Hg, etc., are also known as selective poisons for most catalysts.

The nonprecious metals are particularly vulnerable to selective poisoning by irreversible chemical reaction. The active metal oxides, CuO, CO_2O_3, NiO, etc., react with SO_3 and form catalytically inactive sulfates at operating temperatures.

$CuO + SO_3 \rightarrow CuSO_4$, less than 600°C

It was these reactions which ultimately eliminated the possible use of these oxides for auto exhaust pollution-abatement catalysts.

Nonselective Poisoning

Other components in the pollution-laden gas stream can deposit on the catalyst and react, not with the active components, but with the carrier or each other. Phosphorous compounds present in lubricating oils, used in both stationary and mobile engines, can deposit on an Al_2O_3 carrier and form both aluminum phosphates and metaphosphate dimers and trimers. Ultimately, they affect the surface properties of the carrier, which in turn affect the nature and dispersion of the catalytic components.

Fouling or Masking

Since gas streams pass through piping, it is not surprising that rust or other particulate materials can deposit on the catalyst. This physical deposition is called fouling. It causes deactivation simply by physically masking the active sites from the reactants. Fine particulate matter can deposit on the outside surface of the catalyst and gradually work its way into the pore structure.

At first this leads to increased pore diffusion resistance, but ultimately can block total access to the active sites. In addition to rush, other particles such as SiO_2 present in silicone oils can foul the catalyst.

In the petroleum, petrochemical, and chemical industries, coke deposition is common mode of catalyst fouling. This is relatively uncommon in environmental applications, since operation is usually lean of stoichiometry where coke formation is thermodynamically unfeasible at even modest oxygen concentrations. It is possible, however that due to poor air-to-fuel mixing, some carbon laydown can occur, specifically for those applications which operate near the stoichiometric point. Usually this small amount of carbon does not represent a major masking problem, since, when the proper mixing is resumed, the carbon will burn off in situ. However, a localized hot spot may develop leading to thermal deactivation of the catalyst.

Attrition

Because a catalyst enhances the rate of a reaction, processes can operate with high volumetric flow rates. High linear velocities can lead to fluidization of particulate catalyst beds with subsequent attrition or erosion of the coated catalyst. Thus, the proper choice of materials, preparation procedures, and operating conditions must be found to minimize loss of catalytic material. Honeycomb supports, however, can be effectively used to minimize all practical attrition.

Catalyst Selection Factors

The choice of catalytic components, carriers, particulates vs. honeycombs is obviously dependent on the nature of the pollutants, potential poisons present in the gas stream, the degree of abatement required, the operating conditions, life, and ultimately economics. The decision, therefore, requires a great deal of communication between catalyst manufacturers and users. The catalytic components are almost always Group VIII metals (or metal oxides), however, promoters to enhance activity are often added.

Nonprecious metal oxides, such as Cu, CO, Cr, V, and Mn, are less expensive than the precious metals, Pt, Pd, Rh, etc., but are more susceptible to thermal degradation and selective poisoning by sulfur compounds and are not easy to regenerate. Consequently, the choice depends strongly on the operating efficiency of the facility and quality of the gas stream to be treated. Although more expensive initially precious metals can be recycled after they lose their effectiveness, allowing the user to recover some of his initial cost. By and large, most pollution-abatement catalytic converters utilize precious metals dispersed on high surface area stabilized alumina.

Although Al_2O_3 is used as the carrier, its origin, stabilizer concentrations, and pore structure are often modified dependent on the application. Materials can be selected that will be more resistant to poisons, thermal degradation, etc., but a detailed knowledge of the end use is essential to optimize catalytic materials.

Very briefly, the dominantly used catalysts selected for stationary catalytic incineration are those containing platinum, an alumina, washcoat, and a ceramic honeycomb support structure.

CATALYTIC SYSTEM DESIGN

The major considerations in the design of a catalytic system for abatement of volatile organic compounds (VOC) are:

- Catalyst selection
- Pollutant percent removal requirement
- VOC compound identification
- Pressure drop requirement
- Operating condition constraints
- Catalyst life expected

With knowledge of catalyst technology and chemical reaction engineering, all requirements can be achieved through proper system design.

Commercial Catalyst Systems

A catalyst can be manufactured and used in many forms, such as particulates, woven mesh,

Table 21.1

Materials Available for Honeycombs

Substance	*Formula*	*Composition (parts by weight)*	*Chemical properties - durability*						
			Melting point °C	*Limit of use In oxidizing atmosphere °C*	*Reducing atmosphere*	*Carbon*	*Fluxes acid*	*Fluxes basic*	*Molten metals*
Aluminum oxide	Al_2O_3	100 Al_2O_3	2015	1950	good	moderate	good	good	good
Beryllium oxide	BeO	100 BeO	2550	2400	very good	very good	good	moderate	good
Calcium oxide	CaO	100 CaO	2600	2400	poor	poor	poor	moderate	moderate
Magnesium oxide	MgO	100 MgO	2800	2400	poor	good	poor	good	moderate
Silicon Dioxide (silica glass)	SiO_3	100 SiO_3	—	1200	moderate	good	good	poor	—
Thorsium oxide	ThO_3	100 ThO_3	3300	2700	good	moderate	poor	good	very good
Titanium oxide	TiO_2	100 TiO_2	1840	—	—	—	—		—
Uranium oxide	UO_2	100 UO_2	2878	—	good	good	—	—	poor
Ziconium oxide	Z_2O_2	$93ZrO_2$ 5CaO $2HFO_2$	2677	2500	good	moderate	good	poor	good
Mullite	$3Al_2O_3*2SiO_3$	$72Al_2O_3$ $28SiO_3$	1830 incongruent	1850	moderate	moderate	good	moderate	moderate
Spinel	$MgO*Al_2O_3$	71.8 Al_2O_3 28.2 MgO	2135	1900	—	moderate	moderate	good	—
Forsterite	$2MgO*SiO_3$	57.3 MgO 42.7 SiO_3	1885	1750	good	moderate	poor	good	good
Zircon	ZrO_2*SiO_3	67.1 ZrO_2 32.9 SiO_3	2420 incongruent	1870	moderate	moderate	good	poor	good
Cordierite	2MgO 2 Al_2O_3 5 SiO_2	13 MgO 35 Al_2O_3 50 SiO_2	1485	1200-1300	moderate	very good	poor	moderate	moderate to good

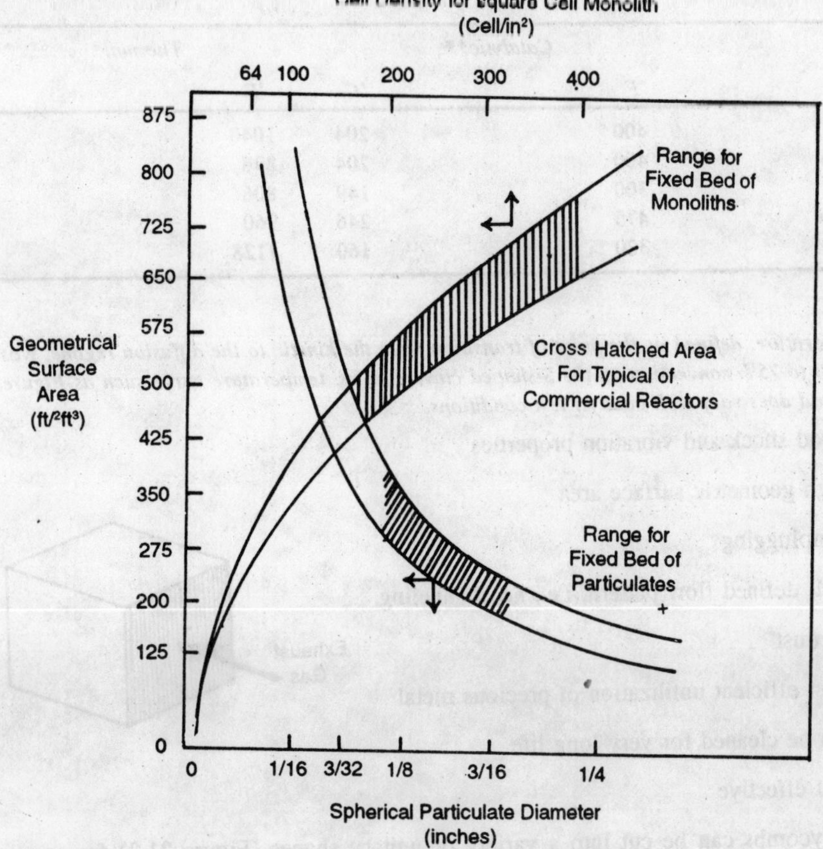

Figure 21.6. Comparison of fixed bed reactors filled with monolith vs. spherical particulates.

screen or honeycomb. The catalytic element is usually placed on high surface area carrier, which in turn is deposited on a mechanical support structure. Catalysts are prepared using various support structures. In recent years, the use of honeycomb supports has been preferred over the conventional particulate support. This preferred support structure contains a high cell density of parallel channels through which exhaust gas passes as shown below:

Monolithic Catalyst Structure

- Major advantages of honeycombs are:
- Low pressure drop
- Fast warm-up
- Lower volume/smaller package/lower weight
- No attrition losses/dimensionally stable reactor beds
- Reactor orientation in any direction

Ignition Temperatures**

	Catalytic**		Thermal*	
	°F	°C	°F	°C
Benzene	400	204	1040	560
Toluene	400	204	896	480
Formaldehyde	300	149	806	430
Methyl Ethyl Ketonw	475	246	960	515
Carbon Monoxide	320	160	1128	609

** NFPA No. 325M*

*** The ignition temperature, defined as the point of transition from the kinetic to the diffusion regime, was taken between 50 to 75% conversion on the S-shaped cinversion vs. temperature curve such as Figures 21.7 and 21.8, and does vary with catalyst test conditions.*

- Good shock and vibration properties
- High geometric surface area
- Nonplugging
- Well defined flow pattern, i.e., no channeling
- No dust
- Most efficient untilization of precious metal
- Can be cleaned for very long life
- Cost effective

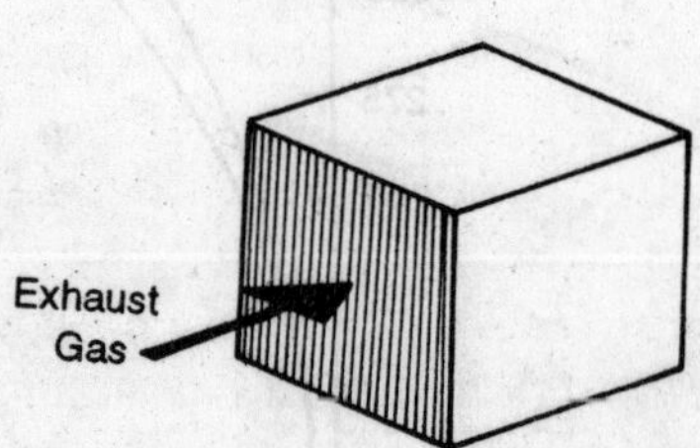

Ceramic honeycombs can be cut into a variety of unitary shapes (Figure 21.2) for convenient

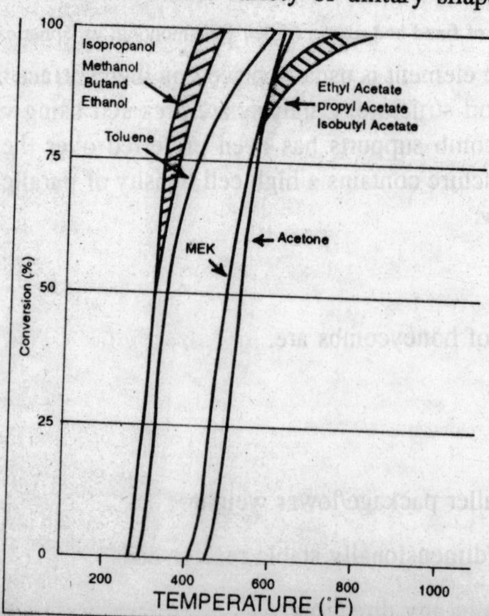

Figure 21.7. Typical compound conversion vs. temperature.

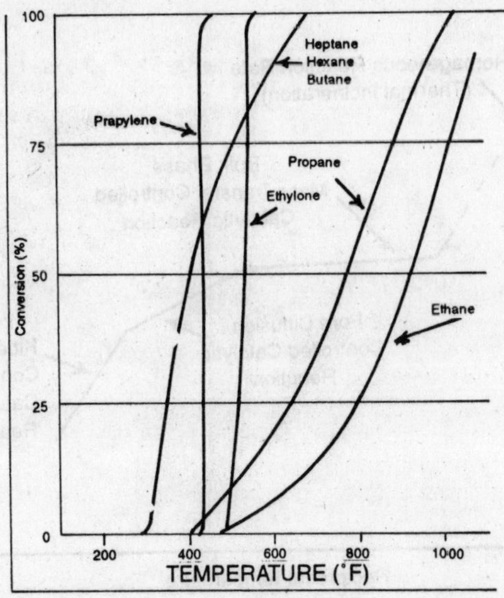

Figure 21.8. Typical compound conversion vs. temperature.

Space Velocity *Hour^{-1}*	*Conversion* *%*
30,000	99.6
50,000	97.8
70,000	96.5

packaging, and many materials are available from honeycomb manufacturers. Table 21.1 lists a number of ceramic candidates (with selected properties) that may be considered. The ceramic selection depends on the operating environment (e.g.,temperature), as well as chemical compatibility with the carrier and catalytic element.

Temp. °C in/out	337/371	285/316	230/260
Delta P, Pa	498	498	373
Gas Components		*Specific Conversion* *% @ 3 months*	
Ethanol	99.1	98	97.6
n-Propanol	99.9	97.3	97.2
n-Propyl acetate	99.6	93	51.0
Heptane (2%of stream)	–	63.8	62.0
Total Organics (G.C. with F.I.D.)	98.0	95.0	85.8

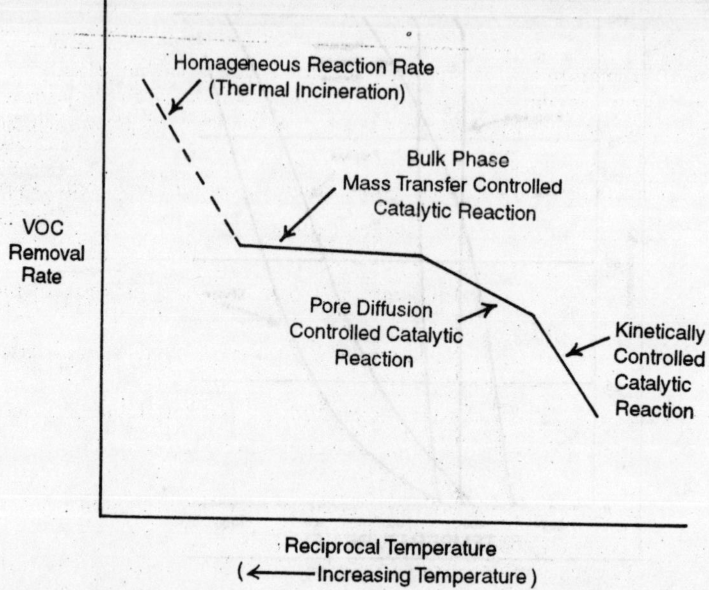

Figure 21.9. VOC removal rate as a function of temperature.

The shape in the individual honeycomb channel can be a circle, square, triangle, etc., and the channel density (or channels per square inch) can be varied. Commercial honeycombs are available from

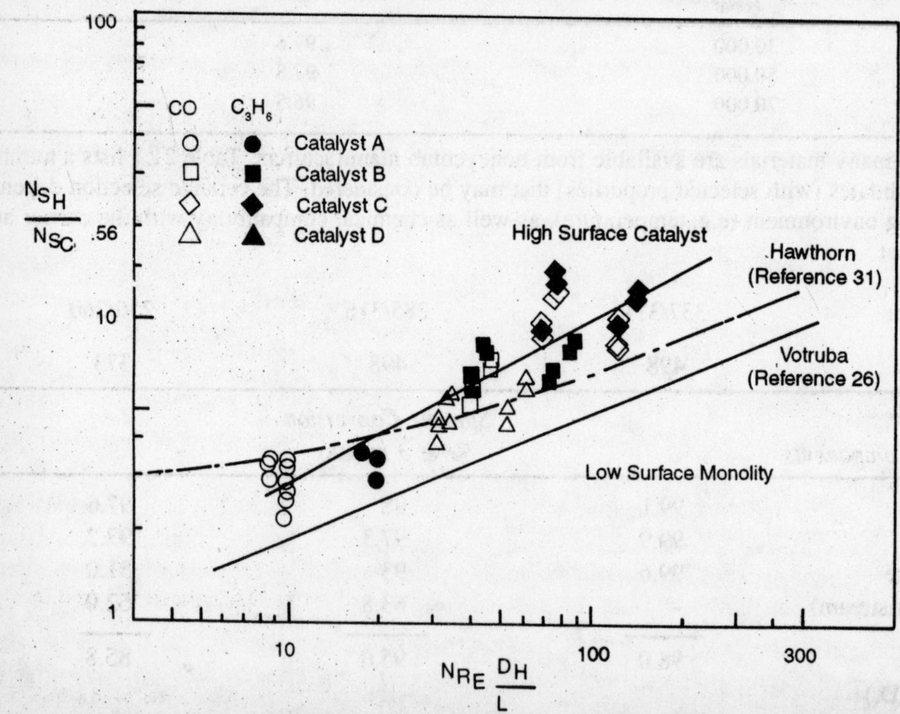

Figure 21.10. Mass transfer conversion vs. reduced Reynold's number.

10 to 600 cells per square inch (çpsi).

Figure 21.6 shows the substantial advantage that honeycombs have over pellets in the geometric surface area available for the catalyst coating. Honeycombs have a high geometric surface area with a low pressure drop. For instance, when comparing a 200 cell/in^2. honeycomb catalyst and particulate catalyst at equal geometric surface area (576 sq ft per cu ft) and high space velocity proven for the honeycomb, the particulate catalyst would be 1/16-in. diameter and have 50 to 100 times higher pressure drop per foot of bed depth. This means that a special reactor need to be designed for this particulate catalyst.

In summary, the catalyzed honeycomb consists of three parts: the precious metal, the carrier, and the ceramic support. Schematically, this is shown in Figure 21.5A.

Operating Temperature Requirement

The catalyst lowers the energy requirements for oxidizing VOCs to harmless carbon dioxide and water. The basic incineration process can be viewed simply whether catalytic or thermal:

Thermal

$$C_xH_y + \left(x + \frac{y}{4}\right) O_2 \xrightarrow{\text{1200-1800°F}} x\,CO_2 + \frac{y}{2} H_2O$$

(hydrocarbon) (oxygen) (carbon dioxide) (water)

Catalytic

$$C_xH_y + \left(x + \frac{y}{4}\right) O_2 \xrightarrow[\text{catalyst}]{\text{400–800°F}} x\,CO_2 + \frac{y}{2} H_2O$$

The thermal energy requirement of the catalytic system is less. Since each hydrocarbon be haves differently, there are various temperature requirements for abatement using both catalytic and thermal processes as shown here:

To properly specify the operating temperature for catalytic abatement, the response of a range of model organic compounds must be determind. Figures 7 and 8 illustrate responses that may be used in specifying the operating temperature requirements using proprietary precious metal catalysts.

Catalyst Volume and Converter Sizing

Some examples of effective space velocities (hence catalyst volume) from slip stream pilot plant studies at a plant printing on plastics are taken from and listed here:

These are for lined out operation between three and six months onstream at an inlet temperature of 315°C. The authors estimated by extrapolation that for the 50,000 hr^{-1} space velocity, design conversion would still be 94-95% after three years on stream with the monolith catalysts used.

Some typical data for the 50,000 hr^{-1} (8.5 cu m/hr flow rate) case are:

Linear velocities can be calculated from the given flow rate and catalyst sample size (1.5-in. dia. by 6.0 in. long).

The response of VOC conversion versus temperature depicted in Figures 21.7 and 21.8 can be used to determine the reaction rate parameters required for catalyst sizing. These data can be used to determine the reaction rate as shown in Figure 21.9.

In good engineering practice, the reactor is designed to operate in bulk phase mass transfer control of the catalytic reaction. Here the reaction rate (and therefore percent abatement) is insensitive to changes in operating temperature until significant aging has occurred.

Table 21.2
Nominal Honeycomb Properties

Honeycomb	d_{ch}	ε	*a*	ρ_B	*t*
9 cpsi	.274	63	110	—	.072
11 cpsi or	.236	64	130	—	.059
	.250	72	138	—	.045
25 cpsi	.157	66	202	—	.036
64 cpsi	.099	70	340	—	.019
100 cpsi	.083	69	398	—	.017
200 cpsi	.059	72	576	31	.012
300 cpsi	.046	65	660	40	.012
400 cpsi	.44	71	852	28	.06

d_{ch} - Hydraulic channel diameter (inch) ρ_B - Bulk density (lbs/ft^3)

ε - Open frontal area (%)   t - Wall thickness (inches)

a - Geometric surface area (ft^2/ft^3)

Another benefit of operating in the mass transfer control region is the ability to size converters using fundamental parameters of mass transfer such as the Sherwood number (N_{Sh}), Schmidt number (N_{Sc}), and Reynolds number (N_{Re}). Correlations developed for mass transfer in honeycomb catalyst under reaction conditions are plotted in Figure 21.10. For use in these mass transfer correlations, the physical parameter of various square-shaped cell honeycombs are given in Table 21.2.

The temperature rise across the catalyst configurations is assumed small for these conditions, and thus a simplistic mass transfer model can apply:

$$\text{Conversion} = 1 - X_{f,out}/X_{f,in} \qquad (1)$$

$$= 1 - \exp \frac{N_{Sh}\, D_{f\text{-}air}\, aL}{V\, D_H} \qquad (2)$$

where Xf = fuel concentration

subscripts *in* and *out* = catalyst inlet and exit

a = geometric total surface-to -volume ratio

D_H = hydraulic diameter

$D_{f\text{-}air}$ = diffusivity of species in air

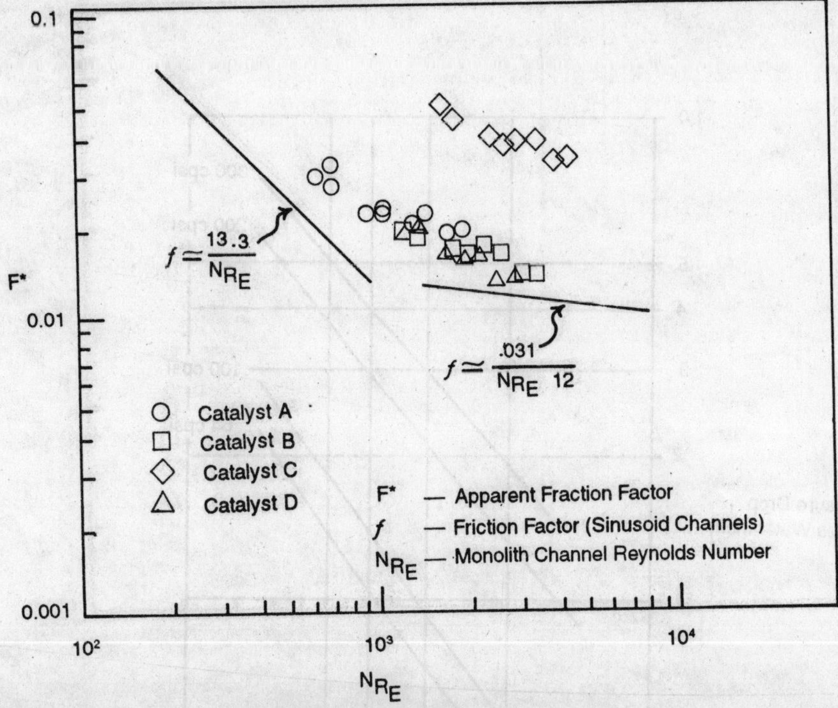

Figure 21.11. Friction factor correlation.

F^* = apparent friction factor

L = catalyst channel length

N_{Re_c} = Reynolds number in catalyst channels

N_{Sh} = Sherwood number (dimensionless mass transfer parameter)

N_{Sc} = Schmidt number (dimensionless mass transfer parameter)

P = pressure

ΔP = pressure drop

T = temperature

V = catalyst face velocity

X = mole fraction of fuel

Physical properties are evaluated at the average temperature in the catalyst. Diffusivities of CO and C_3H_6 were estimated using standard formulas. Calculated values of the dimensionless mass transfer and flow parameters can be used to develop a plot (Figure 21.10) of:

$$\log\left[\frac{N_{Sh}}{N_{Sc}\cdot 56}\right] \text{ vs. } \log\left[N_{Re}\,\frac{D_H}{L}\right] \tag{3}$$

This correlating technique was developed by Votruba for mass transfer in honeycomb catalysts. Data that are in the kinetically controlled regime will not be correlated by this technique because of the

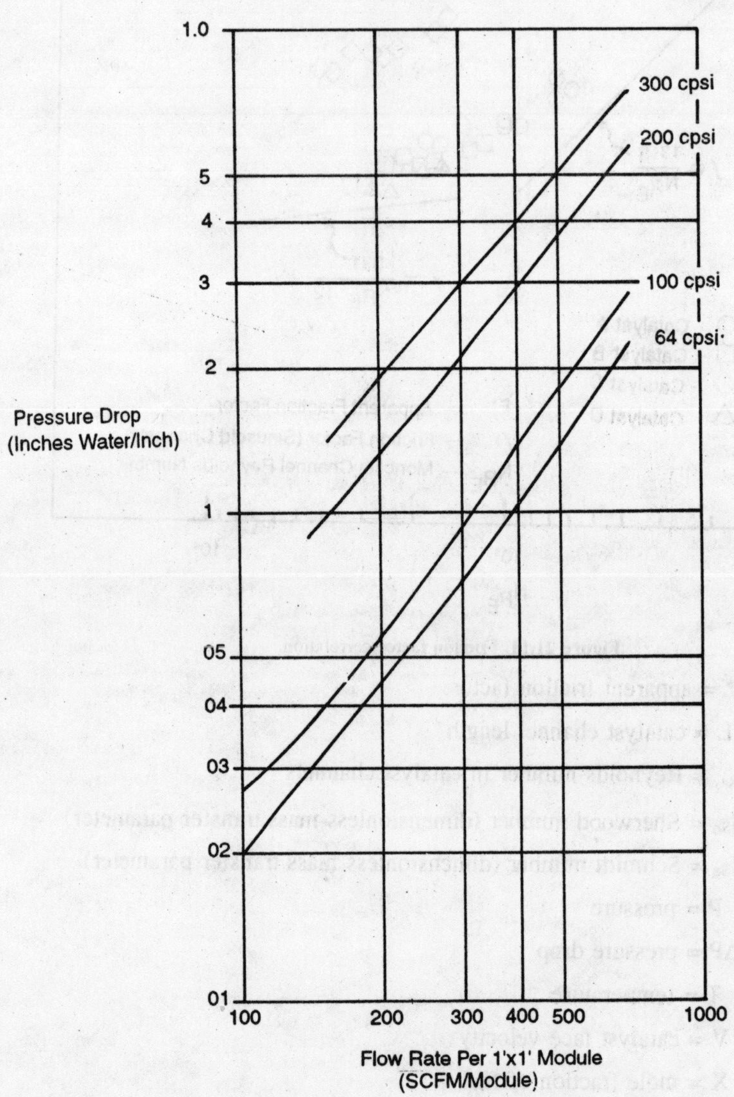

Figure 21.12. Honeycomb pressure drop vs. flow rate.

strong effect of operating temperature in the Arrhenius rate expression. Figure 21.10 shows the experimental results that gave a linear correlation on a log-log plot. Additional proof of mass transfer control is the excellent agreement on this plot for all four different catalyzed honeycombs. Equation 3 or Figure 21.10 is used to calculate the Sherwood number and Equation 2 to calculate the conversion.

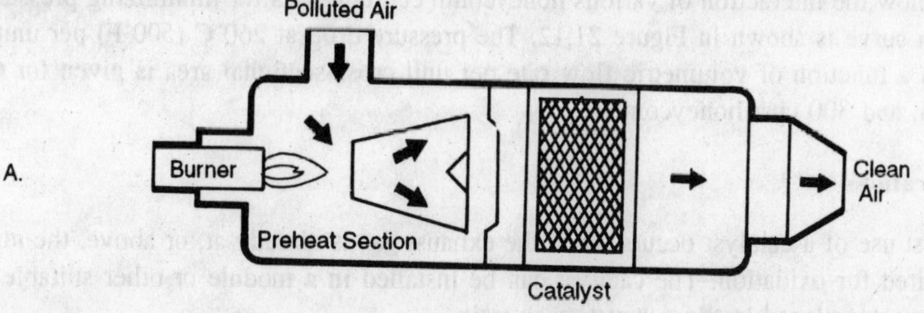

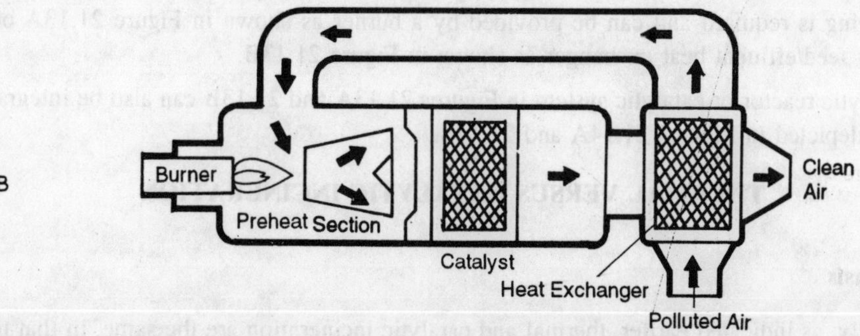

Figure 21.13. System applications.

Mass transfer control performance allows extrapolation to other length configurations (same D_H and a) using the following mass transfer scaling criteria derived from Equation 2, where X_1 and X_2 are the overall conversions for each configuration:

$$\frac{L_1}{L_2} = \frac{\text{In}\,(1 - X_1)V_1(N_{Sh}D_{f-air})2}{\text{In}\,(1 - X_2)V_2(N_{Sh}D_{f-air})1} \qquad (4)$$

Reactor Bed Pressure Drop

Another important design consideration is the pressure drop across the catalytic bed. Correlations have been developed for catalyzed honeycombs relating the apparent Fanning friction factor to the Reynolds number as shown in Figure 21.11, following standard friction-factor correlation procedures.

The free stream face velocity is utilized to calculate the channel velocity from V_{ch} (σ) = V, and therefore:

$$F^* = \frac{1}{2}\,\frac{P}{V_{ch}^2}\,\frac{D_{II}}{L} = \left(\frac{\sigma 2}{2}\right)\left(\frac{\Delta P}{\rho V^2}\right)\frac{D_{II}}{L} \qquad (5)$$

where ρ is the density and σ is the percent open area of the test catalyst. The apparent friction factor includes entrance and exit flow momentum losses, as well as frictional losses, for both developing and fully developed flow, and hence than the conventional pipe flow fricition factor shown as the solid line with its equation in Figure 21.11.

Using the fundamental friction factor and Bernoulli equation for pressure drop,design curves can be generated to show the interaction of various honeycomb cell densities for minimizing pressure drop. One such design curve is shown in Figure 21.12. The pressure drop at 260°C (500°F) per unit length of honeycomb as a function of volumetric flow rate per unit cross-sectional area is given for 64 cpsi, 100 cpsi, 200cpsi, and 300 cpsi honeycombs.

System Considerations

The simplest use of a catalyst occurs when the exhaust gas is already at, or above, the minimum temperature required for oxidation. The catalyst can be installed in a module or other suitable reactor housing and the reactor placed in the exhaust stack train

If the temperature is below the operating temperature listed in Table 21.3 or one established by testing, preheating is required and can be provided by a burner as shown in Figure 21.13A or by the burner and/or a feed/effluent heat exchanger as shown in Figure 21.13B.

The catalytic reactor or catalytic system in Figures 21.13A and 21.13B can also be integrated into the process as depicted in Figures 21.14A and 21.14B.

THERMAL VERSUS CATALYTIC INCINERATION

Theoretical Basis

Chemically, as indicated earlier, thermal and catalytic incineration are the same, in that the basic reactant is oxidized to CO_2 and H_2O when the reaction is carried to 100% completion. For gaseous reactants, thermal incineration is always a homogeneous reaction.

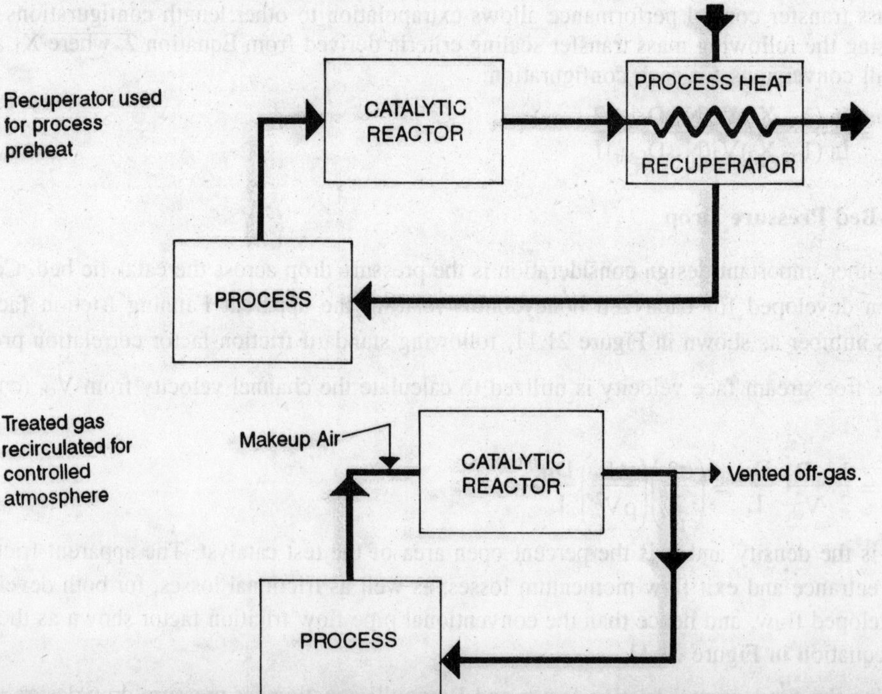

Figure 21.14. Process and system application schomatics.

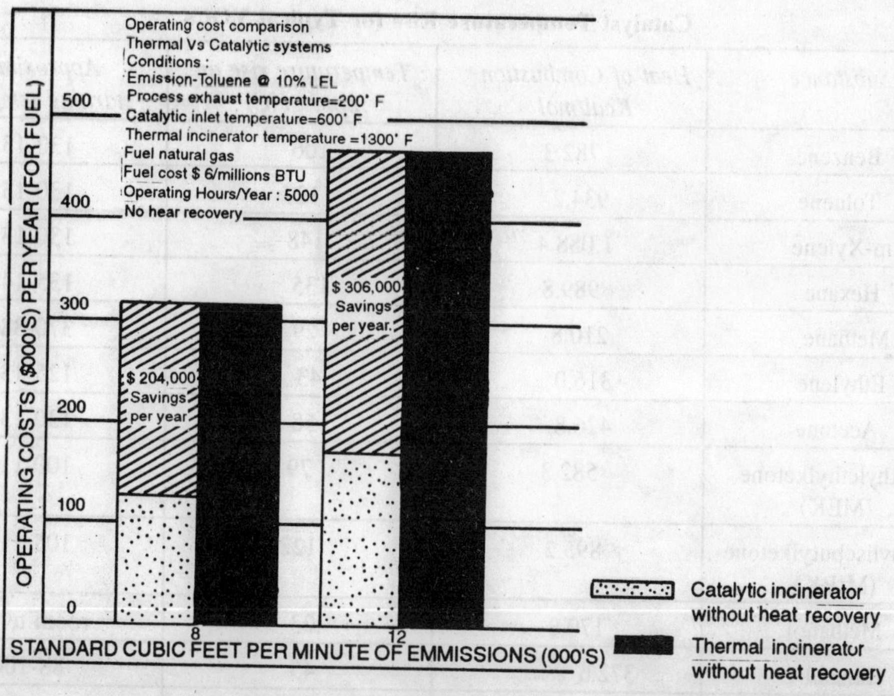

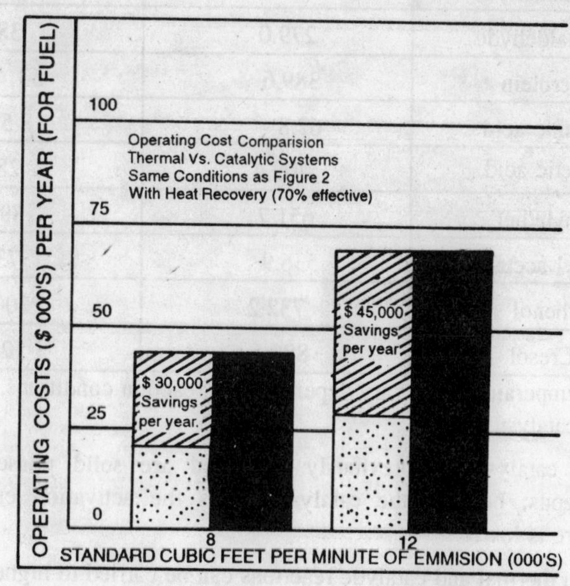

Figure 21.15. Energy savings for catalytic-thermal incineration.

Table 21.3

Catalyst Temperature Rise for Typical VOCs

Substance	*Heat of Combustion Kcal/mol*	*Temperature rise on ppm (°C)*	*Approximate startting temp. (°C)*
Benzene	782.3	106	130-135
Toluene	934.2	127	130-135
m-Xylene	1,088.4	148	130-135
Hexane	989.8	135	135-140
Methane	210.8	29	370-380
Ethylene	316.0	43	125-130
Acetone	426.8	58	130-135
Methylethylketone (MEK)	582.3	79	100-110
Methylisobutylketone (MIBK)	895.2	122	105-115
Methanol	170.9	23	room temp.
Ethanol	372.6	45	88-100
Formaldehyde	134.1	18.2	room temp.
Acetaldehyde	279.0	38	105-110
Acrolein	389.6	53	95-105
Formic acid	62.8	8.5	room temp.
Acetic acid	209.4	28	190-195
Ethylether	651.7	89	140-150
Ethyl acetate	536.9	73	195-205
Phenol	732.2	100	140-150
m-Cresol	880.5	120	145-150

NOTE: Temperature may vary depending on reaction conditions, amount and dispersion of platinum catalyst.

The catalysts as previously described are solid phase and thus, the reaction system is heterogeneous; because the catalyst lowers the activation energy of the process, the operating temperature is lower.

Both thermal and catalytic reactions can be carried to higher degrees of completion by increasing residence time in the reactor, as evidenced calculations using typical first-order kinetics, such as Equation 6, which often can empirically represent the industrial application of concern.

$$\frac{dc}{dt} = kc \qquad (6)$$

where c = concentration of reactant

t = time of reaction

k = k_0 exp (-E/RT) = rate constant

k_0 = pre-exponential factor

E = energy of activation

R = gas constant

T = reaction temperature

At a constant temperature, integration of Equation 6 results in an expression for conversion, Equation 7:

$$X = 1 - e^{-kt} \qquad (7)$$

where X = fractional conversion of reactant

$$X = \frac{C_{in} - C_{out}}{C_{in}}$$

t = elapsed time, residence time, or inverse space velocity

Inspection of this equation shows that as residence time increases, the conversion approaches 1,i.e. 100% The residence time is increased by making the reactor volume larger and, specifically in the case of the catalyst, by adding more catalyst. A detailed kinetic discussion can be found in, EPA-600/7-79-096.

The catalytic reactor volume, however, is much smaller than the thermal reactor volume because of the lower activation energy and higher rate constant resulting from the catalyst selection.

Practical Aspects

These two factors, lower operating temperature and lower reactor volume, can be used to obtain practical economic benefits. First, the smaller reactor volume means less material can be used and the reactor section can be tailored to the process envelope. Since materials such as stainless steel are often chosen, this savings can be appreciable. The smaller size also means less weight and for this reason, roof top installations are viable. Regardless of the materials of construction, the lower operating temperature will usually result in maintenance savings for auxiliary and peripheral equipment such as thermocouples, sample ports, pressure taps, and miscellaneous internals used in both types of reactors.

A major aspect of the lower ignition temperature is a fuel savings, because most pollution-control after-burner syatems are required to operate with a feed stream that cannot sustain combustion. Fuel must be added, and the cost of this additional fuel can increase or decrease drastically over the multi-year life of these incinerators. The use of heat exchangers to recover the heat is applicable for both thermal and catalytic systems.

Generally, the same efficiency can be achieved in either case, but a detailed comparison of efficiency is required for the heat exchanger size and cost, which will usually show that initial investment cost is not greatly impacted.

Table 21.4

Commercial Examples in Japan*

(Catalyst Performance)

Name of constituent	*Chemical formula*	*Formula weight*	*Temp. rise at 1,000 ppm*	*Operating temp.*	*Concentration before treatment*
Styrene	$C_6H_5CHCH_2$	104.15	138°C	250°C	310ppm
Acetaldehyde	CH_3CHO	44.05	35°C	350°C	240ppm
Benzene	C_6H_6	78.12	103°C	210°C	380ppm
Toluene	$C_6H_5CH_3$	92.14	123°C	210°C	320ppm
M-xyiene	$C_6H_4(CH_3)_2$	106.17	143°C	210°C	270ppm
Phenol	C_6H_5OH	94.11	101°C	300°C	380ppm
Formaldehyde	HCHO	30.03	17°C	150°C	410ppm
Acrolein	CH_2CHCHO	56.06	51°C	180°C	500ppm
Acetic acid	CH_3COOH	60.05	26°C	350°C	590ppm
Butyric acid	C_3H_7COOH	88.11	66°C	250°C	370ppm
Acetone	CH_3COCH_3	58.08	57°C	350°C	410ppb
Methyl ethyl ketone	$CH_3COC_2H_5$	72.11	74°C	220°C	380ppm
Methyl isobutyl ketone	$CH_3COC_4H_9$	100.16	116°C	250°C	270ppm
Ethyl acetate	$CH_3COOC_2H_5$	88.11	68°C	350°C	350ppm
Butyl acetate	$CH_3COOC_4H_9$	116.16	108°C	260°C	480ppm
Methyl alcohol	CH_3OH	32.04	21°C	150°C	830ppm
Ethyl alcohol	C_2H_5OH	46.07	44°C	350°C	550ppm
Isopropyl alcohol	C_3H_7OH	60.10	64°C	280°C	230ppm
Butyl alcohol	C_4H_9OH	74.12	84°C	260°C	330pm
Carbon monoxide	CO	28.01	9°C	150°C	4,000ppm
Methyl cellosolve	$HOCH_2CH_2OCH_3$	76.10	55°C	300°C	110ppm
Ethyl cellosolve	$HOCH_2CH_2OC_2H_5$	90.12	76°C	300°C	80ppm
Butyl cellosolve	$HOCH_2CH_2OC_4H_9$	18.18	118°C	300°C	50ppm

NOTE: Temperature may vary depending on reaction conditions, amount and dispersion of platinum cataylst.

Figure 21.15 shows savings realized through lower fuel consumption in one example. The savings were greatest when no form of heat recovery was used, but still substantial after heat exchangers were installed. For retrofit designs, size savings can be anticipated, and for new equipment designs, process engineers should take advantage of all process and system flexibilities to achieve similar benefits.

Table 21.4, although not exhaustive, can be quickly scanned for estimating temperature rises and minimum temperature requirements for VOCs in proven applications. The VOC identification and amount will initially provide a good basis for deciding whether a thermal or a catalytic incinerator will be more suitable or whether a nonincineration, recovery-type process should be applied, followed by a catalyst, for final and more complete clean-up.

The information on gas flow rate and need for a heat exchanger will allow a ± 25% scoping cost estimate to be made within hours.

COMMERCIAL CATALYTIC INCINERATORS

Applications

Over 1,000 commercial applications can be documented for VOC catalytic incineration. These

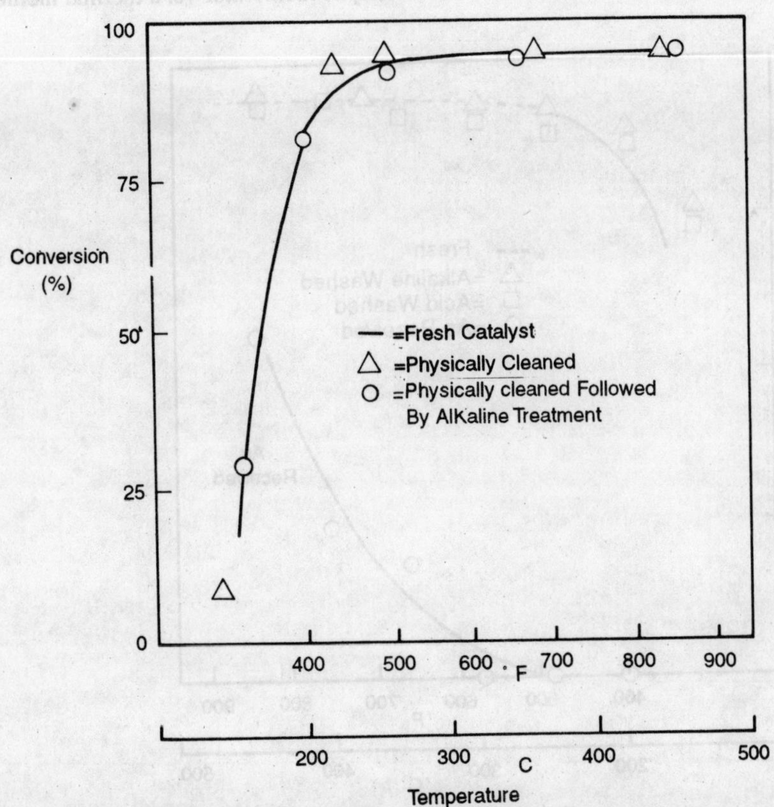

Figure 21.16. Toluene conversion activity for an aged Torvex P-3 catalyst: automotive.

have been discussed in detail in some of the references.

Table 21.4 is from a list of applications in Japan which shows the actual VOC removed and the inlet concentration to the catalyst reactor system. In these cases, design conversion was usually based on odor removal and, in many cases, this utilized designs where less than 1 ppm of the target VOC was found in the exhaust.

Case History Performance

The full-scale unit, was installed in a formaldehyde plant using Reichold's licensed FORMOX™ process for manufacturing formaldehyde from methanol. Recycling is used, but some exhaust gas must be withdrawn from process and vented to remove inerts. The vented gas stream flow rate was 4,500 to 6,000 cu m per hr and entered the catalytic incinerator at 32-43°C. A preheat burner is used to start the system and a heat exchanger used thereafter to maintain catalyst bed exit temperatures in the 425-621°C range sufficient to provide suitable catalyst inlet temperatures 195°C to 271°C. This unit started up in July 1979 and operated with average conversion of the four key components in the 97.9 to 98.5% range without change during a carefully monitored first year.

This unit is still in operation in 1987. NO_x values were measured at 11.0 and 12.3 ppb with a 9.0 ppb background proving the negligible NO_x formation of a catalytic incinerator vs. a thermal incinerator.

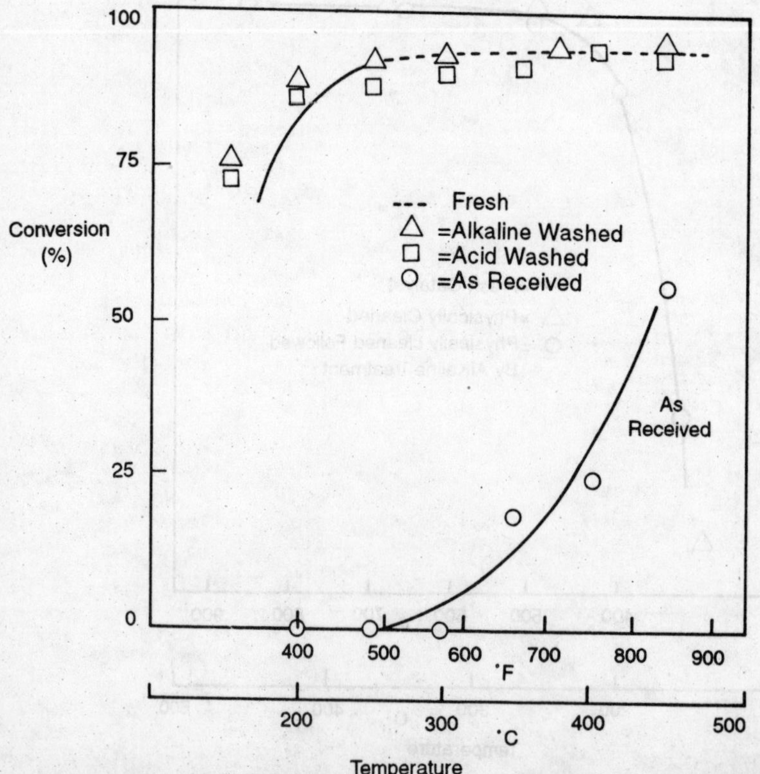

Figure 21.17. Toluene conversion activity for an aged Torvex P-3 catalyst: can coating.

The full-scale unit oxidizes methyl ethyl ketone and dibutyl ketone in a 5000 SCFM gas stream from a rubber fabricating plant. The pressure drop in operation was 6 inches w.c. for the honeycomb catalyst and conversions routinely exceed 90%. The unit was fabricated as a skid-mounted unit weighing 3,000 pounds and shipped to the plant for installation.

A very large system (70,000 SCFM) was installed to treat the off-gas from a phthalic anhydride reactor and operated for six years before catalyst changeout. Both organics and CO were efficiently removed at 95% and 99% respectively, in this large unit. An initial factor in the design selection of a catalytic incinerator was the restriction on natural gas for some uses in 1976, the start-up year. Today,, catalytic incinerators are still often the design choice for phthalic anhydride plants based on energy savings for 1987 costs vs. those in the 1970s. A second incinerator of 45,000 SCFM has been installed for a site expansion and the same conversion efficiency is obtained in both incinerators.

Catalyst Maintenance Practices and Examples

In many respects, catalytic oxidation of organic pollutants in an oxidizing atmosphere is essentially complete (greater than 90%) to carbon dioxide and water. Long life is expected because carbon cannot form on the catalyst surface. In addition, the exhaust gases are usually quite clean because most poisonous materials have been scrubbed for recovery on an economic basis.

Table 21.5
Commercial Examples (1 to 10 Years Operation)

Industry	*Years on Stream*	*VOC Conversion (%)*	
		Before Cleaning	*After Cleaning*
Can Coating	7	66	86
Can Coating	3	60	95
Can Coating	3	65	85
Oil Production	1	50	85
Can Coating	3	70	94
Can Coating	10	85	95
Synthetic Fibers	1	50	99
Can Coating	6	73	90
Plastic Products	2	13	88
Ceramic Products	0.5	—	90
Synthetic Fibers	2	80	90
Rubber Products	1	86	87
Metal Coating	2	75	95
Automotive Paint Bake	2	71	93
Automotive Paint Bake	1	91	95

Automotive Paint Bake	1	—	95
Automotive Paint Bake	10	90	91
Chemical	7	70	90
Metal Coating	6	70	90

Indeed, the selected list of representative process industries shown in Table 21.5 demontstrates usage of the same catalyst for up to 10 years. At the 10-years point, two of the systems are shown to have 85% and 90% VOC destruction efficiency; after cleaning, the 85% case was improved to 95% efficiency. These efficiencies were close to design efficiencies, but were not the maximum efficiency possible for the given application due to some degradation via support sintering and metal agglomeration. In both of these cases, the redesign, either by adding more catalyst (approximately equal amount) or by changing to a high cell density, more active catalyst, could bring the overall VOC conversion to 99% and the same 10-years life can be expected.

Table 21.5 indicates that some of the applications can result in a significantly decreased VOC conversion within one year. However, it is not necessary to change the catalyst. A short shutdown period of a few hours, or occasionally one or two days, can be utilized to clean the catalyst of deposited debris; and, as shown in the specific examples, bring the performance back to 90% conversion, usually the same as the original fresh charge of catalyst. It is known that 10-20 of these regenerative cleanings do not significantly shorten catalyst life before a changeout and expense of a new charge must be incurred. This is determined on a case-by-case basis, from experience and data base of different vendors, the catalyst used, and the specific application.

Reference details various catalyst-cleaning procedures, which can be as simple as blowing clean dry air in reverse flow through the catalyst, then removing the debris and dirt that lightly cover and block or mask the surface. This method is done in place or) when the design contains small 1 foot by 1 foot or 2 foot by 2 Foot individual modules) by removing modules one at a time.

More definitive chemical washing procedures can be used, such as dilute alkaline or acid washing, either at the user's site or at the catalyst vendor's plant. Figure 21.16 reports that physical cleaning by air lancing can restore activity to the initial level the same as an alkaline wash. If this is demonstrated for a given site, a considerable cost savings can result. Typical production plant operators need no special training to clean the catalyst.

A brief description of the automative subassembly plant is given to support Figure 21.16.

A Torvex P-3 (55 cpsi) catalytic incinerator has been used to abate hydrocarbons from an operation which paints automotive plastic subassemblies. The hydrocarbons in paint solvents, such as ketones, toluene, and xylene, together with some solids, are emitted to the exhaust gas. The exhaust gas temperature is maintained at 350°C before entering the incinerator. The temperature rise across the catalyst bed is 72°C.

After 18 months of operation, personnel at the plant noticed a visible emission at the stack and a decline of temperature rise across the catalyst bed. These catalyst modules were subsequently returned for regeneration.

The returned modules were found to contain substantial amounts of deposits. After physically blowing these modules clean, catalyst samples were taken for the standard activity tests. Figure 21.16

shows that this cleaned catalyst gave the same conversion response as a fresh catalyst. Additional treatment with an slkaline solution did not change the catalyst performance.

From these results, it was apparent that the observed performance decline in the commercial installation resulted from dust accumulation in the catalyst modules. These deposits were removed simply by physical cleaning, and chemical treatment was not needed to reactivate the catalyst.

A second example (Figure 21.17) demonstrates that both acid and alkaline washes are applicable in an instance where simple air lancing is not adequate. In this system, Torvex P-3 Catalyst (55 cpsi) was installed downstream of a composition can coating line. After 12 months of servic, performance acitivity tests were requested on the catalyst before and after chemical cleaning. These results were used to determine whether the catalyst should be cleaned or replaced.

The laboratory activity tests on the returned sample indicated very low toluene-conversion activity. Analysis further showed that deposits of silicon and aluminum compounds on the catalyst caused performance loss.These deposits were removed effectively by treating the catalyst with either an acid or an alkaline solution. Figure 21.17 indicates that the cleaned catalyst samples demonstrated the same activity as a fresh catalyst and were free of any other modes of deactivation.

From these results, the alkaline cleaning procedure was recommended.

APPLICATION EVALUATION PROCEDURES

End-User Inquiry Data

The first step in the assessment of any catalytic incineration application is the collection of some very basic information to allow any of the commercial companies to provide a "go/no go" judgment on whether a catalytic approach applies. Inquiry Data Sheet A is one such form which will readily serve this purpose.

The sales application engineer will know immediately in many cases by the process involved whether catalysts apply. For example, formaldehyde plant off-gases and phthalic anhydride off-gases have been catalytically treated and are well- established and in operation for seven or eight years. Most engineering contractors can provide basic designs and costs for the examples listed in Table 21.3.

Engineering Design Options

Major design options are based on pressure drop and energy- recovery constraints. For equipment specification, the designer has to determine whether he needs:

- A preheater
- A blower on inlet or outlet
- A heat exchanger

Design Data Sheets A and are a convenient method for collecting the summary data and preliminary calculations. To use these, the catalyst bed will have been sized for cubic feet of volume, based on the conversion requirement and estimated catalyst life for the given feed stream. For the most part, this volume will not alter as the designer changes the number of modules array to obtain desired drop governed by catalyst bed cross-sectional area and catalyst bed depth, as illustrated previously.

Design Data Sheet C provides a convenient method of tabulating energy and pressure-drop requirements for use in requesting quotes, calculating economics, or reviewing the design at design review meetings.

Economic Evaluation

Each company, project engineer and engineering consultant has specific methods and criteria for evaluating projects on a stand-alone basis or in comparison with alternatives.

Economic Evaluation Sheets A and B are designed specifically to serve a relatively narrow purpose under the basic assumption that a process exists, and a catalytic incinerator retrofit is required. This analysis applies even if a thermal incinerator is in use and high energy costs suggest that a change in pollution control equipment would have a reasonable payback.

The analysis basically allows the project evaluator to computer the initial investment as offset by any depreciation or other tax benefits and then evaluate this return as energy savings. Of course, it is assumed the project justification is essentially mandated by the company's own prior safety practices or legislated code and permit requirements.

However, within this scope, proper design can result in cost savings by implementing the most appropriate pollution control equipment, or by integrating it into the industrial process. This may be in conjunction with the production process design if there is flexibility for preventing pollution formation or minimizing concentration and volume.

As a variation, these sheets can be used to compare the cost/benefit ratios of catalyst X vs. Y, or return on investment for using a heat exchanger even if the original plant did not fully recover all low-level heat sources.

Test Programs

A Test program can maximize results because no two off-gas streams are exactly alike, and a standard solution is not likely to achieve optimum results. Major companies, therefore, often approach air-pollution control with a professional testing program conducted at modest cost. As indicated by a preliminary evaluation, a small-scale test unit is installed and operated under the direction of engineers and technicians and usually in conjunction with a consultant, government agency, or catalyst company. Often, the test unit will be engineered for a specific off-gas stream. Working with data developed from testing, an optimum full-scale system can be designed.

Special abatement problems demand a flexible approach because catalytic systems are not always applicable to gas streams containing significant amounts of chlorine or other halogens, sulfur, or heavy metals. In these cases, a designer may recommend alternate methods of abatement. In evaluating off-gas streams that contain these components, two techniques will be considered:

- Wet scrubbing to remove chlorine and sulfur compounds, followed by cataytlic oxidation to remove combustibles and odors.
- A system of adsorptive techniques combined with catalytic reactors which "polish " primary and stripping gas streams.

INQUIRY DATA SHEET A

CATALYTIC SYSTEMS DATA SHEET FOR THE ABATEMENT OF VOLATILE ORGANIC COMPOUNDS (VOC)

Name____________________ Title____________________

Campany____

Address____

____

Phone__________________________ Date________________

What type of process is producing the emissions ?____________________

____

____

Waste stream flow rate____________________ SCFM, ACFM, Ib./hr.__________

Is this flow rate the minimum quantity for effective operation of process ? Yes________ No________

Waste stream temperature____________________°F

VOC Constituents (type)	Usage or Concentration Ib./hr. or ppm
__________	__________
__________	__________
__________	__________

Does the process emit any of the following catalyst masking agents or deactivators: silicones __________, phosphorous compound__________, heavy metals__________, sulfur compounds particulates__________.

If any of the above items are present, what is concentration of the exhaust ?________________

Fuel available______________________________

Fuel cost________________ $/million BTU

Hours operated per year____________________________________hrs./yr.

Do you currently have any equipment handling these air emissions ?____________________
If yes, please describe____

____

If you have a thermal incinerator, what is the allowable pressure drop in the existing system______ inches w.c.
Does the existing system have a preheat heat exchanger?________________________

____

When will the abatement equipment be required ?________________________
What destruction of voc is required ?______________________________

____

other remarks____

____

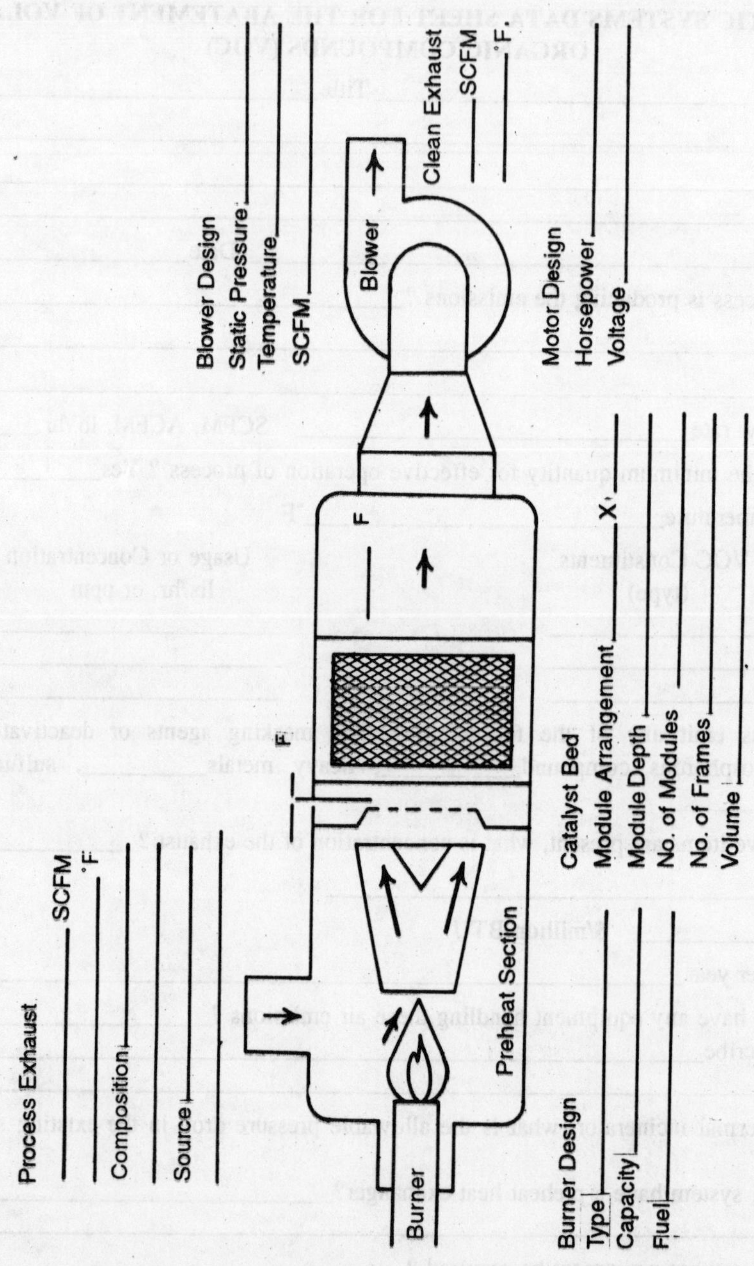

Design Data Sheet A

PROCESS WITHOUT HEAT RECOVERY

Design Data Sheet B

PROCESS WITH HEAT RECOVERY

Burner Design
Type ______
Capacity ______
Feul ______

Blower Design
Static Pressure ______
Temperature ______
SCFM ______

Motor Design
Horsepower ______
Voltage ______

Clean Exhaust
______ SCFM
______ °F

Heat Exchanger
Type ______
Effeciency ______

Process Exhaust
______ SCFM
______ °F
Composition ______
Source ______

Catalyst Bed
Module Arrangement ______ X ______
Module Depth ______
No. of Modules ______
No. of Frames ______
Volume ______

Blower

Preheat Section

Burner

______ F

______ F

______ F

Design Data Sheet B

PROCESS WITH HEAT RECOVERY

DESIGN DATA SHEET C

ENERGY CONSIDERATIONS

- Energy to heat exhaust to catalytic combustion temperature__________
- Solvent Fuel value__________
- Energy saved at__________% recycle__________
- Enery saved through use of Heat Exchanger__________
- Energy to heat burner combustion air to catalytic combustion temperature__________
- Energy required to make up heat losses in the system__________

SYSTEM PRESSURE DROP ESTIMATES, INCHES OF WATER COLUMN

- Catalyst Bed__________
- Furnace Section__________
- Supply and Exhaust Ducts__________
- Heat Exchanger__________
- Other __________

TOTAL __________

Prepared for:__________

at:__________

Prepared by:__________

ECONOMIC EVALUATION SHEET A

ESTIMATED CASH INVESTMENT VS. CASH SAVIGNS ASSOCIATED WITH THE INSTALLATION OF A CATALYTIC REACTOR.

Customer : ______________________________

Operation : ______________________________

I. CASH INVESTMENT :

A. Estimated Reactor Cost: ______________

B. Estimated Catalyst Cost: ______________

C. Estimated Installation Cost: ______________

Total Estimated Cash Investment: (1)

II. CASH FLOW SAVINGS:

A. Estimated First Year Cash Flow savings

Tax Benefit of first year depreciation = _______ % X (A+C)

____

Tax benefit of catalyst cost =_________ % X (B)__________

Other credits ______________________________

Estimated annual energy cash savings (From "Energy Savings Estimate")

Fuel Savings______________________________

TOTAL FIRST YEAR ESTIMATED CASH FLOW SAVINGS:__________ (2)

B. Estimated Second Year Cash Flow Savings:

Tax benefit of second year depreciation = _________ % X (A+C)

____

Estimated annual energy cash savings (From "Energy Savings Estimate")

Fuel Savings______________________________

TOTAL SECOND YEAR ESTIMATED CASH FLOW SAVINGS:__________(3)

C. Estimated Third Year Cash Flow Savings :

Tax benefit of third year depreciation = _________ % X (A+C)

____

Estimated annual energy cash savings (From "Energy Savings Estimated")

TOTAL THIRD YEAR ESTIMATED CASH FLOW SAVINGS :__________(4)

III. INVESTMENT VS. CASH FLOW SAVINGS FIRST THREE YEARS

A. Total Estimated cash investment (1) ______________________________

B. Total estimated cash flow savings over the first three year period (sum of 2, 3 and 4)______________________________

ECONOMIC EVALUATION SHEET B

ENERGY SAVINGS ESTIMATE

CUSTOMER: ____________________

OPERATION : ____________________

PRODUCTION RATE: ____________________

OPERATING SCHEDULE: ____________________

	BTU/HOUR	
	Present Operation	With Catalytic Reactor
Process Heat Load	______	______
Less : Recycle	______	______
Boiler Duty	______	______
Boiler Fuel Consumption (82% Efficiency)	______	______
Reactor	______	______
Total BTU Consumed	______	______
COST SUMMARY	$ /YEAR	
Fuel oil ($________ /gal.)	______	______
Natural Gas (________cents 100 cu. ft.)	______	______
Electricity (________cents KWH)	______	______
Other (______@______ cents million BTU)	______	______
TOTALS	______	______
PRE-TAX SAVINGS		______

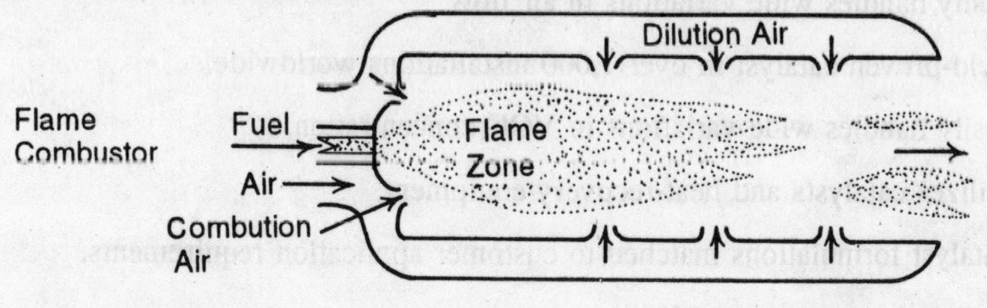

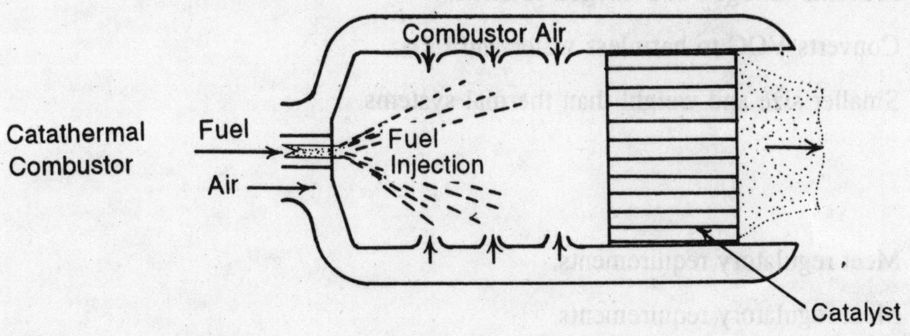

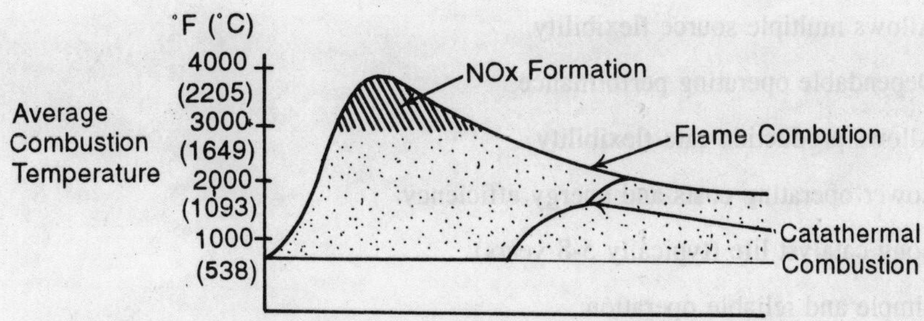

Figure 21.20. CATCOM[TM] **process description and feature.**

For these cases, a careful review of prior art is important and a test program is definitely required in conjunction with guarantees or tests to confirm the effectiveness of the scrubbing process selected.

Features and Benefits

The features and benefits, numbered respectively, of catalytic incinerators are listed in a qualitative manner below in comparison to a thermal incinerator.

Features

1. High percent VOC destruction.
2. Expertly designed catalytic reactors.
3. Easily handles mixtures of VOCs.

4. Easily handles wide variations in air flow.
5. Field-proven catalyst in over 1,000 installations worldwide.
6. Easily handles wide variations in VOC concentration.
7. Utilizes catalysts and heat-recovery equipment.
8. Catalyst formulations matched to customer application requirements.
9. Automatic system controls.
10. Prevents nitrogen and oxygen fixation.
11. Converts VOC to harmless water and CO_2.
12. Smaller size and weight than thermal systems.

Benefits

1. Meet regulatory requirements.
2. Meet regulatory requirements.
3. Allows flexibility in solvent formulations.
4. Allows multiple source flexibility.
5. Dependable operating performance.
6. Allows production rate flexibility .
7. Lower operating costs and energy efficiency.
8. Long catalyst life (typically 5-8 years)
9. Simple and reliable operation.
10. No nitrogen oxides formed from atmospheric nitrogen.
11. No solid or liquid waste disposal.
12. Flexible easy and low-cost installation.

The most important of these features and benefits are numbers 5, 6 and 10, described further.

Proven installations. At least four major catalyst companies have provided platinum catalysts with great success through the world. The most significant geographical centres are the U.S. Japan, Italy, and Germany.

Low concentration of VOCs. This is a very important feature of catalytic oxidation. Essentially, the reactor size and operating temperature are independent of whether the design requires removing 90% of the VOCs from:

10 ppm to 1 ppm
100 ppm to 10 ppm
or 1,000 ppm to 100 ppm

Additional fuel is not required, and catalyst volume and reactor size are the same.

NO_x, is not formed. Nitric oxides are not formed at all over the catalyst. When a preheater is required, some NO_x is formed but this is small at the operating conditions vs. a standard size thermal incinerator. In addition, the pre-heater is usually only used during start-up periods.

For site-specific applications, many of these benefits can be quantified and incorporated into the economic evaluations.

FUTURE DEVELOPMENTS AND CONCLUDING COMMENTS

Special Compounds

At the present time, dioxins, furans, PCBs and other chlorinated compounds are not routinely treated by industrial oxidation catalysts. Some screening tests have been conduced on chlorinated compounds and do indicate conversion can be attained, but activity declines were noted. In these studies, process variables were not optimized and aging runs were not utilized to counteract the activity declines. Further studies are needed in both bench-scale tests and slip-stream tests to establish performance. It would be expected that design temperatures need to be higher, Destruction of the organic portion of the chlorinated compound should free-up HCl and Cl_2; thus their subsequent poisoning effect needs to be quantified. Once increased temperatures provide a desired conversion level. The catalyst volume would be defined and it would be expected that increased volumes may be required to get catalyst life near those for nonhalogenated compounds.

CATCOM™, A Novel Full-Combustion Process

The content of this chapter on catalytic incineration has been restricted to post-treatment of exhausts. However, one process invented around 1970 has not yet found commercialization for a combination of reasons. This process, termed CATCOM™ OR CATATHERMAL™ by its inventors, is scientifically intriguing and has received widespread study by NASA and the EPA.*

Figure 20.18 illustrates that this process replaces a homogeneous burner and operates at a temperature significantly lower than typical flame temperature (such as in a gas turbine combustor) and below the temperature at which NO_x forms. The driving force for commercialization of this technology will occur as result of the following, in order of priority;

1. Requirement for 5 ppm NO_x from all NO_x sources.
2. Need to design smaller, less expensive combustors without exhaustive development trials.
3. Discovery of high-temperature, thermally stable, ceramic supports suitable for catalyst coating.

Concluding Comments

Catalysts and catalytic equipment are in active commercial use for a variety of air pollution control applications. Performance for well-defined and properly engineered systems has been exceptional with

*CATCOM and CATATHERMAL are tradenames of Engelhard Corporation.

catalyst life exceeeding several years. The catalytic devices indeed have economic advantages, especially with regard to energy savings, when incineration is applicable. A strong advantage for catalytic equipment is the proven flexbility to retrofit into existing equipment as well as new plants.

Any prediction of future developments must necessarily be limited to an extrapolation based on most recent past and current status for a work of this scope. These projections seem reasonable:

1. More and more different applications are being added each year as a result of pilot scale tests to confirm some minimum life expectancy for the catalyst.
2. The number of catalytic incinerators will grow as a result of natural industry growth for proven applications and will most certainly grow for applications which can be designed by similarity and because very few catalyst failures have been reported since some early marginal applications in the 1960s.
3. Catalyst costs will decrease relative to other costs, since the character of the industry historically has exhibited two key patterns:
   - Precious metal loadings are decreasing :
   - Honeycomb cell geometry is improving.

Beyond current products and process industries, one can expect some new developments. Catalyst research should result in specific tailoring of activity and selectivity for enhanced conversion of dilute quantities of dioxins, furans, and chlorinated hydrocarbons. On a chemical analogy basis, it is expected that platinum, palladium and their combinations with other materials will soon show good conversion of these compounds in pilot tests with these components. Since solid waste incineration is projected to grow and society's environmental concerns are not expected to diminish, surely pilot testing will proceed, commercialization will take place, and enhanced product development will follow—as it does all industries when a need is established.

22

The Recovery of Waste Organic Solvents

Introduction

For the past eight years, we (see Acknowledgment at end of chapter) have been conducting experiments designed to determined those situations in which the laboratory recovery of organic solvents would be both feasible and economical. These studies have ranged from systems which we consider easily amenable to solvent recovery such as the recovery of liquids of reagent-grade quality, to those in which the purity of the recovered material must be chromatographic or pesticide grade purity or better. In choosing the systems to be studied, we have concentrated on systems in which the waste solvent that was generated fulfilled several requirements:

1. The quanity generated was low enough (a maximum of 50 liters per week) that recovery could be accomplished in-house with an easily operable laboratory distillation unit.
2. The solvent that was to be recovered was not combined in a mixture in which the second component would form an azeotrope with the desired solvent upon distillation. We have found that even in some of these cases, a distilled azeotropic mixture can be either reused directly or chemically treated so that the end solvent is reusable.
3. The recovered solvent was of a purity that was equal to or better than the material used in the original process.
4. There was a significant economical advantage to recovering the used solvent rather than preparing it for waste disposal by commercial means.

The chapter is divided into three sections. First the results of our studies on recovering ordinary laboratory solvents and organic liquids generated in our academic teaching loboratores are presented. The next part deals with the recovery of solvents generated in health care institutions and discusses our results in recovering organic solvents from low-level radioactive liquid scintillation wastes. Finally, we detail those systems in which the recovery of high-purity solvents is feasible. The distillation equipment used in all of this work along with its characteristics is described below. The analytical techniques employed are presented at the end of the chapter.

Equipment

In considering separation techniques that would accomplish the goals of in-house solvent recovery, it was obvious that distillation would be the method of choice. While various commercial equipment is available for conducting distillation, it was necessary to choose equipment in which effective separation of closely boiling mixtures and high throughputs could be achieved. The best prospect involved *spining band distillation,* which had been known for many years to have the best height *equivalent to a theoretical plate* (H.E.T.P) for a laboratory unit. In 1975, B/R Instrument Corp. of Pasadena, Maryland constructed a high-throughput spinning band unit with sufficient eficiency and plates to accomplish our objective. In 1979 , when we first stated out experiments, we chose this type of distillation for our work.

All of the separations described in this work were accomplished on either a B/R Instrument Corp. 8300 or 8400 Spinning Band Distillation unit. While these systems are similar, the differences can be critical in certain separations so they will be described separately.

8300 Spinning Band Distillation Unit

The 8300 spinning band unit, consists of a distillation column 30 cm in length and 21 mm in internal diameter. The column is vacuum-jacketed and silvered in order to maintain adiabatic conditions. This latter feature is necessary so that rapid vapor liquid equilibration can be achieved, thus allowing for maximum still efficiency. The still condenser is a double cooling type with an internal glass coil jacketed by an outside condenser. This construction allows for rapid cooling of vapor and efficient condensation under the high throughput conditions of solvent recovery. The flow of water through the condenser is monitored by a flow valve which in combination with the 8300 microprocessor, described below will shut down the unit if condenser water flow is interrupted. A solenoid-operated reflux valve allows condensate to be removed from the distillation unit at periodic intervals. Vapor-liquid equilibration, and thus separation, is achieved by use of a rotating spinning band. Two PTFE teflon bands are twisted in a helix and attached to a motor-driven teflon shaft. The perforated band is pitched downwards and turned at 2,200 rpm by a sparkless Halleffect DC motor affixed to the band by use of a drive plug. The rotation of the tightly fitted band continually pumps condensate down the column while ascending vapor passes through the descending liquid, substantially increasing effective surface area and thereby providing a high number of theoretical plates. The 8300 system is rated at approximately 20 plates, thereby providing an H.E.T.P (height equivalent to a theoretical plate) of 1.5 cm/plate. This high efficiency for such a short and wide diameter column increases its utility for complete separation of closely boiling mixtures (B.P separations of greater than 10°C) as long as they don't azeotrope.

The distillation unit is controlled by dedicated microprocessor (see Figure 22.1), which allows for the programming and storing of eight distillation protocols. These protocols consist of user-gwnerated information related to the distillation being performed. The information required for each protocol is the following:

1. Equilibration time—the time interval between vapor reaching the distillation head (the condenser) and its removal as condensate. The more difficult the separation, the longer the equilibration time.

2. Open cut temperature—the temperature at which the reflux valve will open and condensate in removed. This marks the beginning of Fraction # 1.

3. Close cut temperature—the temperature at which the reflux valve will close, thereby marking and end of Fraction # 1. Two additional fractions can be programmed in which this close cut temperature is the open cut temperature for the next fraction, and so on.

Figure 22.1. 8500 distillation microprocessor.

4. Reflux ratio—the ratio of condensate returning to the still to that removed to the receiver. In effect, this represents the ratio in which the reflux valve is closed and opened. Thus 1 : 1 reflux ratio indicates that the valve is opened and closed for equal durations of time.

In addition to delineating the parameters for a desired distillation, the 8300 microprocessor controls the entire distillation from start-up to completion. It controls heat input to the heating mantle and shuts the heat off if either water flow to the condenser is interrupted or the temperature of the pot flask reaches 15°C above the close cut temperature of the final programmed fraction. Input to the heating mantle can be varied or preprogrammed protocols can be used. The microprocessor also starts spinning the teflon band at 2,200 rpm once vapor has reached the distillation head and turns the band of upon completion of the run. It will shut the total unit down if any electrical malfunction is sensed. The complete 8300 unit only requires technician attention to fill the pot flash (this system has a maximum pot charge of about 5 liters) and change receiver flasks between fractions (an alarm sounds when this task needs to be performed). The system can be purchased with an automatic changer which even eliminates this task. The completion of a distillation run is essentially completely automatic.

8400 Spinning Band Distillation Unit

The B/R Instrument Corp. 8400 spinning band distillation unit is sightly more flexible than the 8300 and is used when larger volumes of solvents need to be recycled. The unit has a capacity of 12 liters (several 22 L systems are in use in certain laboratories) and is equipped with same microprocessor as the 8300. The distillation unit is longer (61 cm) and can be equipped with stopcocks for vacuum

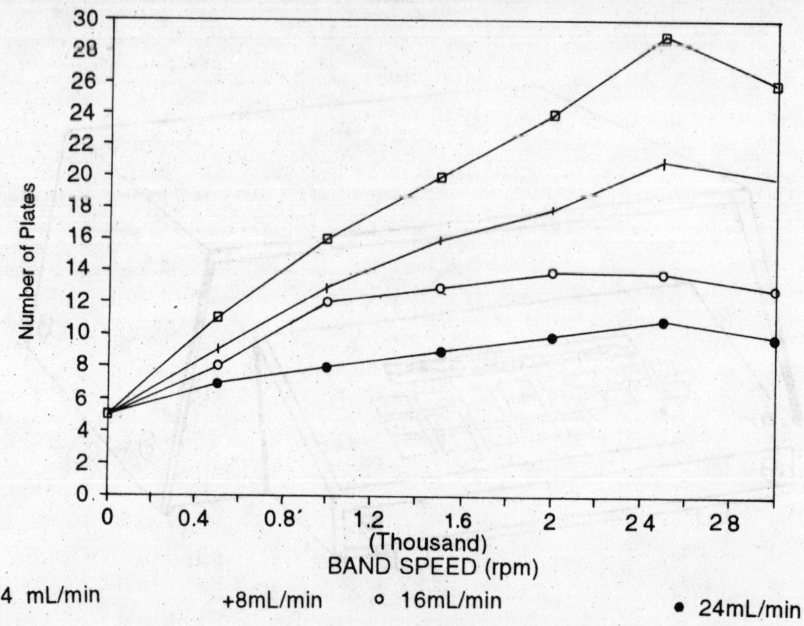

Figure 22.2. Plates versus band speed (40 – T).

distillations or distillation under inert atmospheres. The greater length produces more maximum plates, 30 plates, and an H.E.T.P. of about 2 cm/plate. Boiling point separations of 8°C are possible with this unit.

The fixed parameters of both the 8300 and 8400 systems have been optimized based on experiments done with manual systems. Figures 22.2 and 22.3, for example, show the variation of plate efficiency with band speed and equilibration time of a 61 cm spinning band manual system (the B/R Instrument Corp. 40-T). The maximum number of plates was obtained at about 2,200 rpm and between 1 and 2 hours for vapor-liquid equilibration, regardless of the boil-up rate.

THE RECOVERY OF ORDINARY SOLVENTS AND ORGANIC LIQUIDS FROM AN ACADEMIC TEACHING LABORATORY

John A. Mangravite and Virgil E. Magnuson

There are many situations in industrial and academic laboratories in which large quantities of ordinary organic solvents are generated as waste materials. The laboratory is then faced with the ethical and environmental dilemma of disposing of these wastes. In most cases, if the used solvent is managed correctly rapid recycling can be achieved in which the recovered material is of sufficient purity to be reused in the process in which it was generated or for other ordinary purposes. If the solvent waste is generated in routine procedure, it is advisable that the recycling process be incorporated as the final step in the procedure in which the waste is generated. In this section, we discuss a number of ordinary laboratory solvents that can be routinely recycled and the results of program that we initiated in our First Year College Chemistry laboratory program. Various types of analyses were performed to ascertain

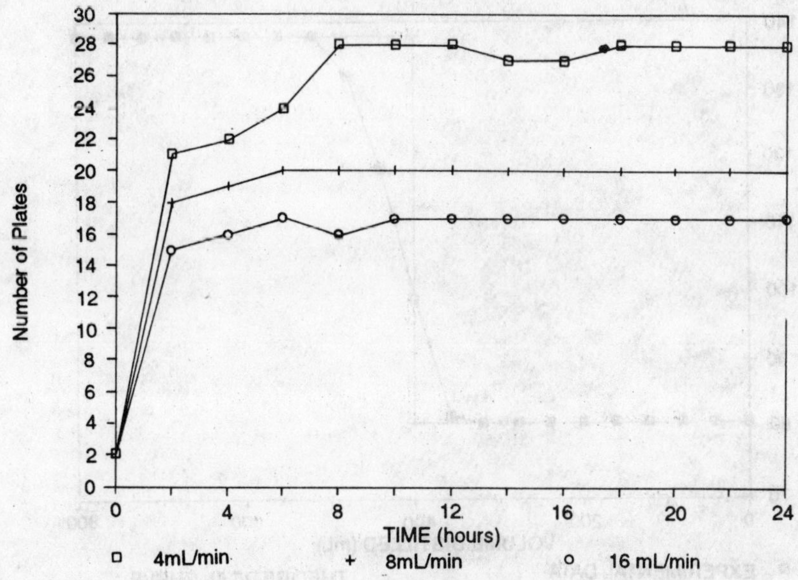

Figure 22.3. Plates versus equilibration time (40 – T).

the purity of the recovered solvent, with gas chromatography as the most conmon method of quality assurance.

Ordinary Solvents

Over the past few years, we have conducted hundreds of distillations of both artificial and real binary and tertiary mixtures utilizing the 8300 and 8400 distillation units. The distillations were monitored by gas chromatography and distillation curves were constructed in which volume distilled was plotted versus boiling temperature. A complete compilation of these results is beyond the scope of this chapter, however typical results are shown in Figures 22.4—22.6. Figure 22..5 illustrates the distillation curve for a 50 : 50 mixture of cyclohexane-xylene in which a perfect separation is achieved. Figure 22.5 shows the excellent separation of a 33 : 33 : 33 mixture of n-hexane, n-heptane, and n-octane, and Figure 22.6 the distillation of a 50 : 50 methanol-toluene mixture in which an azeotrope is collected before pure toluene. In the latter case, distillation was effective in that the composition of the azeotrope (70:30) is constant and distillation proceeds at 63.8°C.

We have chosen five common systems of the many that we have studied to discuss completely in this section. Distillation and analytical data are presented for these systems, all of which were actual wastes generated in specific waste streams.

Acetone

Acetone is a very common organic solvent that is routinely used to rinse glassware or as a degreasing solvent in cleaning and drying precision parts. It is also used as a delusterant for cellulose acetate fibers and as a solvent for potassium iodide and permanganate. As a reagent grade, it is

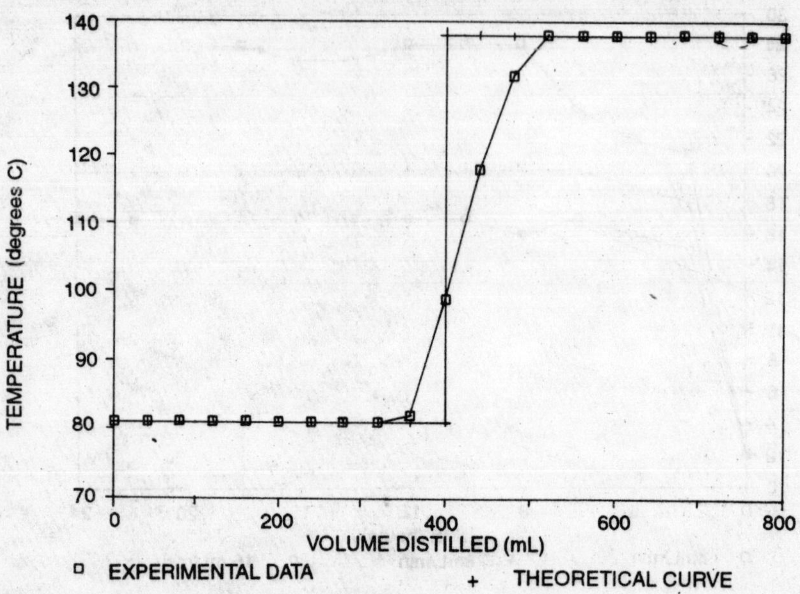

Figure 22.4. Distillation of cyclohexane:xylene.

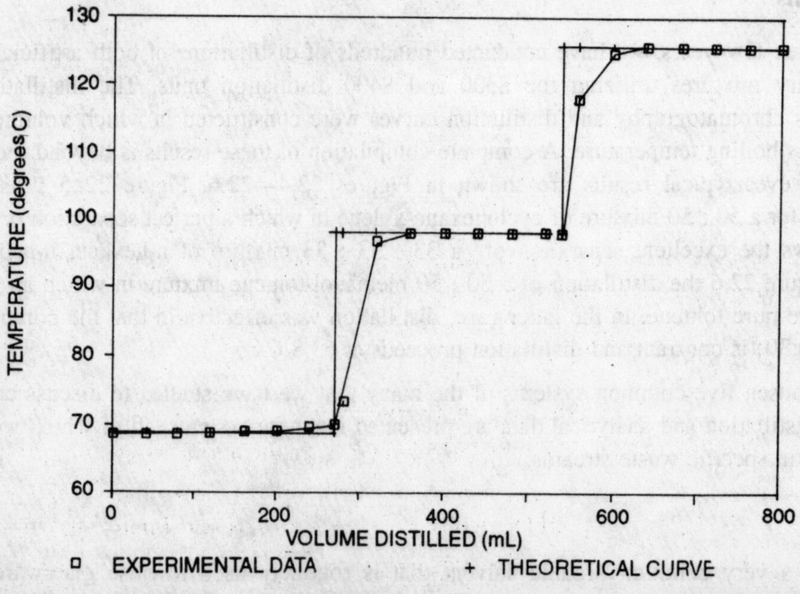

Figure 22.5. Distillation of hexane:heptane:ocatane.

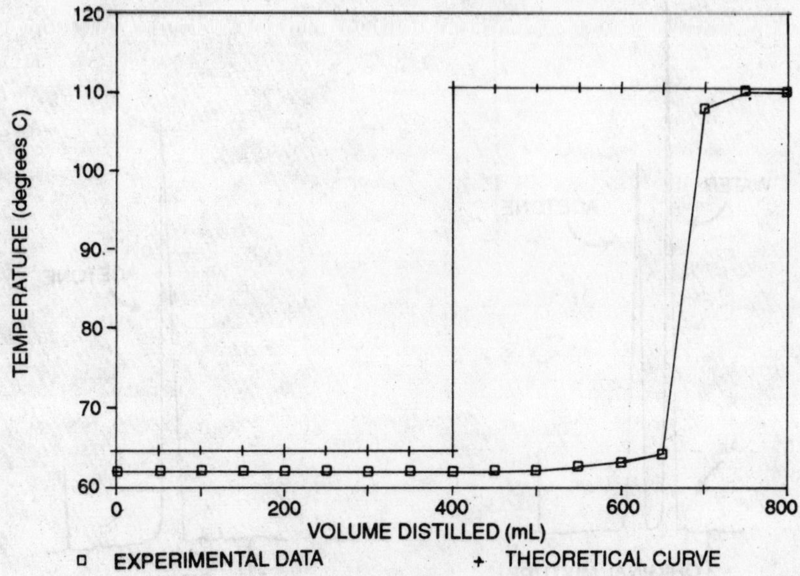

Figure 22.6. Distillation of methanol:-xylene.

commonly used as an extraction solvent. While it is moderately toxic and flammable, in most cases, the water miscibility of the material allows it to be dumped down the drain and rinsed into the local sewage system, as long as it is not contaminated with some hazardous component. An alternative is to collect the used acetone and distill it under high throughput conditions. The major contamination is due to water, with nonvolatile compounds as minor components. Figure 22.7 shows the gas chromatogram (TC detector) of used acetonewater (70 : 30) and acetone recovered on an 8300 unit from an extraction procedure. The appropriate distillation curve is shown in Figure 22.8. Four and a half liters of the mixture were distilled at a 2:1 reflux ratio and 10-minute equilibration time over a four hour period (unattended) leading to a recovery of 3 liters of greater than 99.9% purity acetone. The recovered material was demonstrably reagent grade. It is obvious that on a routine basis, 30 liters of acetone can be easily recycled per week with the resulting product of sufficient purity to be reused in most, if not all, procedures that would be applicable.

Dichloromethane (DCM) and Chloroform

DCM is a moderately toxic solvent that is used as a degreasing agent and as a solvent for pharmaceutical and food extraction. It is incorporated into the methodology of procedures for various drug analyses and as a solvent for cellulose esters and ethers. We have recovered DCM from many of these procedures by conducting distillations at about 1.5-2 liters/hour. By collecting a heart cut of boiling point 40.1-40.2°C, a very pure material is obtained. If completely anhydrous solvent is required, drying of the distilled solvent with a zeolite reduces the water content to less than 0.006%.

Chloroform is a highly toxic, water-immiscible solvent that is used as a general solvent. It reacts slowly with oxygen when exposed to air and light to produce dangerous products. We have used direct distillation to recover chloroform from mixtures with pentane and cyclohexane, removing the heart cut

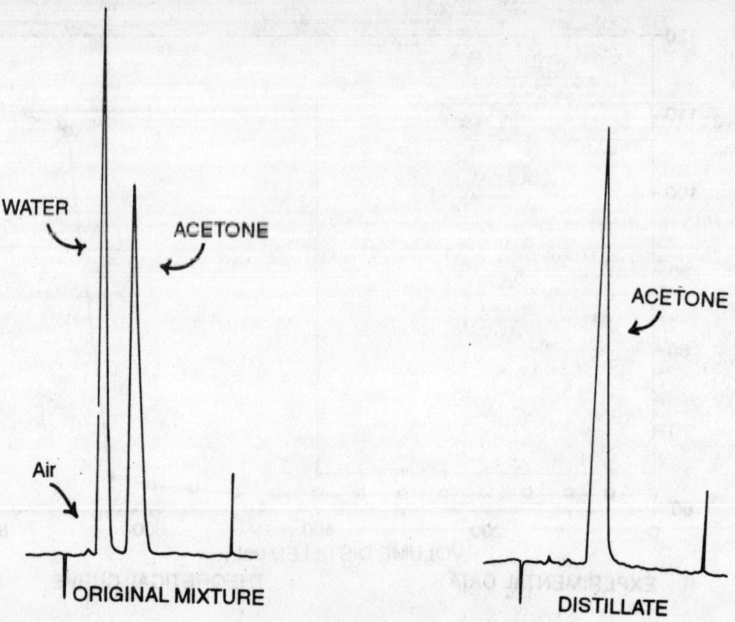

Figure 22.7. Gas chromatogram of acetone: water.

at 61.2°C. In the common mixture of $CHCl_3$ and hexane, a 72 : 28 azeotrope forms making this mixture unsuitable for solvent recovery. A similar azeotrope is formed with $CHCl_3$ and methanol ; however, recovery can be accomplished in this case. The waste mixture containing 50 : 50 $CHCl_3$, and methanol was treated with an equal volume of water leading to the extraction of the water-miscible methanol from chloroform. The separated chloroform was dried over molecular sieves and distilled. Methanol could also be recovered from distillation of the methanol-water extract. It is advisable to add a small amount of methanol to the recovered chloroform to act as a stabilizer if the solvent is to be stored.

N,N-Dimethylformamide (DMF)

DMF is a dipolar aprotic solvent (B.P. 152.8°C) that is flammable and highly toxic. It is an extremely good solvent for vinyl resins and is miscible with water and most organic solvents. The waste streams that we used in our recovery experiments were twofold. In the first case, we carried out experiments in which DMF (90%) was combined with water (10%), while the second involved recovery of DMF from a mixture with water, ethyl acetate and xylene. The initial composition of DMF in this latter case was 73%. Figure 22.9 shows the distillation curve for the DMF-water mixture, while the distillation curve for the latter mixture is shown in Figure 22.10. In both cases, throughputs of about 800 ml/hour were employed, with the resulting recovered material of greater than 99.9% purity.

Tetrahydrofuran (THF)

THF is an extremely useful solvent for natural and synthetic resins and is incorporated as a mobile phase for *high-performance liquid chromatographic* (HPLC) analyses of polymers. In our recovery experiments, 4 liters of THF used in an HPLC procedure for the analysis of polystyrene polymers was

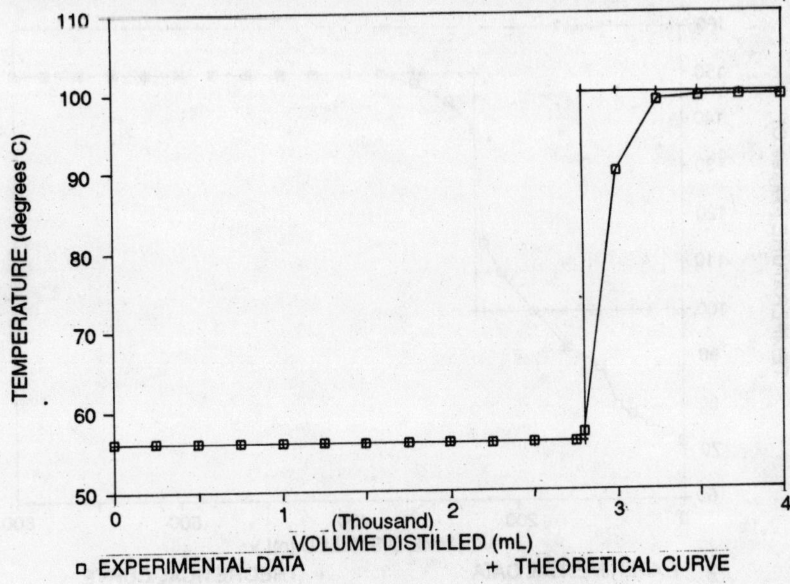

Figure 22.8. Distillation of 70 : 30 acetone:water.

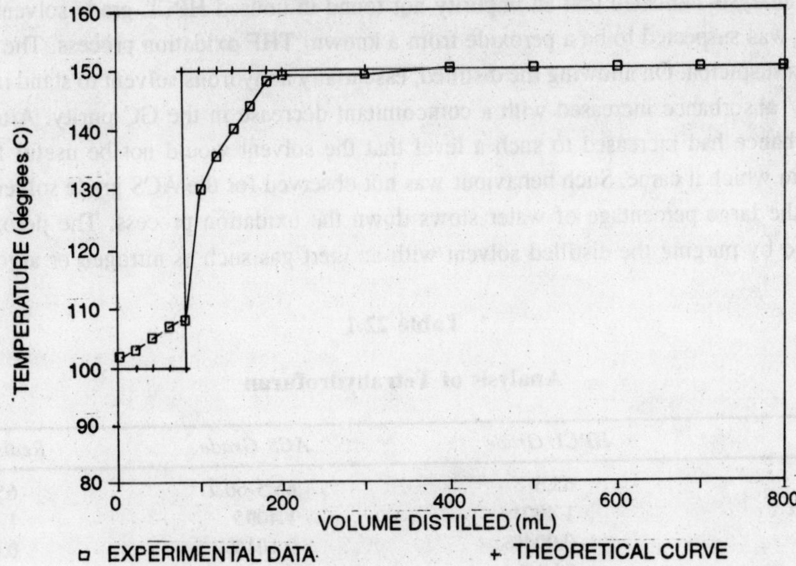

Figure 22.9. Distillation of 90 : 10 DMF water.

distilled at a rate of 1 liter/hour. Table 22.1 compares the recovered solvent with two grades of unused commercial THF (unstabilized HPLC grade and an ACS grade). In addition, the ultraviolet spectrum is seen in Figure 22.11. Results from the analysis demonstrate the high purity of recovered solvent,

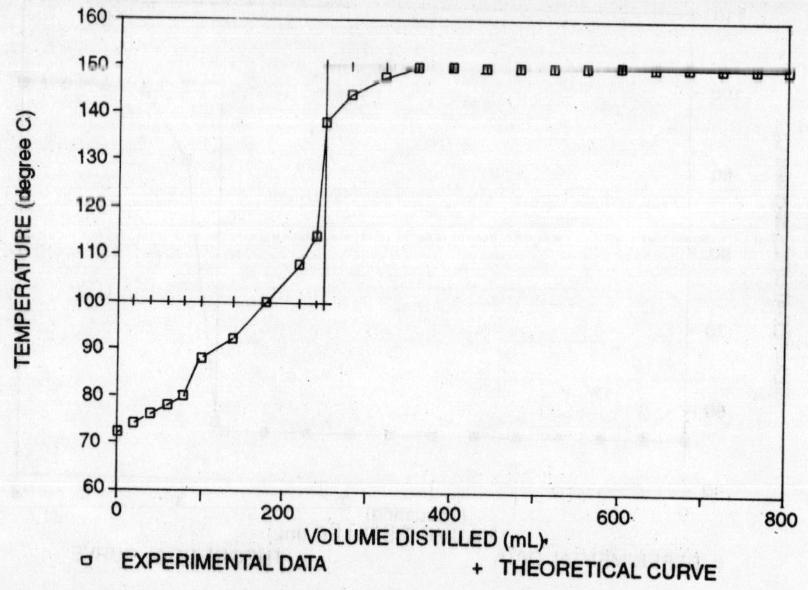

Figure 22.10. Distillation of 73 : 27 DMF:water.

although GC analysis indicated that an impurity not found in unused HPCL grade solvent was present. This impurity was suspected to be a peroxide from a known THF oxidation process. The UV spectrum confirmed this suspicion. On allowing the distilled, essentially anhydrous solvent to stand in the presence of air, the UV absorbance increased with a comcomitant decrease in the GC purity. After three days, the UV absorbance had increased to such a level that the solvent would not be useful for the HPLC procedure from which it came. Such behaviour was not observed for the ACS grade solvent in this time frame, since the large percentage of water slows down the oxidation process. The peroxide impurity can be avoided by purging the distilled solvent with an inert gas such as nitrogen or argon.

Table 22.1

Analysis of Tetrahydrofuran

		HPCL Grade	*ACS Grade*	*Redistilled*
Boiling Range		65.6	65.5-66.7	65.5
Refractive Index		1.4001	1.4095	1.4001
Water		0.004%	0.101%	0.005%
GC Purity		99.9%	98.2%	99.8%
Fluorescence				
(EX 310 nm):	320 nm	1%	5%	1%
	360 nm	1%	26%	0.5%
	400 nm	1.5%	54%	1.5%
	440 nm	1%	15%	0.5%

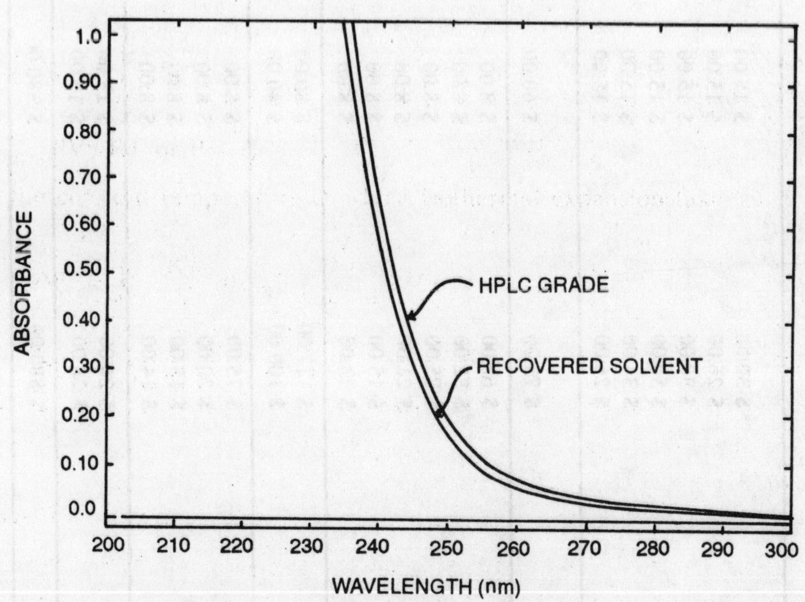

Figure 22.11. UV spectrum of tetrahydrofuran.

Organic Solvents Generated in a Teaching Laboratory

West Chester University is a moderately sized state university (10,000 students) located in southeastern Pennsylvania. The Chemistry department offers several B.S. degrees (160 majors) and has a moderate M.S. program. Each year, approximately 600 students enroll in the Freshman Chemistry laboratory in which 24 experiments (12 per semester) are conducted. The text used in this course is *Chemical Principles in the Laboratory*. Among these experiments, six involve the heavy use of organic solvents. Both the cost of these solvents and the financial burden of their disposal prompted us to initiate a program of solvent recovery in these labs. This program will be implemented in all of our teaching labs in the near future, considerably extending the cost savings to the institution.

Table 22.2 lists the materials that we have recovered in this program, the amount and initial cost of the solvents, and the estimated cost of disposal. In implementing such a program, it is important that the waste materials be managed carefully; therefore, all wastes were analyzed by GC both before and after distillation. The resultant recovered solvent, in all cases, was of sufficient purity to be reused. The total estimated savings per academic year in this one teaching laboratory was $1,260.00, an amount that surely justifies the incorporation of solvent recovery into the management of such a program. We have now added this duty to the undergraduate and graduate teaching assistants, who accomplish the task efficiently with a minimum of extra time.

Table 22.2

Organic Liquids Recovery from CRL 103 and 104 (General Chem. I and II) During Academic Year 1985-86

	Recovered Material	*Quantity*	*Solvent Cost*	*Disposal Cost*
Experiment 1—Density of Liquids	95% Ethanol	2 Liters	$ 30.00	$ 15.00
	1,1,1,-Trichloroethane	2 Liters	$ 26.00	$ 15.00
	2-Propanol	2 Liters	$ 30.00	$ 15.00
	n-Heptane	2 Liters	$ 50.00	$ 15.00
	Cyclohexane	2 Liters	$ 34.00	$ 15.00
	Toluene	2 Liters	$ 26.00	$ 15.00
Experiment 3—Paper Chromatography	Eluting solvent (1-Butanol as base)	8 Liters	$ 73.00	$ 60.00
Experiment 8— Molar Mass of a Liquid	1-Bromopropane	1 Liter	$ 60.00	$ 8.00
	1-Chlorobutane	1 Liter	$ 32.00	$ 8.00
	n-Hexane	1 Liter	$ 75.00	$ 8.00
	2-Butanone	1 Liter	$ 21.00	$ 8.00
	2-Propanol	1 Liter	$ 15.00	$ 8.00
	1,1,1,-Trichloroethane	1 Liter	$ 13.00	$ 8.00.
Experiment 15—Freezing Point Depression	Paradichlorobenzene	10 Liters	$ 117.00	$ 80.00
	Hexanes	5 Liters	$ 100.00	$ 40.00
Experiment 33—Aspirin	n-Hexane	1 Liter	$ 75.00	$ 8.00
	Ethyl Acetate	1 Liter	$ 20.00	$ 8.00
	Ethanol	1 Liter	$ 15.00	$ 8.00
	Acetone	1 Liter	$ 14.00	$ 8.00
Experiment 34 Resins	Toluene	2 Liters	$ 26.00	$ 15.00
	Acetone	2 Liters	$ 28.00	$ 15.00
		TOTAL	$ 880.00	$ 380.00

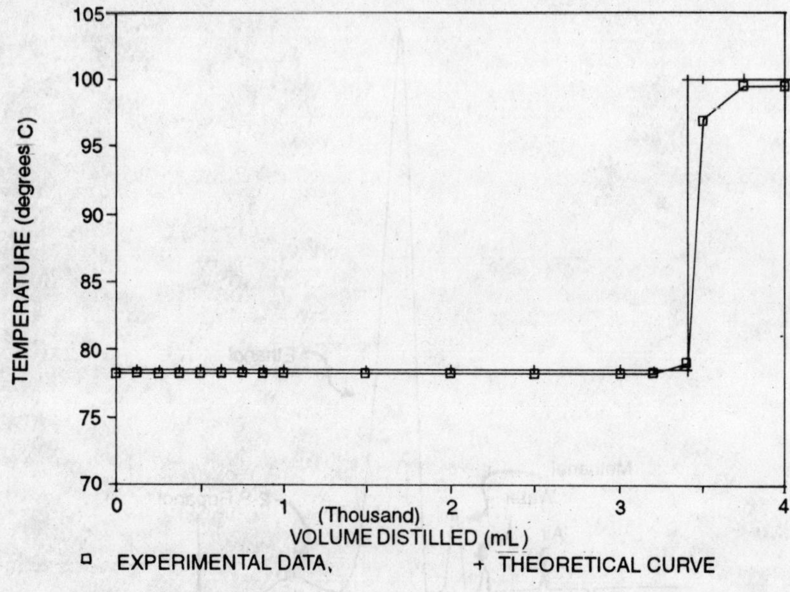

Figure 22.12. Batch distillation of 85 : 15 EtOH:water.

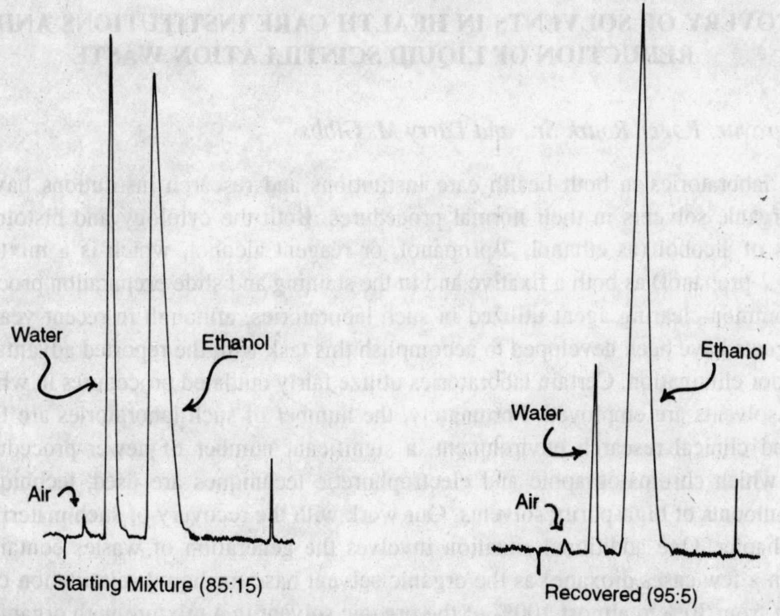

Figure 22.13. Gas chromatogram of ethanol/water mixtures.

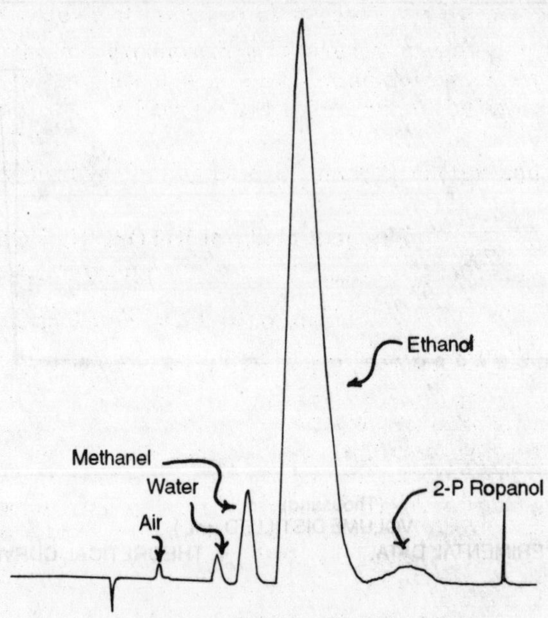

Figure 22.14. Gas chromatogram of reagent alcohol.

THE RECOVERY OF SOLVENTS IN HEALTH CARE INSTITUTIONS AND VOLUME REDUCTION OF LIQUID SCINTILLATION WASTE

John A. Mangravite, Roger Roark Sr., and Larry M. Gibbs

Clinical laboratories in both health care institutions and research institutions have a very large demand for organic solvents in their normal procedures. Both the cytology and histology labs utilize large amounts of alcohol (as ethanol, 2-propanol, or reagent alcohol, which is a mixture of ethanol, methanol, and 2-propanol) as both a fixative and in the staining and slide preparation procedures. Xylene is the most common clearing agent utilized in such laboratories, although in recent years, a variety of substitute reagents have been developed to accomplish this task with the reported advantages of reduced toxicity and odor elimination. Certain laboratories utilize fairly outdated procedures in which chloroform or other such solvents are employed. Fortunately, the number of such laboratories are few. Within the health care and clinical research environment, a significant number of newer procedures have been developed in which chromatographic and electrophoretic techniques are used, techniques employing considerable amounts of high-purity solvents. Our work with the recovery of such materials is discussed later in this chapter. One additional situation involves the generation of wastes containing xylene or toluene (and in a few cases dioxane) as the organic solvent base for liquid scintillation counting. These wastes contain from 30% to almost 100% of the organic solvent in a mixture with organic fluors, water, detergents, and nonvolatile organic compounds which have been tagged with a low-level radioactive atom (usually tritium or C-14). In this section, we discuss our work in this area.

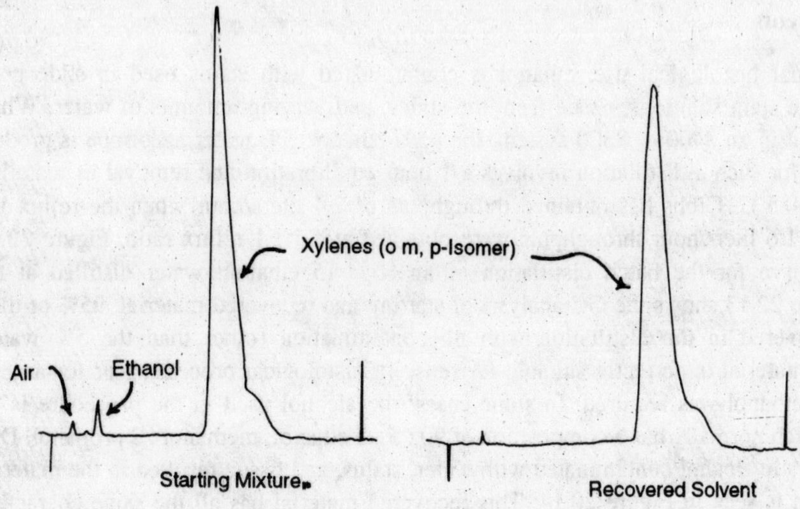

Figure 22.15. Gas chromatogram of xylene.

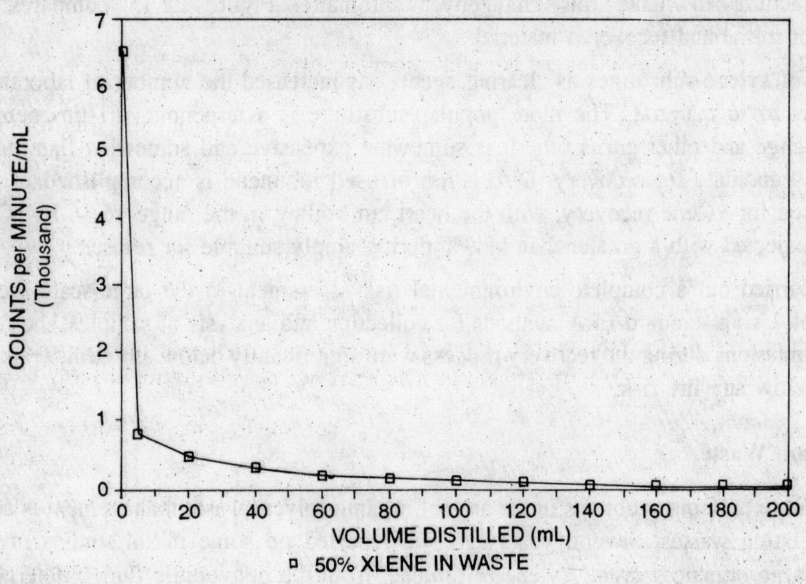

Figure 22.16. Distillation of xylene-based LS waste.

Histological Solvents

Under normal histological use, ethanol is contaminated with stains used in slide preparation, methanol from the stain solutions, tissue from the slides, and varying amounts of water. When such a mixture is distilled in an 8400 or 8300 system, the 95% ethanol: 5% water azeotrope is produced. The normal protocol for such a distillation involves a 1 hour equilibration and removal of material boiling in the range 77-80.5°C. Gibbs has obtained throughputs of 2.4 liters/hour when the reflux valve was kept open, while 1.6 liters/hour throughputs were obtained at a 1 : 1 reflux ratio. Figure 22.12 shows the distillation curve for the batch distillation of an 85 : 15 ethanol: water distilled at the lower throughput. Figure 22.13 shows the GC analysis of starting and recovered material. 95% of the original ethanol was recovered in the distillation with no contamination (other than the 5% water of the azeotrope). This material is perfectly suitable for reuse in histological processing or for any other use in which a 95% ethanol was required. In some cases, the alcohol used in the procedure is a reagent grade alcohol, which normally has a composition of 90 : 5 : 5 ethanol: methanol : 2 propanol. Distillation of used reagent grade alcohol contaminated with water, stains, and tissue resulted in the material whose gas chromatogram is seen in Figure 20.14. This recovered material has all the same characteristics of the 95 : 5 ethanol; water azeotrope and is easily recycled in the histological procedures.

The recovery of xylene can be accomplished at higher throughputs of up to 5 liters/hour. Two fractions are collected. Fraction # 1 is removed in the boiling range of 60°to 138°C and contains various impurities such as alcohols and water carryover into the xylene waste. At 138°C, fraction #2, which is pure xylene (all three positional isomers, orthometa-, and para-, are present), is collected. An alarm sounds to inform the operator to change receivers to collect this pure fraction. An option is available from the manufacturer to make this changeover automatic. Figure 22.15 compares the gas chromatograms for initial and recovered material.

The advent of xylene substitutes as clearing agents has increased the number of laboratories that no longer use this toxic material. The more popular substitute is a terpenoid, d,1-limonene, that is extracted from orange and other citrus oils. It is somewhat expensive and somewhat flammable, thus making it a good candidate for recovery. Distillation of used ilmonene is accomplished in the same manner as that used for xylene recovery, with the heart cut boiling in the range 161—164°C. A 90% recovery can be expected with a greater than 99.9% purity, amply suitable for re-use.

Gibbs has carried out a complete environmental risk assessment in the large-scale recovery of xylene and alcohol. Using standard EPA methods for collection and analysis of samples, he found that levels of solvent emissions during the recovery process were significantly below any health risk standard and magnitudes below any fire risk.

Liquid Scintillation Waste

One of the more pressing problems in the area of organic solvent waste management is associated with liquid scintillation wastes. Several years ago, we reported on some initial studies involved in separating the volatile organic solvent, xylene or toluene, from the nonvolatile fluors, detergents, and radioactive labeled material, thereby reducing the volume of waste requiring disposal. We observed that distillation of xylene- and toluene-based liquid scintillation cocktails that had been spiked with tritiated compounds (0.1 μCi/ml methyl-(^{3}H)-thymidine or 8-(^{3}H)- caffeine) successfully removes all of the

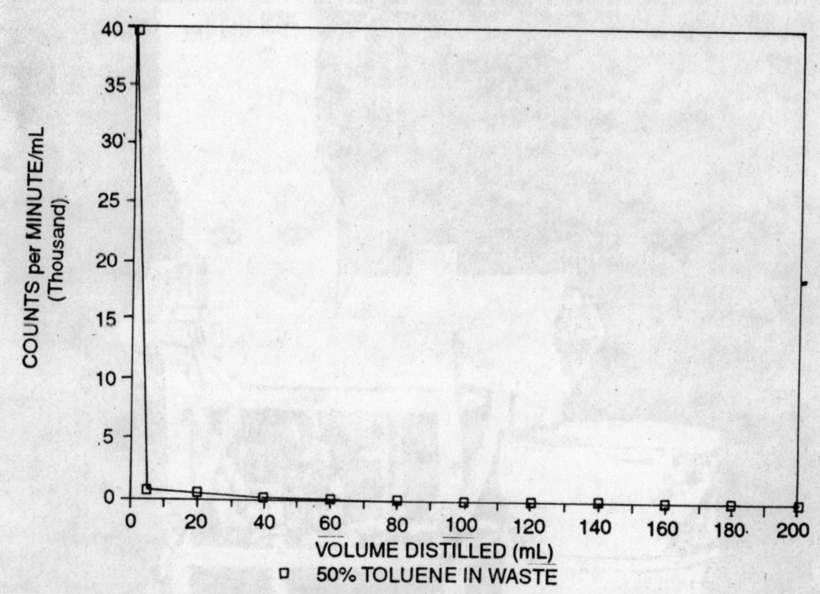

Figure 22.17. Distillation of toluene-based LS waste.

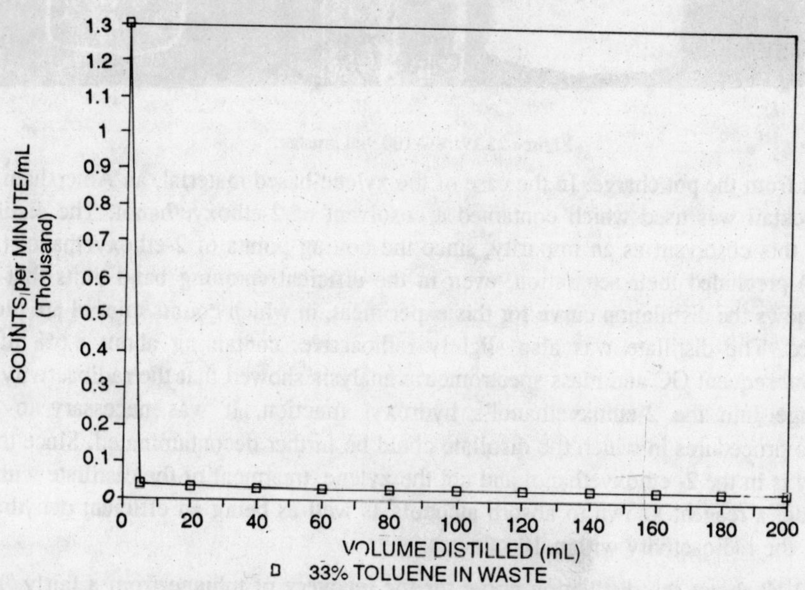

Figure 22.18. Distillation of toluene-based LS waste.

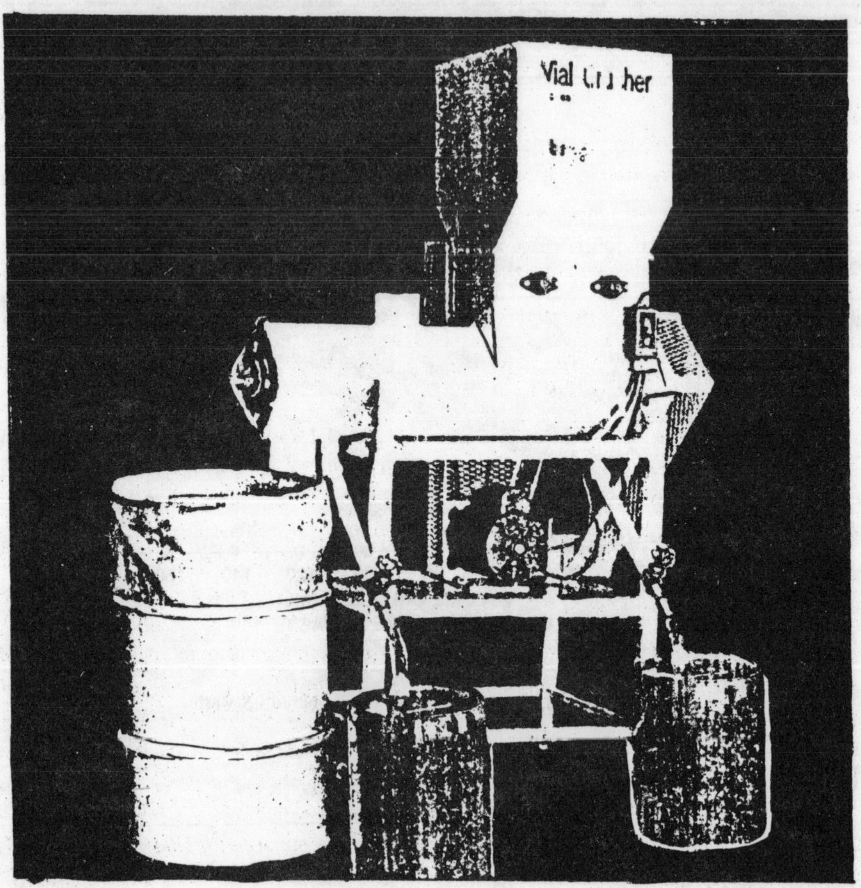

Figure 22.19. VC-100 vial crusher.

organic solvent from the pot charge. In the case of the xylene-based material, an Amersham PCS liquid scintillation cocktail was used which contained a cosolvent of 2-ethoxyethanol. The distillate in this case contained this cosolvent as an impurity, since the boiling points of 2-ethoxyethanol (135°C) and xylene (138°C) precluded their separation, even in the efficient spinning band units that were used. Figure 20.16 shows the distillation curve for this experiment, in which counts/min/ml are plotted versus volume distilled. The distillate was also slightly radioactive, containing about 8.6% of the initial radioactivity. Subsequent GC and mass spectrometric analysis showed that the radioactivity was due to tritium exchange into the 2-ethoxyethanol's hydroxyl function. It was necessary to extend our methodology to procedures in which the distillate could be further decontaminated. Since the source of the radioactivty is in the 2- ethoxyethanol and not the xylene, treatment of the distillate with anhydrous calcium chloride, a reagent known to absorb alcohols as well as being an efficient dehydrating agent, removed all of the radioactivity within 10 minutes.

Figure 22.17 shows the distillation curve for the recovery of toluene from a fairly "hot" sample containing 1.0 μCi/ml of tritiated thymidine. Again, all of the organic solvent is successfully distilled from the pot charge, but is also slightly contaminated (0.9% of the original radioactivity). In this case, exchange has taken place with water which co-distills with toluene as the known [6c] azeotrope. Once again, treatment with anhydrous calcium chloride removed this radioactive material.

In both these cases, the distillate was radioactive due to the exchange of the isotope from an easily exchangeable position of the labeled material to a molecule (2-ethoxyethanol or water) that was capable of accepting the exchange. In neither case was there any radioactivity in the recovered organic solvent—a solvent that proved to be a reagent grade material that could be re-used in any appropriate procedure. If the radioactive label is positioned in a molecule that does not undergo easy exchange, the distillate does not contain any radioactivity. For example, Figure 22.18 shows the distillation curve for actual low level radioactive waste obtained from the University of Alabama Medical Center. This material was a toluene-based waste (approximately 33% toluene) that contained tritiated steriods in addition to the fluors, detergents and water. The label was situated in a non-exchangeable position in the steroid, so that no radioactivity was observed in the toluene : water azeotrope that distilled.

Gibbs has followed up on our work by examining the potential environmental risk associated with the recovery process. He has conducted environmental air sampling which demonstrated that no radioactivity is emitted during the distillation procedure. In addition, he has developed a simple and risk-free procedure for decontaminating the very small amounts of radioactivity that remain in the distillation unit after use.

Recently, an ancillary piece of equipment that greatly increases the appeal for volume reduction of liquid scintillation wastes through organic solvent recovery has been developed. Since the majority of LS wastes are found in 10 ml glass or plastic vials, it is necessary to remove the liquid from the vials before distillation can be accomplished. One means that could be used is the manual removal of the vial contents to the pot flask. This, of course, is totally unacceptable when thousands of vials are involved. Figure 22.19 shows a picture of the commercially available vial crusher (the VC-100) manufactured by B/R Instrument Corp. that effectively automates this task. The vial crusher is simply operated by loading scintillation vials in the hopper through the top. A continuous feed mechanism crushes thousands of glass and plastic vials per hour. The crushed vials are collected in a 120 liter drum lined with a special polypropylene woven bag. This bag will hold the crushed vials and allow any residual liquid to drain off. The bag can then be easily removed and prepared for solid waste disposal. The radioactive liquid waste is collected in two separate 20 liter drums and can be pumped or poured into the still for volume reduction and organic solvent recovery. The VC-100 is a safe piece of equipment with all operating parts fully enclosed. The system can be easily vented to a hood through an attached 4 inch vent hose assembled to a side port. By incorporating these techniques into clinical procedures and research experiments involving low-level radioactive materials, we propose a proven procedure for solving a major environmental problem.

THE RECOVERY AND RE-USE OF HIGH-PURITY SOLVENTS

by John A. Managravite and Sr. R. M. Katusz

Our initial efforts in the investigation of in-house laboratory recycling of used solvents was in the recovery of high-purity solvents. We were prompted to study this area because of the rising costs of high-purity solvents used in chromatographic and other analytical procedures. Our original work involved the recovery of acetonitrile from an HPCL mobile phase containing this organic solvent mixed with an aqueous buffer. We have extended this work to the recovery from mobile phases containing

Table 22. 3
Analytical Data for Acetonitrile

Analysis	*HPLC Grade*	*Recovered Pure Solvent*	*Recovered Azeotrope*
Boiling Range	81-82	82	77-78
Refractive Index	1.3466	1.3465	—
Water (Karl Fischer)	0.07%	0.08%	15.6%
GC Purity	99.9%	99.9%	99.7%
UV Cutoff (nm)	190	190	190

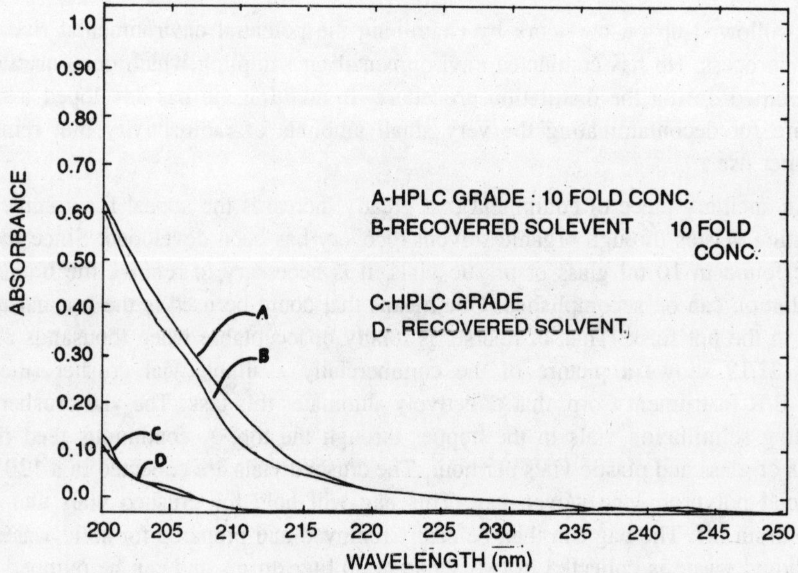

Figure 22.20. UV spectrum of acetonitrile.

acetonitrile, methanol, and buffer and to other types of mobile phases, both polar and nonpolar. We report on these results in this section. A report on the recovery of deuterated NMR solvents by micro spinning band distillation is presented elsewhere. The details of quality assessment are found in the "Analytical Procedures" section at the end of this chapter.

Acetonitrile

From aqueous acetonitrile mixtures. The waste solvents that were used in our studies were either an acetonitrile-4% aqueous ammonia solution or an acetonitrile-water-acetate buffer used in theophylin or tricyclic antidepressant analyses, or the analysis of opium derivatives. These solvent mixtures were equilibrated for 2 hours and distilled at a throughput of 250 ml/hour. The lower throughput than our other recoveries was found to be necessary in order to produce a high-purity distillate. The first 10% of distillate was discarded in all distillations and a pure acetonitrile, b.p. 81-82°C, was recovered when

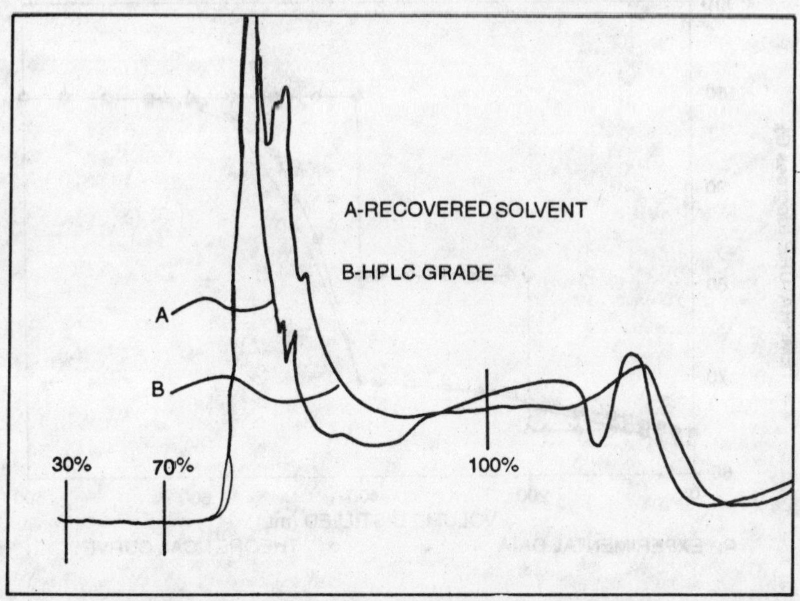

Figure 22.21. Dynamic HPLC chromatogram—acetonitrile.

the starting pot charge contained the lower percentage of water. When the recovered solvent contained an inital high concentration of water, the expected azeotrope, b.p. 77-78°C, containing 15.7 mole % water, was obtained. Solvent purity, in both cases, was determined from boiling range, refractive index, Karl Fisher titration for water, UV absorbance, and fluorescence spectral data, as well as chromatographic data. The solvent specifications are seen in Table 22.3. It is obvious from the data in this table that the recovered solvent has the same gross specifications as a high-purity acetonitrile. In addition to these results, we conducted two experiments which provided a better assessment of solvent purity. In the first experiment, samples of recovered acetonitrile were preconcentrated by reducing 100 ml portions to 10 ml by flash evaporation at room temperature. The ten-fold preconcentrates were then spectrophotometrically compared to samples of unused HPLC grade acetonitrile prepared in the same manner. A comparison of the UV spectra is seen in Figure 22.20 While the static spectra show nearly

Table 22.4

Analytical Data for Methanol

Analysis	*HPLC Grade*	*Recovered Pure Solvent*
Boiling Range	64-65	64-65
Refractive Index	1.3311	1.3310
Water (Karl Fischer)	0.09%	0.17%
GC Purity	99.8%	99.8%
UV Cutoff(nm)	205	205

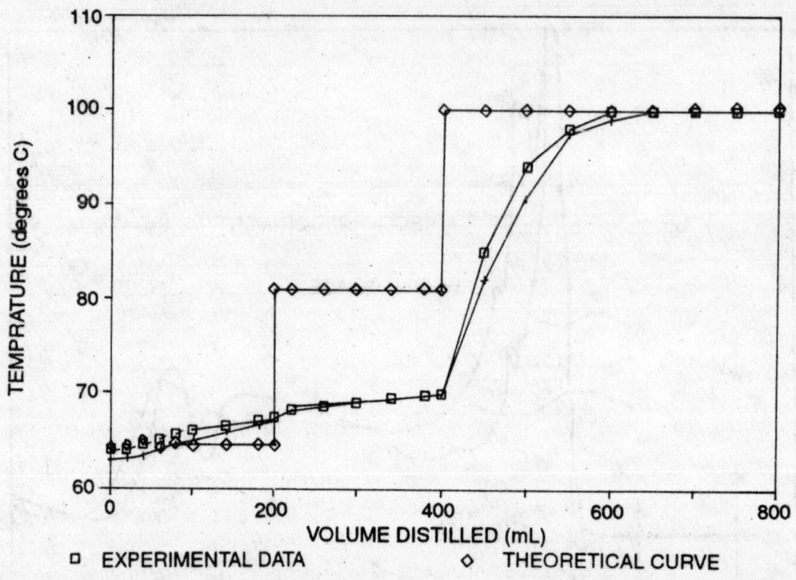

Figure 22.22. Distillation of 25 : 25 : 50 ACN:MeOH:water.

identical absorbance, the preconcentrated samples show that the unused HPLC grade has higher absorbance in the 210-225 nm region. A second type of experiment that also demonstrated the high purity of our recovered solvent (either recovered or unused) were compared after 60 min. of on-column HPLC enrichment using acetonitrile-water, 30-70, mobile phase passed over a C-18 reversed phase column at 3 ml/min. Impurities accumulating on the column were stripped off by using a step gradient of acetonitrile-water, 70 : 30 for 15 min., followed by a step gradient of 100% acetonitrile. The results

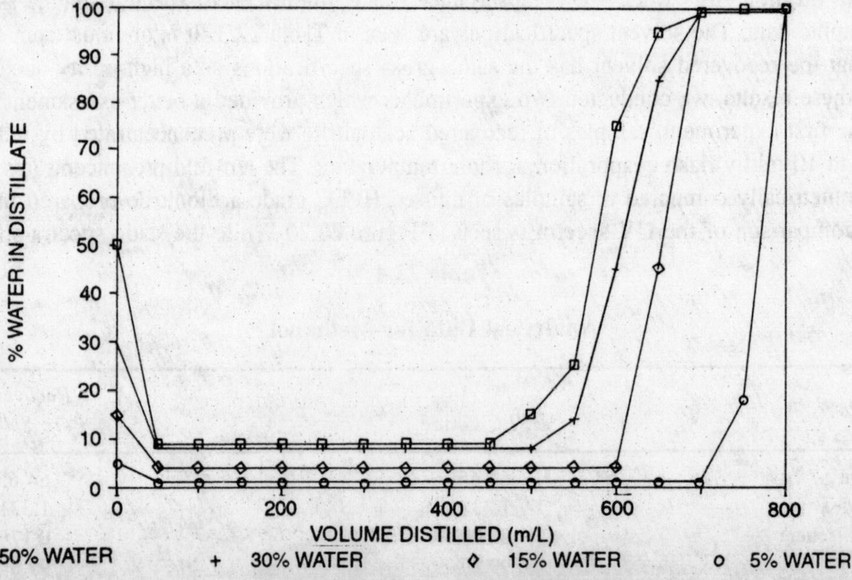

Figure 22.23. Distillation of methanol-water mixtures.

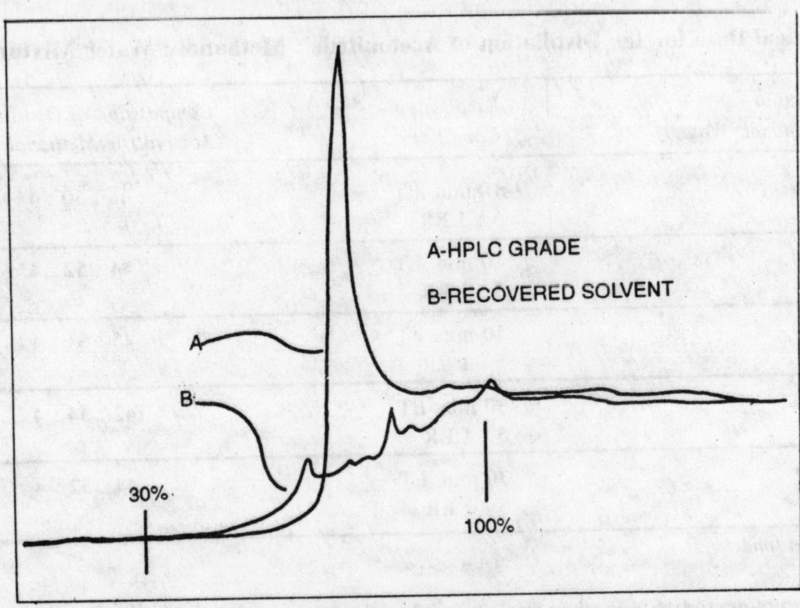

Figure 22.24. Dynamic HPLC chromatogram—methanol.

are shown in Figure 22.21. The lower level of impurities of the recovered solvent compared to unused HPLC grade is amply demonstrated when this chromatogram is followed at 210 nm.

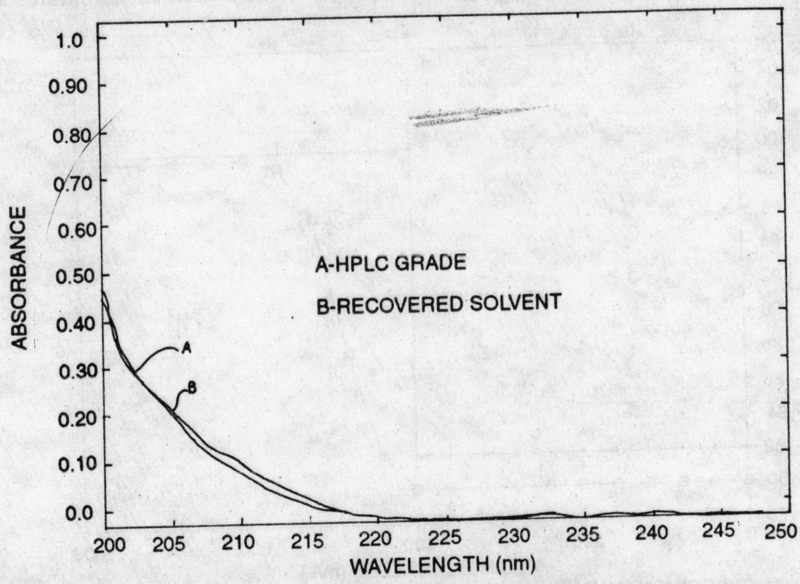

Figure 22.25. Preconcentrated UV spectrum—methanol.

Table 22.5

Analytical Data for the Distillation of Acetonitrile : Methanol : Water Mixtures

Starting Composition (Acetonitrile: Methanol: Water)	*Distillation Conditions*	*Composition of Distillate (Acetonitrile:Methanol:Water)*
30:30:40	10 min .ET 5 : 1 RR	46 : 50 : 4
40:40:20	10 min. ET 5 : 1 RR	44 : 52 : 4
25:25:50	10 min. ET 5 : 1 RR	45 : 51 : 4
25:25:50	30 min. ET 5 : 1 RR	42 : 54 : 4
25:25:50	30 min. ET 5 : 1 RR	44 : 52 : 4

ET = Equilibration time
RR = Reflux ratio

From aqueous-acetonitrile-methanol mixtures. In recent years, anumber of routine HPLC analyses have been developed which involve the use of a mobile phase that contains acetonitrile and aqueous buffer along with various amounts of methanol. Recovery of acetonitrile from such mixtures poses a difficult separation problem, in that the distillation profile consists of several constant boiling mixtures. Thus, acetonitrile cannot be scccessfully separated from methanol nor from water. It turns out, however, that regardless of the composition of the starting mobile phase, if one distills such mixtures and collects

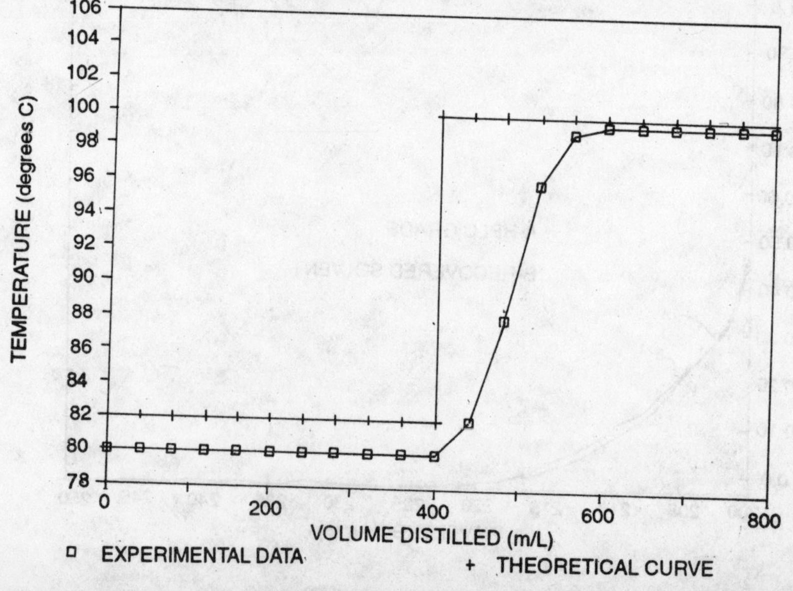

Figure 22.26. Distillation of 2-propanol : water.

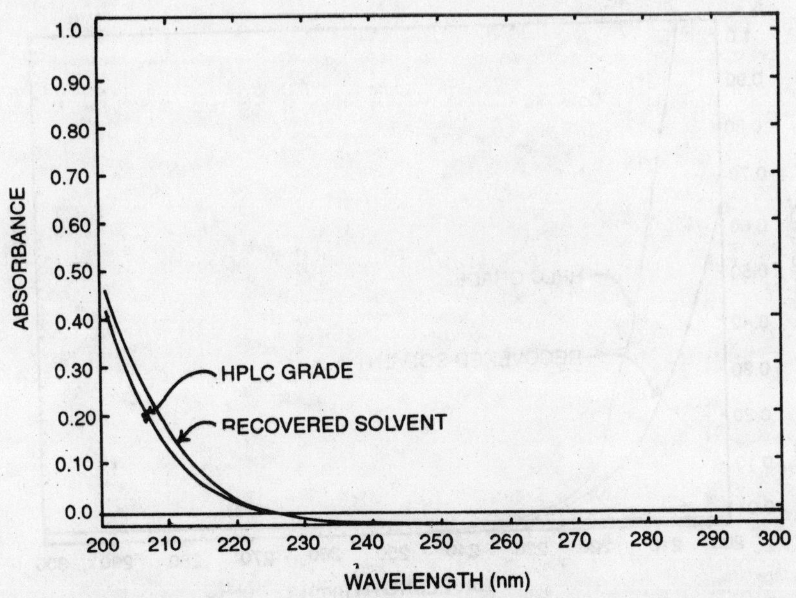

Figure 22.27. UV spectrum of hexane.

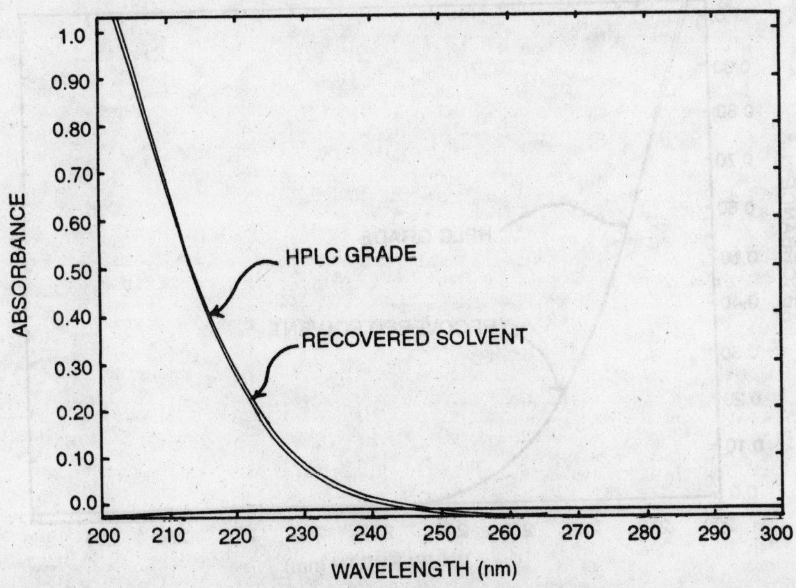

Figure 22.28. UV spectrum of heptane.

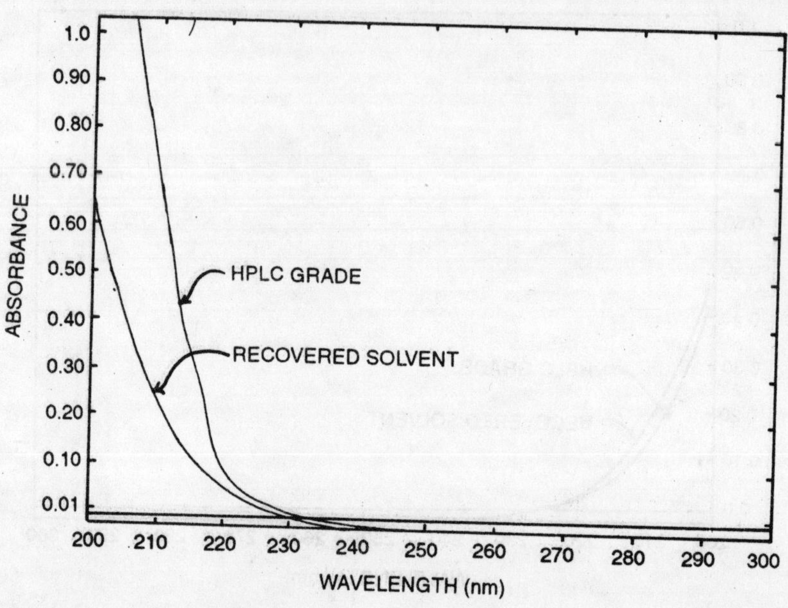

Figure 22.29. UV spectrum of iso-octane.

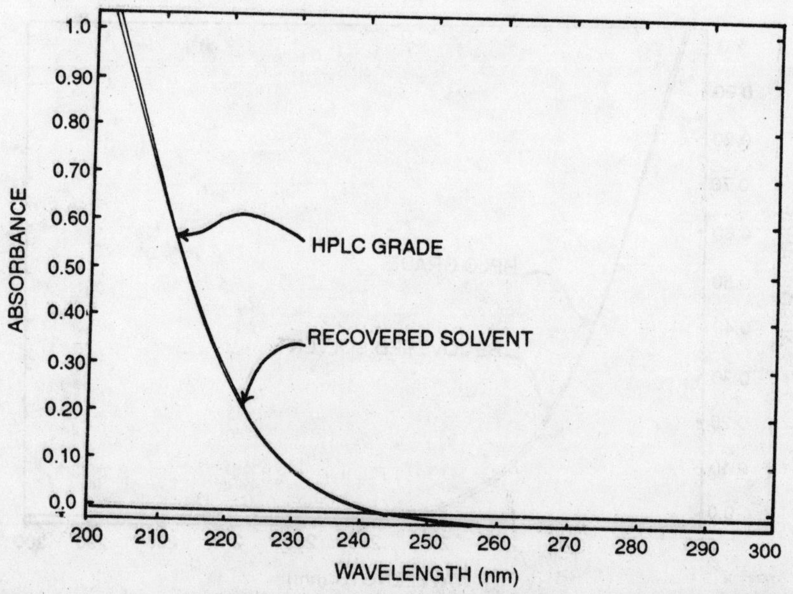

Figure 32.30 UV spectrum of cyclohexane.

Table 22.6

Analytical Data for Hexane

Analysis	*HPLC Grade*	*Recovered Pure Solvent*
Boiling Range	68-69	68-69
Refractive Index	1.3755	1.3751
Water (Karl Fischer)	0.01%	0.01%
GC Purity	99.9%	99.9%
UV Cutoff (nm)	195	195

Table 22.7

Analytical Data for Heptane

Analysis	*HPLC Grade*	*Recovered Pure Solvent*
Boiling Range	98-99	98
Refractive Index	1.3878	1.3972
Water (Karl Fischer)	0.01%	0.01%
GC Purity	99.9%	99.9%
UV Cutoff (nm)	200	199

Table 22.8

Analytical Data for Iso-Octane

Analysis	*HPLC Grade*	*Recovered Pure Solvent*
Boiling Range	99-100	99
Refractive Index	1.3915	1.3918
Water (Karl Fischer)	0.03%	0.04%
GC Purity	99.7%	99.8%
UV Cutoff (nm)	215	215

Table 22.8

Analytical Data for Cyclohexane

Analysis	*HPLC Grade*	*Recovered Pure Solvent*
Boiling Range	80-81	81
Refractive Index	1.4258	1.4258
Water (Karl Fischer)	0.02%	0.02%
GC Purity	99.9%	99.8%
UV Cutoff (nm)	200	202

material boiling up to 70°C, the distillate composition is constant. Table 22.4 shows the results of a number of distillations conducted with different percentages of starting components. In all cases, the material recovered below 70°C is of an approximate composition 42% acetonitrile: 54% methanol: 4% water. The distillation curves for several runs are shown in Figure 22.22. The purity of these mixtures was greater than 99.9% and is of sufficient purity to be re-used providing that the experimenter adds the necessary buffers and makeup organic solvents. While this recovery of acetonitrile is certainly more involved than recoveries from mobile phases that do not contain methanol, it is certainly justified by the high cost of such solvents.

Methanol

Aqueous methanol mobile phases are commonly used in reverse phase HPLC analyses including analysis of aromatics, quantitation of pigments, and analysis of phenols and weak acid drugs. Used samples were distilled at throughputs of 300 ml/hour and the recovered solvent subjected to a full screening. Figure 22.23 shows the distillation curves for the recovery from several different starting mixtures. In this display, the % water in the distillate is plotted versus volume distilled. It is obvious that under the conditions of the distillations, it is difficult to obtain an anhydrous methanol. Even when longer equilibration times and slower takeoffs are employed, water is present in the distillate. Such a fact has been noted elsewhere.

Table 22.5 compares the gross analytical properties of recovered solvent and an HPLC grade. More importantly, the dynamic HPLC run shown in Figure 22.24 and the preconcentrated UV spectra in Figure 22.25 emphatically demonstrate the high degree of purity of the recovered solvent. While the cost of high-purity methanol is much less than acetonitrile, the fact that it can be easily recovered mitigates against its disposal.

Isopropanol

Isopropanol is employed to a much lesser extent than methanol and acetonitrile; however, it is still the solvent of choice in certain analyses. We have recovered isopropanol from used mobile phases utilizing the same distillation conditions as those employed with methanol and acetonitrile. As in the case of the latter solvent, distillation of aqueous isopropanol solutions lead to the recovery of an azeotrope (see Figure 22.26) of 13% water content. The purity of the recovered solvent is identical to that of the initial material.

Nonpolar Solvents: Hexane, Heptane, Iso-octane, and Cyclohexane

When normal phase HPLC analyses are employed, nonpolar solvents must be used as the mobile phase. Hexane is used in the analysis of hydroxylated aromatics and azo dyes on normal phase porous silica, while cyclohexane is the solvent of choice when alkylamino metabolites are determined by ion pair chromatograhy. Similarly, heptane and iso-octane are routinely used in HPLC analyses and the latter is important in pesticide analysis by gas chromatographic techniques. There are certain mobile phase mixtures (for example, hexane, methylene chloride, and THF combinations) that do not lend themselves to solvent recovery since their vapor compositions are too complex. However, in most cases we have been able to recover a high grade of solvent for re-use. Figures 22.27–22.30 show the preconcentrated UV special comparisons for all four of these solvents. Analytical data are provided in

Tables 22.6–22.9. We are currently working on solving some problems associated with the dynamic HPLC analyses on normal phase silica columns and will report on such at a later time.

We believe that our work demonstrates the feasibility of recovering high-purity solvents for reuse in HPLC analyses as well as other procedures in which solvent purity is critical. The static and dynamic evaluations of acetonitrile and methanol amply demonstrate the high quality of the recovered materials. In the case of acetonitrile, the distilled solvent is of a greater purity than that reported by Kiesele. Our results with the recovery of nonpolar solvent by direct spinning band distillation are extremely promising; however, initial experiments in the recovery of high-purity THF, DMF, DMSO, and dichloromethane, as well as freons, indicate that pre- and/or post-chemical treatments may be necessary. In some cases, recovery and re-use will not be successful. When the initial mixtures azeotrope, the distillate may or may not be useful. In the cases of acetonitrile and isopropanol, the azeotropes can be re-used by the addition of the appropriate amount of makeup solvent. Distillation of hexane (b.p. 68°C)/ethyl acetate (b.p.77°C) mixtures give an azeotrope (b.p.65°C). The boing point of this azeotrope is so close to that of pure hexane that separation is not possible on a 20 or 30 plate distillation unit.The result is that a material of uneven composition is obtained from experiment to experiment. Another situation in which recovery by distillation is not feasible is when thermal reactions occur. Recovery of methanol from a methanol/water/acetic acid mixture was not successful in that esterification occurred under the distillation conditions leading to the formation of ethyl acetate. This could be prevented by neutralization before distillation; however, the added time and effort was deemed no longer economically feasible. In any event, solvent recovery can be an important factor when high-purity solvents are employed, and we trust that greater involvement in this area will be forthcoming.

Conclusion

The analytical data that we have presented in this chapter amply demonstrate the feasibility of in-house solvent recovery in a variety of areas. Recovery of ordinary solvents is simple and should be incorporated into routine procedures in all cases. Recycling of histology solvents should also be commonly carried out and, in fact, is now being accomplished in humdreds of hospitals and research laboratories. With the development of an inexpensive vial crusher, a solution to the problem of liquid scintillation waste reduction has now been solved. Finally, our work on the recovery of high-purity solvents shows the important economical impact and environmental promise in this high-volume area.

ANALYTICAL PROCEDURES

Materials

All solvents were either Baker Reagent grade solvents or B and J HPLC grade. Radioisotopes for the liquid scintillation studies were purchased from ICI Radiochemicals, Inc., while the cocktails were obtained from Amersham Co.

Equipment

The following equipment was used in all of the work described in this chapter.

Gas Chromatographic Analysis

GC analysis was carried out using one of the following instruments:

- A Perkin-Elmer 3920B all glass dual Flame Ionization (FID) detection system with a Carbowax 20M or Chromosorb 101 packed column.
- A Perkin-Elmer Sigma 3B capillary chromatograph with FID detection system and a 50 m DBI fused silica column.
- A Hewlett-Packard 5880 chromatograph fitted with a Thermal Conductivity (TC) detector and a packed Carbowax 20M column. Water analysis was carried out using the method of Hogan.

Ultraviolet Spectral Analysis

UV analysis was carried out using one of the following instruments:

- A Cary 219 recording UV-Visible spectrophotometer.
- A Hewlett-Packard 1485 diode array spectrophotometer.

In addition to obtaining direct UV spectra, a preconcentration method was also used. Preconcentration was achieved by reducing 100 ml samples of the solvent to 10 ml by flash evaporation. The concentrated sample was analyzed by obtaining its UV spectrum and compared to a commercial HPLC grade solvent that was similarly treated.

Fluorescence Spectral Analysis

Flurescence spectra were measured using a Perkin-Elmer MPF-2A fluorescence spectrophotometer with excitation wavelengths determined experimentally.

Liquid Chromatographic Analysis

HPLC analyses were performed on one of the following instruments:

- A Jasco Familic 100 micro HPLC system equipped with a UVIDEC 100-II variable wavelength detector and a reversed phase (C-18) column for polar solvents or a silica column for nonpolar solvents, was used for base line studies.
- A Perkin-Elmer Series 4 HPLC system fitted with a Whatman Partisil (5% C-18) or an Alltech (18% C-18) column was used for the dynamic HPLC analysis.

The dynamic HPLC analysis is a technique that involves enriching an HPLC column with an aqueous solution of the solvent being tested at a high flow rate for a considerable length of time. Following enrichment, the column is stripped of any accumulated compounds with a more concentrated solvent mixture. This procedure is an attempt to imitate, in a condensed time frame, the conditions for continual solvent use in an HPLC analysis. The presence of any impurities detected in a dynamic run is indicative of the purity level of the solvent. Such impurities would result in the formation of "ghost peaks" during normal solvent usage, thereby interfering with the analysis. There are two ways to conduct the test. One, used by some solvent manufacturers, is tc run a gradient of 4% per minute until the final composition is reached. This procedure, however, is not the more stringent test since a gradual washing is employed, marking the magnitude of the impurity concentrations. The second procedure of step gradients is more critical and was used in our work. In the case of acetonitrile, step gradients from 30% to 70% to 100% were employed. With methanol, the steps were 30% to 100%.

Refractive Index

Refractive Indices were measured using a Bausch and Lomb refractometer.

23

Application of Dry Flue Gas Scrubbing to Hazardous Waste Incineration

Introduction

Historically, the most common method of removing acid gases and particulate from hazardous waste incineration flue has been a wet scrubbing process. This typically involves an adiabatic quench for saturating the gases with water, followed by an inertial scrubber for particulate removal and a packed-bed scrubber for acid gas removal.

An important operation in this process is the treatment of the resultant scrubber effluent. Whereas this may not be a significant concern for a large chemical plant complex with an adequate facility wastewater treatment system, it can be a very substantial cost and source of problems for a commercial hazardous waste incinerator. Since a dry scrubbing process has only a solid residue discharge, it has become a viable alternative to wet scrubbing for many incineration applications..

Other advantages of dry scrubbing systems include:

1. Lime can be used as the neutralizing reagent. On a molar equivalent basis, the caustic used in wet systems is approximately 8.5 times more expensive than pebble lime, and thus reagent costs will be lower.

2. Exit temperature of gas leaving the dry system baghouse will be above the saturation temperature, and therefore there will not be visible stack plume except during extreme weather conditions.

3. A significant reduction of heavy metals contained in the flue gases is achieved by this system. The high degree of atomization of the lime slurry produces particles of essentially all sizes up to 100 microns. This enormous amount of surface acts to adsorb the bulk of the carried over in the gas stream.

   In addition recent performance testing on solid-waste incinerator flue gas has shown that significant amounts of products of incomplete combustion are also adsorbed by the lime slurry.

4. The use of a baghouse for particulate matter remoal provides a means of accommodating sustained particulate overloading (with shorter cleaning cycles) and a means of dampening particulate excursions that might occur due to a momentary high ash loading. Also, the cake that forms on the bags contains excess lime, which will continue to react with any residual

acid gas that passes through the spary dryer absorber and provide a means of dampening acid gas excursions that occasionally might occur.

History

In the mid-1970, developmental work was initiated by several firms to apply the dry scrubbing process to industrial and utility boilers for flue gas desulfurization (FGD). Several contracts were awarded in the late1970 and the first FGD dry scrubbing systems came online in 1981. Many FGD systems have since been successfully operating with the dry scrubbing process in this country and abroad.

The Niro Atomizer Company first applied the dry scrubbing process to hazardous waste incineration in Europe in 1982 at the Kommunekemi facility in Denmark. The process was then applied at the SAKAB facility in Sweden and at the Ruhimaki, Finland facility. Another hazardous waste incinerator at Biebesheim, Germany utilizes a conventional wet scrubber, but the effluent is injected into a spray dryer upstream of the wet scrubber in order to obtain a dry residue discharge. All of the above facilities are regional, non-commercial incinerators with capacities in the range of 100-130 mm Btu/hr.

The first application of dry scrubbing to hazardous waste incineration in North America was at the Tricil facility in Sarnia. Ontario, Canada. This large commercial liquid injection incinerator utilizes an alkaline waste stream as the reagent in the spray dryer absorber to remove at least 75% of the HCI and 50% of the SO_2 in the flue gas. The Tricil-designed spray dryer utilizes three dual-fluid nozzles to inject the reagent and cool the gases from 800°F to 400°F with an average outlet volume of 65,000 acfm at 400°F The gases then flow to a pulse-jet baghouse, ID fan, and stack.

The first contract for dry scrubbing on a hazardous waste incinerator in the U.S was awarded to Research-Cottrell for the U.S Army base at Tooele, Utah. It is a very small system (5,000 acfmat 400°F exit temperature) with dual-fluid nozzle atomization and came onstream June 1986. The first commercial hazardous waste incinerator to utilize dry scrubbing is Trade Waste Incineration (TWI) in Sauget, Illinois. This is a somewhat larger system (14,000 acfm at 450°F exit temperature) with a rotary atomizer provided by Joy/Niro, and came onstream in November 1986.

Performance Requirements

As with wet systems dry scrubbing systems on hazardous waste incinerators in this country must fulfill the process requirements of the Code of Federal Regulations in 40 CFR 264 (RCRA). More specifically, 40 CFR 264.343 requires a maximum stack emission of 4 lb/hr of HCI or 1% of the HCl in the flue gas prior to entering the dry scrubbing system, whichever is greater. In addition particulate matter emissions must be restricted to 0.08 gr/dscf, corrected to an oxygen level of 7% on a dry volume basis.

State particulate emission standards may be more stringent than the RCRA standard. Furthermore, state air quality standards for SO_2 and NO_x may be a factor in permitting. Since hazardous waste incinerators are typically 130 mm Btu/hr or smaller, and their waste is generally low in sulfur, the SO_2 standards are not normally a major constraint.

PROCESS CONSIDERATIONS

General

The basic elements of the dry scrubbing process for hazardous waste incineration are shown in Figure 23.1. We will assume for this discussion that the primary acid gas to be removed is HCl, and that HF, HBr, H_3PO_4, SO_2 and NO_x are minor constituents.

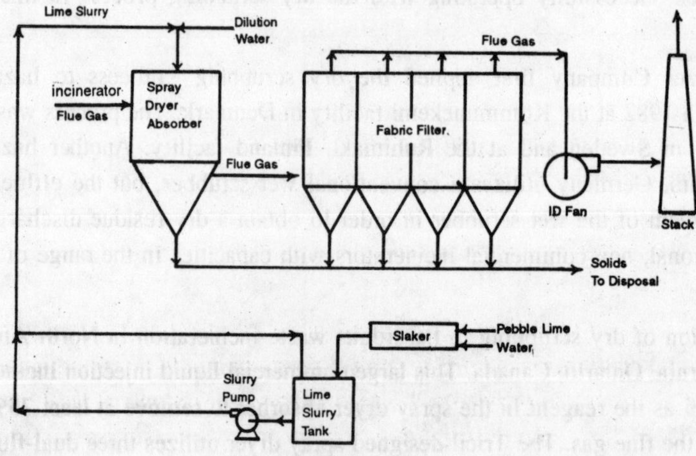

Figure 23.1. Dry scrubbing process.

Combustion Gas Flow

Prior to entering the spray dryer absorber (SDA), the flue gas from the combustion chamber can alternately be cooled by enthalpy extraction in a waste heat boiler or by an evaporative cooling chamber utilizing finely atomized water sprays, or it can be introduced to the SDA directly at elevated temperatures. In all utility, solid waste, European hazardous waste, and most industrial applications, a waste heat boiler is used to cool the gases to 400-500°F prior to entering the SDA. The hazardous waste incinerator for the U.S. Army at Tooele, Utah, uses an evaporative cooler with air-atomized water sprays to cool the gases from 2000 to 600°F prior to SDA. At the TWI incinerator, gases will be brought directly into the SDA at 1700-2000°F.

The use of an evaporative cooler upstream of the SDA results in essentially two "dryers" in series and is thus a somewhat complicated approach. Also, the addition of water into the gas stream in the cooler increases the SDA inlet humidity and thus decreases the evaporation driving force in the SDA. Therefore, it may seem more desirable to bring the gases directly into the SDA at elevated temperatures. In fact, this creates other problems due to the necessity of dispersing the gas into the SDA through a refractory-lined duct.

Since the gases from a waste heat boiler or evaporative cooler enter the SDA at 800°F or less, they can be introduced around the atomizer through a highly efficient, carbon steel gas dispersion mechanism. However, the gases entering the SDA at elevated temperatures must be introduced through a refractory-lined with little provision for gas dispersion. This results in a larger SDA chamber, greater lime usage, and the need for higher SDA outlet temperatures to ensure complete drying.

Another advantage of cooling the gases below 800°F prior to the SDA is the simultaneous cooling of any molten particulates below their fusion point so they can be easily removed. Molten particulate has historically been a source of operating problems for commercial and industrial incinerators.

It would appear that the use of a waste heat boiler upstream of the SDA would be the best approach if the resultant steam can be utilized at the facility. The boiler will result in the smallest dry scrubbing system of the three alternatives due to the lower volume of evaporated water and more efficient gas dispersion. However, the boiler should be designed with a flue gas recirculation system to allow for greater control of the SDA inlet temperature, to compensate for load changes; as well as a large section to cool molten particulate below its fusion point prior to the convective section.

Spray Dryer Absorber

The combustion gas from the incinerator enters the spray dryer absorber through a roofmounted gas disperser and is distributed symmetrically around one or more atomizers. The finely atomized absorbent (lime slurry) reacts essentially quantitatively with the acid gas, although the actual reagent feed will be at least 50% excess lime to assure high removal efficiencies. The principal reaction that occurs is:

$$Ca(OH)_2 + 2HCl \rightarrow CaCl_2 + 2H_2O$$

To a lesser extent, when SO_2 is present, the following reaction will occur:

$$Ca(OH)_2 + SO_2 \rightarrow CaSO_3 + H_2O$$

Some carbonation of lime also occurs due to the high carbon dioxide contet of the flue gas:

$$Ca(OH)_2 + CO_2 \rightarrow CaCO_3 + H_2O$$

Lime usage is generally defined in terms of SDA inlet molar equivalence ratios (MER) where:

$$MER = \frac{\text{mole–equivalents of active } Ca(OH)_2}{\text{mole–equivalents of' HCl and } SO_2}$$

The SDA chamber geometry is determined by the type of slurry atomization and gas-flow distribution employed (see Figure 23.2). In all cases, a nominal 10-12 seconds of gas residence time in the active SDA volume is allowed to ensure complete drying of the absorbent without wetting the chamber walls. Although the chamber geometries may differ, the gas flow is normally downflow and cocurrent and cocurrent with the reagent flow.

The SDA inlet temperature is normally maintained at 450-800°F, depending on the acid gas concentration. The maximum practical temperature of 800°F is determined by the working limits of carbon steel construction for the SDA. This in turn limits the amount of lime slurry which can be dried in the chamber, and therefore, the maximum acid gas concentration which can be sccommodated by the dry scrubbing process is limited to approximately 15,000 ppm of HCl by volume in the SDA inlet. This is difficult to translate into a maximum weight percent of chlorine in the waste feed because of the influence of combustion gas, excess air and humidity.

The SDA outlet temperature is normally maintained as low as possible in order to keep the droplets

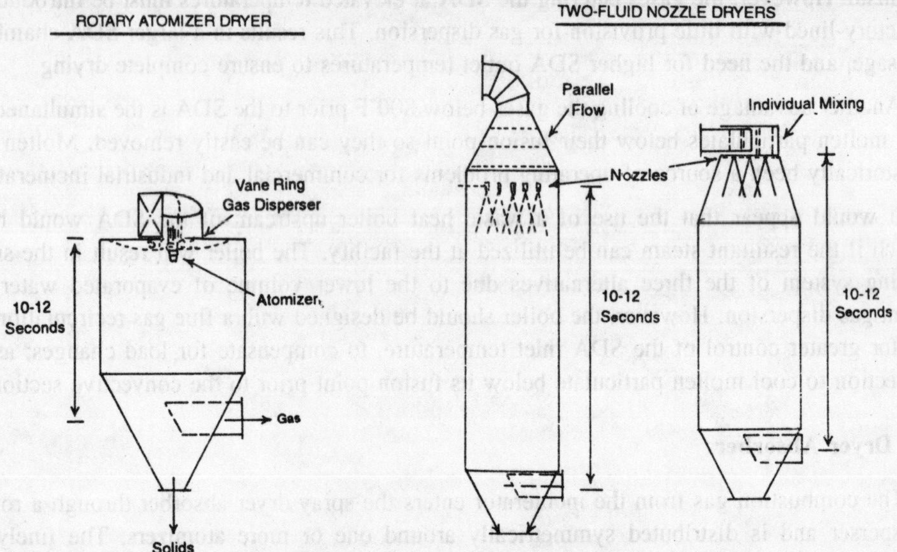

Figure 23.2. Spray dryer arrangements.

moist as long as possible, and thereby increase the reaction rate. Moist particles will remove HCl through absorption, whereas dry particles will only use a slower adsorption mechanism.

$CaCl_2$ exists as an anhydride above 392°F. Below this temperature, hydration at the rate of 2 moles of water per mole of anhydride will occur. At 86°F, an additional four moles of water per mole of anhydride are added. The high water vapor partial pressure in the SDA outlet gas will enhance the formation of hydrates. Since only a fraction of the dried product is $CaCl_2$, the outlet temperature can be as low as 270-300°F without problems. Below this temperature, however, the hygroscopic $CaCl_2$ can be extremely corrosive to carbon steel equipment.

Atomization Systems

Slurry atomization and gas dispersion are the two most critical factors in the performance of the SDA. Two basic atomization methods are utilized in the dry scrubbing process: rotary atomization and two-fluid nozzles.

Rotary atomizers typically consist of a 3,600 rpm motor coupled to a step-up gearbox which provides the required rotational speed (10,000-20,000 rpm) for the atomizer wheel. Slurry is introduced into the center of the wheel and extends over the rotating surface as a thin film. Centrifugal forces accelerate the slurry at the perimeter of the wheel where shear forces disintegrate the film into droplets ranging from 30-100 microns in diameter. The droplet trajectory produces a wide spray pattern that partially determines the chamber diameter.

Dual-fluid nozzles use 70-90 psig compressed air to provide the energy to disintegrate the slurry into fine droplets (70-200 micron). The spray created by the nozzle is a narrower droplet cloud than that created by a rotary atomizer, and, as such, results in a smaller chamber diameter.

Both types of atomizers require the use of abrasion-resistant, replaceable inserts due to the action of the high-velocity slurry stream. The following comparisons can be made between the two atomizer types:

1. Rotary atomizers provide smaller droplets and a narrower droplet size distribution than two-fluid nozzles and therefore may provide higher removal efficiencies.

2. A rotary atomizer is a highly sophisticated, high-speed machine requiring well- trained maintenance staff with special tools and equipment. A dual-fluid nozzle can be maintained with common tools and is more simple in design.

3. A rotary atomizer has higher maintenance and capital costs, but affords somewhat lower reagent costs due to the higher efficiency provided by smaller droplet size.

4. Experience has shown that for larger systems (above 100,000 acfm), dual-fluid nozzle maintenance and capital costs and power requirements approach those of rotary atomizers.

5. Gas flow turndown is much lower on a rotary atomizer (2.5 : 1) versus a dual-fluid nozzle (20 : 1), but reagent flow turndown is much higher on a rotary atomizer (20 : 1) versus a dual-fluid nozzle (3 : 1).

Lime Slurry System

High-calcium pebble lime from a storage silo and water are fed into a paste slaker, which produces a 25%-solids lime slurry. A smaller facility would probably use the more costly hydrated lime instead, to avoid the capital and maintenance costs associated with slakers. This slurry can be diluted to as low as 5% solids in the surge tank in order to accommodate major acid gas load concentration changes in the flue gas, which may result from switching waste blend tanks or some other step change. Since there is a substantial amount of deadtime in the process due to the slurry tank volume, and additional control mechanism must be employed for faster control response. As indicated in Figure 23.3, a small (30-gallon) dilution tank can be employed directly upstream of the atomizer for this purpose.

The control system shown in Figure 23.3 utilizes a feedback signal from an HCl analyzer mounted

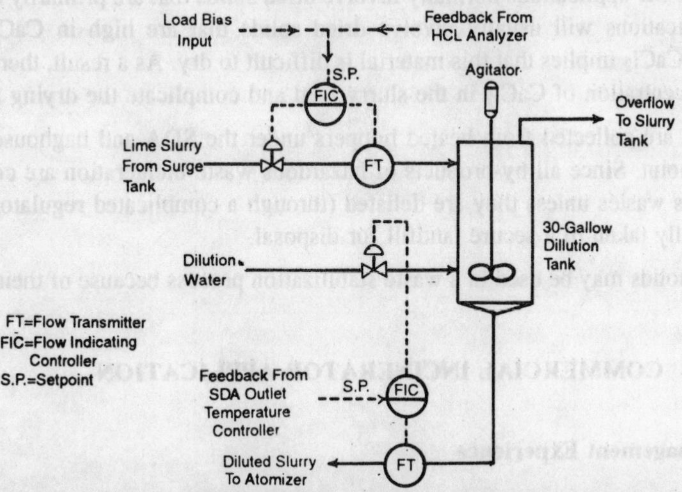

Figure 23.3. Slurry flow control system.

in the stack and a bias signal from a combustion gas flowmeter to adjust the setpoint of the lime slurry flow controller. The spray dryer outlet temperature is used to adjust the setpoint of the total dilution slurry flow controller, which modulates a dilution water control valve.

Fabric Filter

The combustion gases leaving the SDA enter a multi- compartment, pulse- jet fabric filter for particulate removal. The dried solids in the gas stream form a cake on the bag surface and provide additional air/solids contact. Due to the excess lime in the solids, 15-20% of the acid gas is removed at this point. This cake also acts as reserve system alkalinity to smooth out any sudden acid gas load changes.

An electrostatic precipitator would not have this dried solids cake layer, and would thus require all acid gas to be removed in the SDA. For this reason, a fabric filter is preferred for dry scrubbing systems for hazardous waste incineration. Since most of these incinerators are 130 mm Btu/hr or smaller, a pulse-jet filter is normally used.

The baghouse nomally consists of separate modules to facilitate insulation for corrosion prevention. The net air-to-cloth ratio with one module off-line should be in the range of 3 : 1, or lower, in order to provide low velocity and high solids residence time.

Dried Solids System

The process depicted in Figure 23.1 is notably different from most dry FGD systems in that there is no recycle of dried solids to the slurry systems, Most FGD systems use recycle to increase reagent utilization and inprove slurry drying characteristics. However, hazardous waste systems would not normally utilize recycle of solids for the following reasons:

As stated previously, hazardous waste incinerators are generally much smaller than dry FGD applications, and, as such, the additional capital cost for a recycle system cannot normally be justified.

In addition, dry FGD applications normally involve dried solids that are primarily $CaSO_3$, whereas hazardous waste applications will usually involve dried solids that are high in $CaCl_2$ content. The hygroscopic nature of $CaCl_2$ implies that this material is difficult to dry. As a result, there is little reason to create a higher concentration of $CaCl_2$ in the slurry feed and complicate the drying process

The dried solids are collected from heated hoppers under the SDA and baghouse and conveyed to a single discharge point. Since all by-products of hazardous waste incineration are considered under RCRA to be hazardous wastes unless they are delisted (through a complicated regulatory process), the dried solids are normally taken to a secure landfill for disposal.

Alternately, the solids may be used in a waste stabilization process because of their high alkalinity and flyash content.

COMMERCIAL INCINERATOR APPLICATION

Chemical Waste Management Experience

Chemical Waste Managenment (CWM) has two existing hazardous waste incinerators with wet scrubbing systems: the SCA Chemical Services unit in Chicago and the TWI incinerator in Sauget,

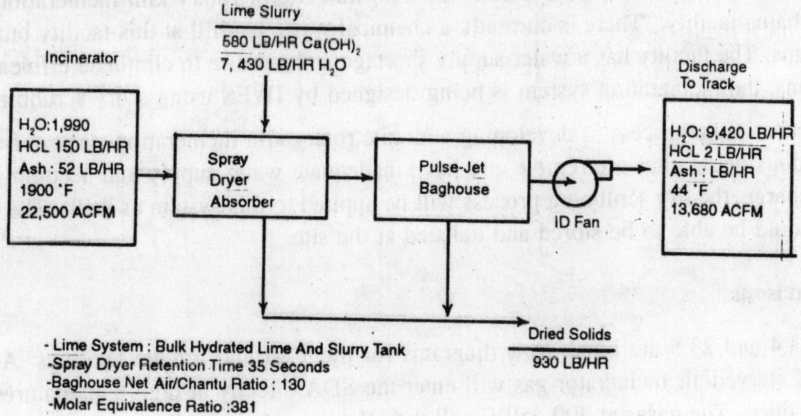

Figure 23.4 TWI gas cleaning flow diagram.

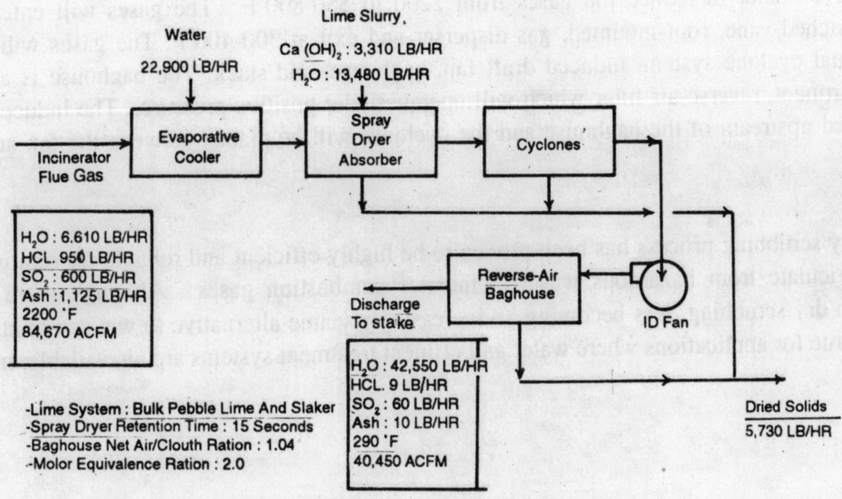

Figure 23.5. Emelle gas cleaning flow diagram.

Illinois.

In order for the SCA commercial PCB incinerator to discharge its wet scrubber effluent into the municipal sanitary district sewer, it must maintain limits on the pH, temperature, suspended and dissolved solids, and heavy metals concentration. Similarly, the TWI facility must control heavy metals concentration in its effluent discharge to the municipal sewer.

TWI obtained a permit from the Illinois EPA to construct two new 15 mm Btu/hr incinerators in February 1985. An economic and process evaluation by International Waste Energy Systems (IWES) showed that the capital and operating costs of wet scrubbing systems, including effluent treatment, for these incinerators was much greater than that for dry scrubbing systems.

CWM has filed an RCRA Part B permit for a 50 mm Btu/hr rotary kiln incineration system for its Emelle, Alabama facility. There is currently a chemical waste landfill at this facility but no effluent treatment systems. The facility has a water supply shortage and a desire to eliminate effluent discharge. For these reasons, the incineration system is being designed by IWES using a dry scrubbing process.

IWES is also in the process of developing a mobile rotary kiln incineration system. Because most remedial hazardous waste sites are remote and have inadequate water supply and accommodations for an effluent discharge, the dry scrubbing process will be applied to this system as well. The dried solids and kiln ash should be able to be stored and delisted at the site.

System Comparisons

Figures 23.4 and 23.5 are block flow diagrams for the TWI and Emelle facilities. As discussed earlier, the TWI starved-air incinerator gas will enter the SDA directly at high temperatures through a refractory-lined duct. The gases at 400-450°F will then flow to a four-module, pulse-jet filter and then to the fan and stack.

In comparison, the Emelle rotary kiln incinerator gas will first be conditioned by water sprays in an evaporative cooler to reduce the gases from 2200 to 550-800°F. The gases will enter the SDA through a pitched-vane, roof-mounted, gas disperser and exit at 300-400°F. The gases will then flow through a dual cyclone system, induced draft fan, baghouse, and stack. The baghouse is an existing, eight-compartment, reverse-air filter which will operate under positive pressure. The induced draft fan will be placed upstream of the baghouse and the cyclones will be placed between the fan and SDA.

Conclusion

The dry scrubbing process has been proven to be highly efficient and reliable for removal of acid gas and particulate from hazardous waste incinerator combustion gases. Although there are some limitations to dry scrubbing, it is becoming an increasingly viable alternative to wet scrubbing. This is particularly true for applications where water and effluent treatment systems are unavailable, inadequate, or costly.

24

Wet Air Oxidation of Hazardous Wastes

Technology Overview

Wet oxidation is the liquid-phase oxidation of organics or oxidizable inorganic components at elevated temperatures and pressures. Oxidation is brought about by combining the waste-water with a gaseous source of oxygen (usually air) at temperatures in the range from about 175-327°C (350-620°F). Typical pressures range from about 2,069-20, 690 kPa (300-3,000 psig) for the above-noted temperature range. The enhanced solubility of oxygen in aqueous solutions at elevated temperatures provides a strong driving force for oxidation.

Wet oxidation has been demonstrated in full-scale, pilot-scale, and bench-scale tests to break down hazardous compounds into carbon dioxide and innocuous end products. While total conversion to carbon dioxide is not achieved, the characteristics of the wastewater are changed. Final treatment of the wet oxidation effluent via biological means is often employed.

In other applications, such as the treatment of organically contaminated brines, wet oxidation again serves to significantly reduce the organic level and, thereby, reduce the hazardous nature of the wastewater. Thus for hazardous wastewaters, wet oxidation functions as an effective pretreatment process.

Process Description

The general processing steps for wet oxidation are:

- Wastewater pressurization/air compression
- Preheating
- Reaction
- Cooling
- Depressurization/Gas Separation
- Off-gas Treatment

The flowschemes used are shown in Figure 24.1 and. 24.2 and differ only in terms of heat-exchanger arrangement.

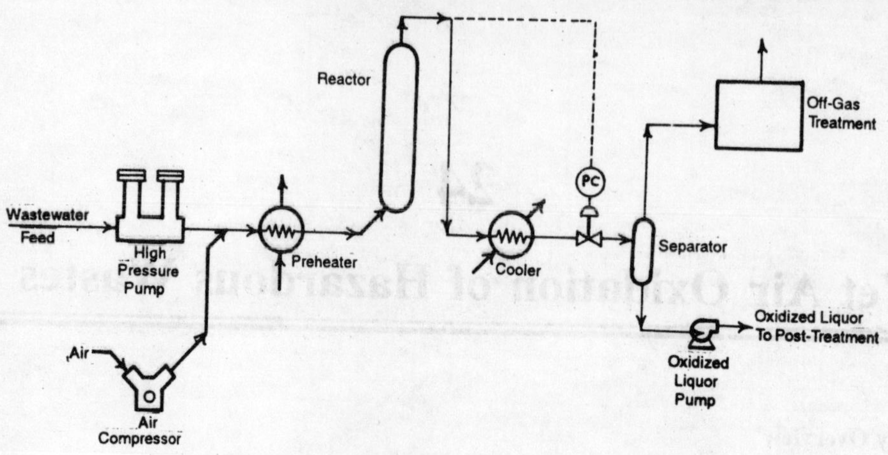

Figure 24.1. A general wet air oxidation flow scheme where wastewater is preheated using an external heat source.

The wastewater liquid or slurry is brought to system pressure using a high-pressure pump. Air from a compressor is added to the pressurized wastewater after the high-pressure pump.

Preheating is necessary to raise the temperature of the wastewater-air mixture such that when the mixture enters the reactor vessel, the exothermic heat of reaction will raise the mixture temperature to the desired maximum. Preheating can be accomplished using an external source of heat (not shown) as in Figure 24.1 or using the reactor effluent as shown in Figure 24.2. Startup energy is provided by that same external heat source to the preheater in Figure 24.1 or to the auxiliary heater in Figure 24.2.

The reactor provides the necessary residence time for the oxidation reaction. The temperature of the wastewater-air mixture rises as the reaction occurs.

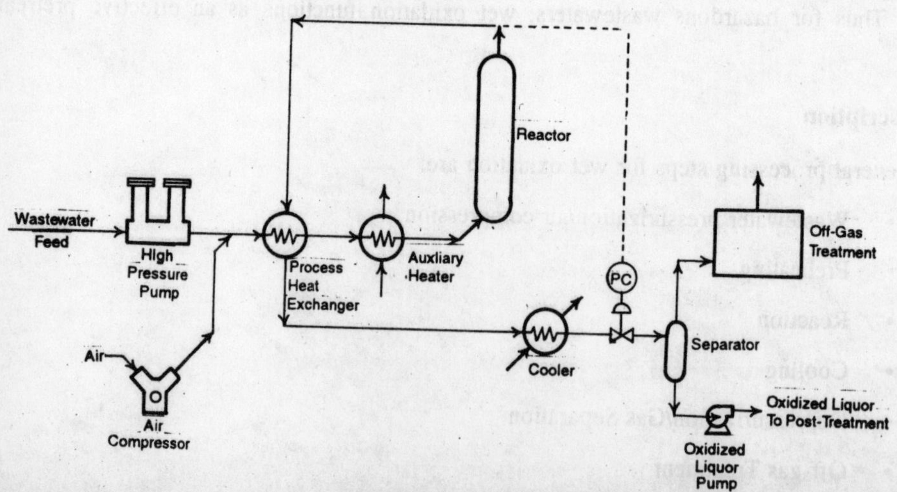

Figure 24.2. A general wet air oxidation flow scheme where wastewater is preheated using oxidized effluent from the reactor. Supplemental preheating is provided by an auxiliary heater.

The reactor effluent is cooled against cooling water as shown in Figure 24.1 or against the wastewater-air mixture as shown in Figure 24.2. Cooling is usually to about 35-57°C (95-135°F).

The pressure of the oxidized liquor-spent air mixture is reduced through a control valve. The gas phase is disengaged from the liquid phase in a separator vessel.

The off-gas from wet oxidation systems is usually treated to reduce the concentration of hydrocarbons. Water scrubbing, which is commonly used to cool the gas stream, results in some removal of hydrocarbons. Adsorption columns using activated carbon provide more organic emissions removal. Afterburning provides the most complete removal of organic emissions.

DEVELOPMENT STATUS

History

A major impetus for the development of wet air oxidation as a commerical process came in the early 1960s with application to the recovery of pulping chemicals from waste liquors. This application required the upper range of wet air oxidation temperature, 300-327°C (572-620°F). At about this same time, wet oxidation was applied for the complete oxidation of sewage sludge.

The work with sewage sludge led to the discovery of a new application of wet oxidation, that of conditioning the sludge for easy dewatering, utilizing the lower range of temperature, 177-204°C (350-400°F). Within 10 years, the utilization of intermediate temperatures, 200-240°C (392-464°F), was adapted to such applications as spent activated carbon regeneration.

Commercial applications in the treatment of hazardous waste appeared in the early 1970s with several installations in Japan. Later in the 1970s and into the early 1980s, the first North American systems were installed. Information on all these installations is given later, in the section "Experience"

Laboratory/Pilot Capability

Bench-scale wet air oxidation studies can be conducted in laboratory autoclaves (capacities up to 1.0 gallon) which are used for batch processing. Continuous-flow pilot-plant wet air oxidation units with capacities up to 1.0 gallon per minute (0.063 l/sec) are also available. Batch and continuous-flow wet air oxidation mobile units are available that can be transported to waste generator sites for wet air oxidation testing.

HAZARDOUS WASTE APPLICATION

Fate of Components

The wet air oxidation process can be used for treatment of a variety of hazardous wastes. The applicable waste stream must contain oxidizable materials, either of an organic or inorganic nature, which are soluble, colloidal, or suspended in an aqueous medium.

Most organic compounds are stoichiometrically oxidized in the wet air oxidation process, the organic carbon oxidized to carbon dioxide, the hydrogen going to water, any halogen to inorganic halides, any sulfur to inorganic sulfate, any phosphorus to phosphate, and any nitrogen to ammonia, nitrate, or elemental nitrogen. At higher processing temperatures (above 260°C or 500°F), where higher oxygen demand reductions are achieved, the remaining organic matter is usually comprised of low-molecular-weight oxygenated compounds, predominantly carboxylic acids. As such, the effluent from the wet air oxidation process is readily treatable by biological wastewater treatment processes.

Halogenated aromatic compounds such as chlorobenzene, PCB's, etc., are the most resistant to wet air oxidation. Even at temperatures of 320°C (608°F), typical destruction levels are only approximately 70 per cent. Most other organic compounds are destroyed in excess of 99 per cent.

The applicability of wet air oxidation for treatment of F and K listed RCRA wastes is indicated in Table 24.1. In addition, there are 122 individual compounds in the P classification and 239 compounds

Table 24.1

Application of Wet Oxidation to EPA F and K Wastes

EPA Hazardous Waste Classification	*Type*	*Wet Air Oxidation*	
		Applicable	*Nonapplicable*
F001-F016	Spent solvent	All	—
K001-K008	Wood preservative and pigments	K001	K002-008
K009-k029	Organic chemicals	All others	K021
K030-K047	Pesticides/explosives	All others	K031
K048-K052	Petroleum refining	All	
K053-K059	Leather tanning	All others	K055
K060-K069	Iron, steel and copper	K065	All others

in the U classification. In excess of 90 per cent of these compounds are judged applicable for wet air oxidation treatment. Finally, actual full-scale treatment results for the destruction of cyanide wastes by wet air oxidation are reported in Table 24.2.

Pre-treatment/Post-treatment Requirements

Most wastewaters can be treated by wet air oxidation without pre-treatment as long as the waste stream is pumpable. Other considerations relate to the concentration of organics and inorganics in the wastewater. Dilution may be necessary to ensure that excessive evaporation does not occur.

The wet air oxidation process should be viewed as a detoxification stage in the overall treatment of hazardous wastes. The effluents from the wet air oxidation process can be post-treated by biological wastewater treatment technologies. Biological treatment can be accomplished by dedicated biological treatment units or by discharge to a publicly owned treatment works (POTW).

The effluent from a wet oxidation unit, which is processing a hazardous waste, is ideally suited to post-treatment by the PACT™ process. The PACT™ process is a wastewater treatment technology which combines carbon adsorption, using powdered activated carbon, with a biological activated sludge system. Many unique wastewater treatment advantages are obtained from this proprietary combination of physical adsorption with biological treatment.

Process Limitations

Treatment of hazardous wastes by wet air oxidation is limited to wastewaters containing oxidizable organic and inorganic compounds. The oxidizable compounds can be soluble or suspended in the

Table 24.2

Wet Oxidation of Cyanide Wastewaters, Casmalia Resources

Description	Oxidation Temp/Time °C/Min	Study	COD,mg/l Influent/Effluent	COD,mg/l Percent Removal	Cyanide, mg/l Ifluent/Effluent	Cyanide, mg/l Percent Removal
Caustic cyanide waste	280/60	A	47,600/6,550	86.2	67,900/229	99.7
	280/60	B	1,030/230	77.7	91.5/0.63	99.3
	280/60	C	13,700/3.010	78.0	2,635/25.6	99.0
	280/60	D	12,750/2,600	79.6	31,250/286	99.1
Metal plating waste	280/60	A	29,000/4,600	84.1	19,010/53	99.7
	280/60	B	14,300/7,270	49.2	19,678/276	99.6
	280/60	C	4,250/930	78.1	449/9	98.0
	270/60	D	39,900/9,600	75.9	26,285/199	99.2
Metal stripper waste	280/60	A	24,400/3,000	87.7	21,100/21	99.9
	280/60	B	34,900/6,230	82.1	6,527/66	99.0
	270/60	C	42,020/12,740	69.7	17,220/259	98.5
Photo reproduction waste	280/60	A	38,200/2.460	93.6	1,880/22	99.8
Precious metal recycle waste	280/60	A	15,600/4,800	69.2	11,340/131	98.8

wastewater. The desired oxidation or destruction efficiency should be obtained within the temperature range of 175-327°C (350-620°F). Reduced destruction levels of oxidation-resistant compounds, e.g., PCB's and chlorinated aromatics, in the wet air oxidation process may limit the application of this technology. However, waste streams which are resistant to conventional wet air oxidation may be amenable to treatment by this technology using a catalyst or by wet air oxidation at higher temperatures, e.g., supercritical wet air oxidation at temperatures above 374°C. Homogeneous metal catalysts have been used in wet air oxidation systems to improve the degree of destruction for some oxidation-resistant compounds.

In general, wet air oxidation technology has few limitations in treating hazardous wastes. Dissolved or suspended solids (up to pumpable limitations) do not limit the application of wet air oxidation.

Environmental Impact

Hazardous wastes are treated by wet air oxidation for the purpose of detoxification. Ultimately two effluent streams, offgas and oxidized liquid effluent, exit from the wet oxidation process. The liquid effluent can be further separated in some cases into a liquid fraction, with all soluble components, and insoluble solids. The environmental impact of these effluent phase—gas, liquid, and solid—must be addressed when considering wet air oxidation for hazardous waste treatment.

Offgas Components and Treatment Systems

Wet air oxidation treatment of hazardous waste is usually carried out to the extent that the toxic organic compounds are nearly completely destroyed. However, the oxidation products are not entirely carbon dioxide and water. Some low-molecular-weight organic compounds, such as aldehydes, ketones, hydrocarbons, etc., are also formed as oxidation products. These organic compounds are distributed between the process off-gas phase and the oxidized liquid phase. Volatile organic compounds (VOC's) can also be stripped from certain hazardous wastes. In general, however, due to the bubble-column design of wet air oxidation reactors, the volatile organic compounds (VOC's) in the offgas are present at the parts per billion (volume/volume) concentration level. The total concentration of organics in the wet air oxidation process offgas, however, typically ranges from 10 to 1,000 parts per million (volume/volume) measured as total hydrocarbons expressed as methane. Thus volatile organic compounds generally contribute less than 0.1% of the total hydrocarbons in wet air oxidation offgases.

The organic components that are emitted from the wet oxidation process in the process offgas can be controlled by a variety of technologies, including scrubbing techniques, granular carbon adsorption, and fume incineration. The specific technology used is selected on a case-by-case basis.

Composition and Disposal of Liquid Effluents

The liquid phase effluent from the wet air oxidation treatment of hazardous wastes usually contains measurable concentrations of low-molecular-weight compounds, which are predominantly carboxylic acid and other oxygen-containing compounds. These soluble organic compounds are readily treated by biological wastewater treatment processes, or a combination of biological and physical adsorption processes. The combination of wet oxidation followed by biological and physical adsorption treatment can produce a high-quality effluent water that is suitable for direct discharge or reuse.

Solids Residuals

The effluent from a wet air oxidation system will contain suspended solids if the influent waste has a high metal content, an ash component, or dissolved salts which are precipitated during the wet oxidation process. Whenever, the level of wet air oxidation required to detoxify hazardous wastes produces an effluent that contains an insoluble ash, the ash material is usually comprised of metal oxides and other insoluble salts such as sulfates, phosphates, and silicates. The insoluble ash can usually be dewatered and subsequently land-disposed. In most instances, the dewatered solid residue will easily meet the requirements of the Extraction Procedure (EP) toxicity test and will be classified as a nonhazardous residue.

DESIGN CRITERIA

Standards

Wet air oxidation treatment systems are designed to treat hazardous wastes and accomplish a desired treatment objective. The desired treatment objective may be the destruction of specific components, of the waste, e.g., cyanide; conversion of elemental forms, e.g., sulfide to sulfate; or detoxification of the waste. To obtain the desired treatment objective, the wet air oxidation unit is designed to process the waste at a given processing rate (gallons per minute), at a maximum temperature (usually within the range of 150-325°C or 302-617°F), and at a maximum system pressure (usually within the range of 2,069-20,690 KPa or 300-3,000 psig).

Since most of the wet air oxidation treatment system is composed of unified pressure vessels, pressure piping, and associated pressurized process equipment, the general engineering design standards of the American Society of Mechanical Engineers (ASME) and American National Standards Institute (ANSI) apply.

Materials of construction for the wet air oxidation process (i.e., vessels, piping, pipe fittings, valves, process equipment, instrumentation, etc.) must be selected on the basis of waste compatibility with respect to general corrosion rates at normal operating conditions, as well as the potential for localized corrosion problems such as stress cracking, pitting, erosion, etc. Materials of construction for pressure vessels and pressure pipings/fittings, etc., should be of ASME-approved materials. The selection of materials of construction for all components of the WAO system that come in contact with the waste being processed should be conducted by a competent corrosion expert in accordance with the recommedations of the National Association of Corrosion Engineers (NACE).

Ancillary process equipment, such as high-pressure pumps, air compressors, steam boilers, and instrumentation, should be selected on the basis of sound engineering principles with respect to materials of construction and design capabilities.

All process equipment mounting structures and pipe supports should be constructed in accordance with the applicable requirements of the American Institute of Steel Construction (AISC) and the American Concrete Institute (ACI).

Critical Operating Parameters

The critical operating parameters for treating hazardous waste by wet air oxidation are temperature, reactor residence time, off-gas oxygen concentration, and system pressure. All of these parameters are monitored in the operating systems and typically all are controlled, automatically, between operator set limits. If any parameters deviates from the set limits, the system alarm is activated. The operator must then take corrective action.

All four of these operating parameters are important to the overall treatment of hazardous waste. The effect of each parameter is addressed as follows.

Temperature

Wet air oxidation reaction rates of the substrate compounds in the hazardous waste will slow with decreasing temperature resulting in lower destruction efficiencies of the substrate compounds. The overall oxygen demand reduction in general will decrease with decreasing operating temperature.

Reactor Residence Time

When the reactor residence time is reduced by increasing the pumping rate to the wet air oxidation system, lower destruction efficiencies of substrate compounds and lower overall oxygen demand reduction in general will result.

Off-gas Oxygen Concentration

Wet air oxidation systems are rarely oxygen-transfer limited in the temperature range used for processing hazardous wastes. However, if the oxygen concentration in the off-gas is depleted, the system will be starved for oxygen. Again, lower destruction efficiencies for substrate compounds and lower degrees of oxygen demand reduction in general will result.

System Pressure

The wet air oxidation system pressure is chosen and maintained to limit the degree of evaporation of liquid water at the corresponding operating temperature. A small variation in system pressure will not generally affect the oxidation efficiencies in wet air oxidation.

Energy Requirements

Wet air oxidation systems consume electrical energy primarily for air compression and high-pressure liquid pumping. The other major energy uses are preheating and off-gas afterburning. The energy required for preheating depends on the flowscheme chosen and the amount of oxidation occurring. The energy required for afterburning is directly proportional to the air flow rate to the wet oxidation unit.

The electrical power for air compression corresponds to about 18-22 kW per pounds of oxygen consumed per minute (0.136-0.167 kW/kgO_2/sec). Wet oxidation systems consume approximately the amount of oxygen equivalent to the oxygen demand reduction for the treated wastewater.

The electrical power for high-pressure liquid pumping is generally about ten per cent of that for the air compressor.

The energy required for preheating is inversely proportional to the amount of oxidation occurring. The more heat that is released in oxidation, the less heat is required for preheating. If the oxidation occurring corresponds to 15-20 g/l of oxygen demand reduction, then no external energy would need to be provided from the auxiliary heater shown in Figure 24.2. All of the preheating requirements could be met with the process heat exchanger. If the oxidation results in less than 15-20 g/l of oxygen demand reduction, auxiliary energy addition would be necessary. When a process heat exchanger is used, maximum auxiliary energy consumption is in the 500-600 Btu/gallon (139-167 MJ/m^3) range.

When a process heat exchanger is not used, all preheating is provided by external means. This preheating will range from not required for highly concentrated wastes to about 4,000 Btu/gallon (1,115 MJ/m^3) for very dilute wastewaters.

The energy required for afterburning is approximately 1,000 Btu/pound (326 kJ/kg) of air fed to the wet oxidation unit.

COSTS

Capital

Capital costs for wet air oxidation systems depend on several factors, such as capacity of the system, oxygen demand reduction of the wastewater, severity of the oxidation conditions required to meet the treatment objectives, and the materials of construction of the wet oxidation system. A range of installed capital costs versus wet oxidation capacity is shown in Figure 24.3. In preparing this cost range, it was assumed that the wet oxidation unit would operate at maximum conditions of 280(536°F), 13,800 kPa (2,000 $lb/in.^2$), and would accomplish a 50 g/l COD reduction. The costs do not include site preparation, a building, or any ancillary storage tankage.

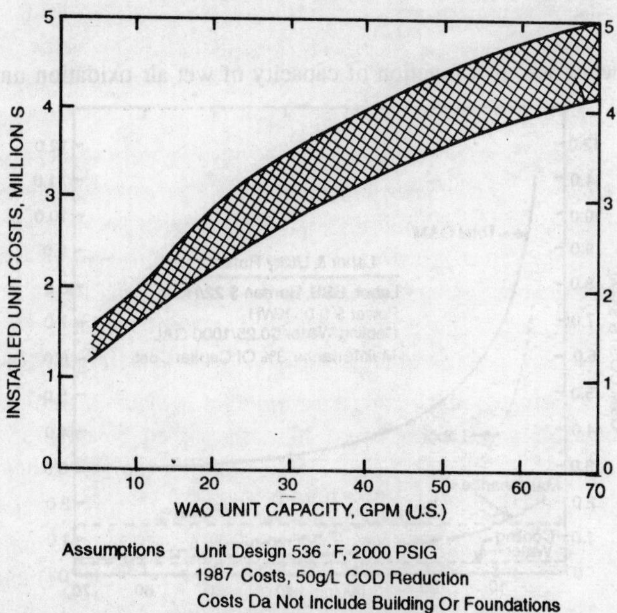

Figure 24.3. High pressure wet air oxidation installed unit: capital cost vs. wet air oxidation unit capacity. The relationship between system cost and system capacity is shown graphically.

Nominal Plant Life

The nominal plant life for a wet air oxidation installation is 20 years. During this nominal plant life, all process equipment may require repair and rotating equipment (pumps, compressors, etc.) will require periodic maintenance and perhaps replacement.

Operating

Operating costs for wet oxidation as a function of the capacity of the unit are shown in Figure 24.4. These operating costs again are made on the basis that the wet oxidation unit would operate at maximum conditions of 280°C (536°F), 13,800 kPa (2,000 ib/in.^{2} gauge), and would accomplish a 50 g/l COD reduction.

Total operating costs are greatly affected by the labour element at lower unit capacities. Operating experience at hazardous waste treatment facilities indicates that one dedicated operator is required for the operation of a single wet oxidation system within the capacity range of 10 to 70 gal/min (0.63 to 4.4 liters/sec.).

The utilities shown in Figure 24.4 are limited to cooling water and electric power. For most applications of wet air oxidation, operation will be autothermal (i.e., no external fuel requirement). For more dilute wastewaters (less than 20 g/l oxygen demand) oxidized in a system according to the Figure 24.1 flowscheme (no energy recovery), a maximum of about 4,000 Btu/gal (1,115 MJ/m^3) will be required. also, where an afterburner is employed for off-gas treatment, a fuel requirement of approximately 3,000 Btu/gal (836 MJ/m^3) of waste processed will be encountered.

Life Cycle

Life cycle treatment costs as a function of capacity of wet air oxidation unit are shown in Table

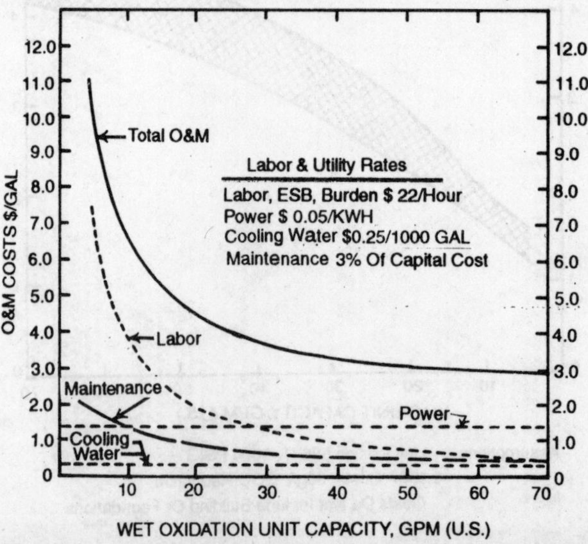

Figure 24.4. Hazardous waste wet oxidation O & M costs vs. flow rate. Operating costs for wet oxidation as a function of capacity of the unit are shown graphically.

24.3. The total cost for treatment of hazardous wastes by wet air oxidation ranges from approximately 2.7 to 8.4 cents per gallon (0.7 to 2.2 cents per liter), depending on the capacity of the wet air oxidation unit.

EXPERIENCE

North American Facilities

The following list of installations use the wet air oxidation process for treatment of hazardous wastes:

Facility	*Location*
1. Casmalia Resources	Santa Barbara County Casmalia, California, USA
2. Bofors-Nobel Inc.	Muskegon, Michigan, USA
3. Dominion Foundries and Steel Company (DOFASCO)	Hamilton, Ontario, Canada
4. Northern Petrochemical Company	Morris, Illinois, USA
5. Uniroyal Chemicals Ltd.	Elmira, Ontario, Canada

A brief summary of the operation of the wet air oxidation facility at each installation follows.

Casmalia Resources

Casmalia Resources operates a Class I hazardous waste disposal facility in Casmalia, California. Casmalia Resources installed a 10 gal/min wet oxidation unit in 1983. The wet oxidation unit was provided by Zimpro/Passavant Inc., Rothschild, Wisconsin. The wet air oxidation unit was permitted by the State of California Department of Health Services to treat the following classes of hazardous wastes: cyanide wastes, sulfide wastes, phenolic wastes, pesticide wastes (halogenated aromatic pesticides excluded), solvent still bottoms, and general organic wastewaters.

Wet air oxidation treatment effectiveness for each of these categories of wastes was determined by a series of demonstration tests. Each demonstration test consisted of a continuous eight-hour wet air oxidation run. During each run, composite samples of waste feed and oxidized effluent were obtained and analyzed to determine the destruction levels for the specific hazardous component. These wet air oxidation treatment results, along with the corresponding process conditions, are summarized in Tables 24.4 and 24.5.

The liquid wet oxidation effluents at the Casmalia Resources site are discharged to evaporation ponds for post-treatment.

Casmalia Resources continues to use the wet oxidation system for commercial treatment of a variety of hazardous wastes, including pesticide wastewaters, waste from petrochemical processes, propellant wastewaters, solvent recycling wastes, cyanide- containing metal plating wastes, sulfide and

Table 24.3

Life Cycle Treatment Costs for Wet Air Oxidation Treatment for Hazardous Wastes

Capacity(gpm)	5	10	15	20	30	40	50
Capital Cost* ($\$ \times 10^6$)	1.5	2.0	2.5	2.9	3.6	4.2	5.0
O & M Cost ($/yr)**	180,000	234,000	280,000	324,000	399,600	475,200	558,000
Present Worth (20 yr at 10%)($\$ \times 10^6$)	3.03	3.89	4.89	5.66	7.00	8.25	9.75
Life Cycle Average Cost, cents/gal	8.42	5.41	4.53	3.92	3.24	2.86	2.71

**Wet air oxidation facitlity including building, storage tanks, etc.*

***250 operating days per year.*

Table 24.4

Wet Air Oxidation Demonstration Program Summary of Demonstration Period Wet Air Oxidation Conditions*

Wastewater Class	*Phenolic and Organic Sulfur*	*General Organic*	*Cyanide*	*Pesticide*	*Solvent Still Bottoms*
Date Procesesd	May 19,1983	July 28,1983	July 29,1983 August 18, 1983	March 28,1984	March 29,1984
Oxidation Temperature, °C	268	277	257	281	268
°F	515	531	495	537	514
Nominal Residence Time, min	113	120	80	182	118
Waste Flow Rate, gpm	5.3	5.0	7.5	3.3	5.1
Compressed Air Flow Rate, scfm	190	190	190	190	190
Reactor Pressure,psig	1610	1515	1220	1600	1550
Residual Oxygen Concentration,%	3.7	4.1	7.1	2.3	3.8

**Average values determined from hourly readings recorded during each demonstration period.*

Table 24.5

Summary of Casmalia Resources Wet Oxidation Treatment Performance

Wastewater Class	*Phenolic and Organic Sulfur*			*General Organic*		*Cyanide*			*Pesticide*			*Solvent Still Bottoms*	
Sample Description	*Influent*	*Effluent*		*Influent*	*Effluent*	*Influent*	*Effluent*		*Influent*	*Effluent*		*Influent*	*Effluent*
COD, g/l		108.1	11.6		76.0	2.5	37.4	4.2		110	5.15		166 45.9
DOC, g/l		3.68		20.83	0.69	14.71	1.71		26..60	1.04		15.20	7.99
BOD5, mg/l	—	—		29,680	880	—	603		41,000	2,530		83,000	21,900
pH	13.0	8.3		1.9	3.3	12.6	9.0		2.0	.4.4		12.5	8.2
Total phenols, mg/l	15,510	36		—	—	—	—		—	—		—	—
Organic sulfur, mg/l*	3,010	180		—	—	—	—		—	—		—	—
Cyanide, mg/l	—	—		—	—	25,390	82		—	—		—	—
Pesticides, mg/l †	—	—		—	—	—	—		169.0	0.93		—	—
Total solids, g/l	88.6	59.7		10.0	3.1	135.1	91.2		8.1	0.7		38.6	30.0
Total ash, g/l	57.1	50.2		0.7	1.1	112.9	77.4		1.1	0.4		25.5	19.8
Volatile solids, g/l	31.5	9..5		9.3	2.0	22.4	13.8		7.0	0.3		13.1	10.2

**Organic sulfur assumed to be equal to total sulfur minus sulfur (inorganic sulfide concentration is nil)*
**Sum of Dinoseb, methoxyclor, carbaryl, and malathion concentrations*

phenolic spent caustic wastewaters, and general organic chemical production wastewaters.

Bofors-Nobel Inc.

Bofors-Nobel, through Environmental Systems Corporation, a joint venture company formed specifically to treat wastes from chemical manufacturing and an abandoned landfill site, uses wet air oxidation as an integral part of its hazardous waste treatment system. A flow diagram of the hazardous waste treatment system is shown in Figure 24.5. Wet air oxidation is used to detoxify several chemical-production wastewater streams which contain toxic organic components in the concentration range of 600 to 1,200 mg/l and total organic concentrations in excess of 50 g/l (Chemical Oxygen Demand). Typical wet oxidation treatment results are summarized in Table 24.6.

Table 24.6
Bofors-Nobel Wet Air Oxidation Performance

	Feed	Effluent	Percent Reduction
Toxic Compounds†	600-1200mg/l	2-9 mg/l	99.8%
COD	70-80 g/l	30-40 g/l	50 + %

* *The Wet Air Oxidation unit is typically operated at 5.5 gpm (0.35 l/sec), 520°F (271°C), and 1800 psig (12,414 kPu) reactor pressure.*

† *Organic priority pollutants.*

Effluents from the wet oxidation of some of the process wastewater contain substantial amounts of ammonium sulfate. These oxidized effluents are directed to a crystalline ammonium-sulfate recovery process.

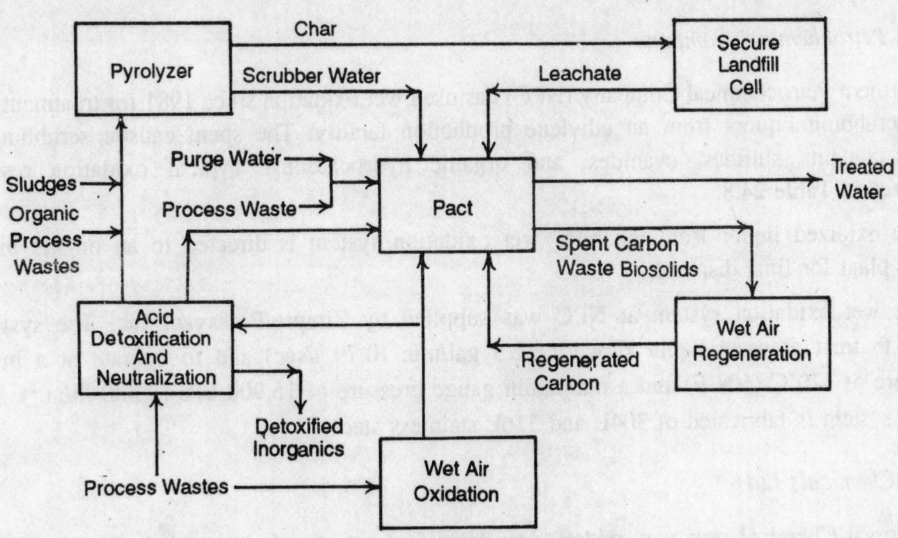

Figure 24.5. A general flow diagram for all wastes handled at Bofors Nobel facilities. Muskegon, Michigan.

A second wet oxidation system is employed by Bofors-Nobel to regenerated powdered activated carbon from the powdered activated carbon treatment (PACT™ unit involves the controlled addition of powdered activated carbon to the aeration basin of a biologcal wastewater treatment system. This latter treatment system is used to process the wet oxidation effluent, certain low-strength process wastewaters, and groundwater from the contaminated site. Both wet oxidation systems at Bofors-Nobel are 10 gpm (0.63 l/sec) skidmounted units supplied by Zimpro/Passavant Inc. Both units are fabricated from titanium for maximum compatibility with a variety of production wastes. Each system is designed to operate at a maximum temperature of 280°C (536°F) and a maximum gauge pressure of 13,800 kPa (2,000 ib/in.^{2}).

Dominion Foundries and Steel Company

Dominion Foundries and Steel Company (DOFASCO) uses wet air oxidation to detoxify spent coke oven gas (COG) scrubbing liquors. Saleable products are recovered from the treated stream. The COG liquor contains cyanides, phenols, thiocyanates, thiosulfates, and other organic constituents. In wet oxidation, cyanides are converted to carbon dioxide and ammonia, thiocyanates are converted to sulfates and ammonia and thiosulfates are converted to sulfates. In the DOFASCO wet oxidation process, hot gases and vapors are separated from the liquid stream in the reactor and are used to preheat the incoming waste. The liquid stream contains approximately 35 percent ammonium sulfate and a small amount of acid. Typical analyses of waste feed and oxidized brine are reported in Table 24.7. High-quality ammonium sulfate crystals are recovered from wet oxidized brine and are sold as fertilizer.

The wet oxidation system at DOFASCO was supplied by Zimpro/Passävant Inc. It was designed to treat a process flow stream of 18 gal/min (1.135 l/sec) and to operate at a maximum temperature of 280°C (536°F) and a maximum gauge pressure of 11,000 kPa(1,600 ib/in.^{2}). The wet oxidation unit is fabricated from various grades of unalloyed titanium.

Northern Petrochemical Company

Northern Petrochemical Company (NPC) has used wet oxidation since 1981 for treatment of spent caustic scrubbing liquors from an ethylene production facility. The spent caustic scrubbing liquor typically contains sulfides, cyanides, and organic hydrocarbons. Typical oxidation results are summarized in Table 24.8.

The oxidized liquor from the NPC wet oxidation system is directed to an on-site biological treatment plant for final disposal.

The wet oxidation system at NPC was supplied by Zimpro/Passavant Inc. The system was designed to treat a waste liquor flow of 12.5 gal/min (0.79 l/sec) and to operate at a maximum temperature of 320°C (608°F) and a maximum gauge pressure of 15,900 kPa (2,300 lb/in.^{2}). The wet oxidation system is fabricated of 304L and 316L stainless steels.

Uniroyal Chemicals Ltd

Uniroyal Chemical uses wet oxidation to detoxify a variety of chemical production wastewater streams. The detoxified wet oxidation effluents are then directed to a biological treatment facility for final disposal. The wet oxidation system was supplied by WetCom Engineering, a firm that previously licensed the Wetox™ wet oxidation system configuration which is based on a cascade of completely stirred tank reactors. The system is designed to treat a waste stream flow of 10 gal/min (0.63 l/sec) at

Table 24.7
Wet Oxidation of COG Liquor at DOFASCO

Component	*Waste Feed*	*Oxidized Brine*
pH	8.4	1.9
COD, g/l	100.1	1.0
NH_4SCN,g/l	94.8	0.008
$(NH_4)_2$ S_2O_3 g/l	24.7	—
Total S, g/l	44.7	130.6
Sulfate—S, g/l	0.9	130.3
NH_3 g/l	25.2	116.9

Oxidation temperature = 270°C (518°F)
Reactor gauge pressure = 9050 kPa (1300 lb/in.^{2})
Waste feed flow = 18.25 gal/min (1.15 l/sec)

Table 24. 8
Wet Oxidation of NPC Spent Caustic Scrubbing Liquor

Component	*Waste Liquor*	*Oxidized Liquor*
pH	1.3	8.1
COD, g/l	32.6	0.7
Sulfifde, g/l	12.3	0.0006

Oxidation temperature = 304°C (580°F)
Reactor gauge pressure = 13,000 kPa (1900 lb/in.^{2})
Waste liquor flow = 8.0 gal/min(0.50 l/sec)

a maximum temperature of 250°C (482°F) and amaximum gauge pressure of 16,600 kPa (2,400 lb/in.^{2}). The reactor and heat exchanges are fabricated from various materials and alloys, including titanium and Hastelloy.

Other Installations

Several wet air oxidiation units have been installed in Japan for the treatment of hazardous wastes. Two of the application are similar to the hazardous waste applications described under North American installations for the treatment of coke oven gas scrubbing liquors and for spent caustic liquors. In addition, systems have been built for five firms to treat the wastewaters from the production of acrylonitrile. These wastewaters contain several different organic cyanides (acrylonitrile, acetonitrile, hydrocyanic acid, and acrylamide). Wet air oxidation has been demonstrated to reduce cyanides from several thousand mg/l to levels of less than five mg/l.

25

Hazardous Waste Treatment Capabilities of Mobile Infrared Treatment Units

Demonstration Unit Process and Equipment Description

Following is a description of the mobile demonstration unit as currently employed by Shirco for the purpose of performing treatability/feasibility studies, either on-site or at the Shirco testing center in Dallas, Texas.

The Portable Demonstration Unit is designed to demonstrate the performance of the Shirco Infrared Furnace in a variety of thermal processing applications. It consists of a feed system, infrared primary chamber, combination gas-fired/infrared secondary chamber, off-gas handling, data acquisition and control systems, heating-element power centers (HEPCs), and emission control systems. All are enclosed in a 45 ft van trailer. The system processes up to 150 lb/hr of waste.

The system was operational in 1980, and was used to conduct on-site incineration tests of industrial wastes. Permit approvals have been granted for numerous hazardous waste tests since July, 1985.

The present system is contained in a single 45 ft trailer and can be transported without special permits. Once the trailer is positioned at the test site, the infrared heating elements and exhaust stack are installed, and utilities are hooked up. On-site utilities required are 30 KVA/480 volt and 15 amp/120 volt electrical, 150 SCFH natural gas or propane, and 20 gpm water with drain. A two-man crew can place the unit in operation in four to six hours from arrival at the test site when utilities are available.

The primary chamber consists of a rectangular cross section "box" of carbon steel lined with layers of lightweight, efficient ceramic fiber blanket. The material to be processed is conveyed through the furnace on a woven wire belt. A friction drive system is used to pull the belt through the furnace. When the material reaches the discharge end of the furnace, it drops off of the belt into a hopper.

Infrared energy is provided by silicon carbide resistance heating elements, grouped into two control zones, with each zone being powered by a 10 KVA heating element power center (HEPC).

The primary chamber has nominal external dimensions of 2.5 ft wide × 9.0 ft long × 7.0 ft high, and an installed weight of 3,000 pounds. Its process capabilities include 500-1850°F process temperature, with material residence time continuously variable between ten and one hundred eighty minutes (10-180 min.). Oxidizing, reducing, or neutral atmosphere can be provided.

The secondary chamber consists of a rectangular carbon steel box, also lined with ceramic fiber blanket insulation. A propane fired burner is used to ignite combustible gases present in the exhaust, and burn them at a predetermined setpoint temperature. An array of silicon carbide heating elements and a 10 KVA power center will be installed at a later date, to allow the chamber to serve as an electric infrared afterburner.

The chamber has external dimensions of 3.0 ft wide × 9.0 ft long × 3.0 ft high, a weight of 1,500 pounds, and process capabilities of 2300°F, two seconds gas residence time, and 0-100% excess air. A continuous pilot and electronic flame monitoring devices are provided.

Combustion air for the primary and secondary chambers is supplied by a blower with manual adjustment of airflow rate and distribution. In the primary chamber, air is injected at various points along the length of the chamber through a manifold system. In the secondary chamber, air is injected through the chamber top, directed at the interface of the burner flame pattern and exhaust gas inlet flow Adjustment of flowrate is by means of a blast gate valve and the burner air registers.

Exhaust gases from the secondary chamber pass through a venturi scrubber with a sump tank. This system consists of an adjustable venturi section containing water sprays, and a separator tower. A manually adjustable cone damper is located in the venturi section to control scrubber pressure drop. In addition to removing particulate, the scrubber cools the gases from their incoming temperature (1000-2300°F depending on system configuration) to saturation temperature (usually about 180°F).

An induced draft exhaust blower is located on the discharger side of the scrubber. The blower is capable of exhausting the primary and secondary chambers, producing a slight (0.01 in. WC) draft on the system, while overcoming the scrubber pressure drop. A removable exhaust stack is installed through the trailer roof. The stack extends 10 feet above the trailer roof, and is equipped with two standard EPA sampling ports. Access to the sampling ports is provided by scaffolding or a lift platform installed alongside the trailer. Standard EPA sampling ports are also provided in the primary and secondary chamber exhaust ducts for emissions testing and off-gas quality sampling.

A control panel contains the instruments required for control of primary chamber zone temperatures, mechanical components, and primary chamber belt speed. A multi-point temperature recorder, digital thermocouple reading display, and power consumption totalizers are included.

Full-scale Mobile Unit

An 80-250 TPD system was constructed and delivered in November, 1986, and utilizes the same operating and process test unit described above. The primary chamber will be permanently mounted on one 67 ft wheel-mounted chassis, which can be rapidly leveled at the work site. The secondary chamber and off-gas handling system are permanently mounted on two other trailers, with a fifth trailer for the control room, laboratory, and administrative offices. A sixth trailer is needed for miscellaneous ancillary equipment.

The primary chamber has dimensions of 9 ft wide × 61 ft long when erected, and an installed weight of 90,000 pounds. Once the chassis is leveled, the infrared heating elements and furnace belt are installed. The secondary chamber has nominal dimensions of 9 ft wide × 11 ft high × 72 ft long, and weighs 80,000 pounds including trailer. A six-man crew can complete the mating and leveling, system interconnection, and component installation in less than one week.

No site preparation beyond a graded area for trailer parking is required. Leveling of trailers is performed using hydraulic outrigger jacks. Large, removable concrete pads can be placed under the jacks to prevent settling of the chassis. A lightweight, removable roof can be erected over the system to provide protection from weather for operators and maintenance personnel.

Site utilities requirements consist of 1,700 KVA/480 volt and 15 amp/120 volt-electric power supply, and makeup scrubbing liquor (amount varies, depending on material being processed). A four-man crew operates the system, with additional personnel, as required, for feed preparation and transport to the incinerator.

Primary Chamber

The Primary Chamber is built up of six powered, one feed, and one discharge module constructed of mild carbon steel. These modules are bolted together and mounted on a skid, which has a removable "goose neck" and transportation "dolly" attached for towing to each designated site.

Each module is insulated with a one in thick layer of ceramic fiber blanket and 3 mil. stainless steel vapor barrier next to the steel shell with additional "Z Block" fiber insulation added as the innermost temperature barrier. The interior steel surface of each module is sprayed with stilastic before the insulation is installed to further protect the shell from corrosive volatiles which might penetrate the insulation at process temperatures. The exterior shell is primed and painted with high-temperature-resistant paint to provide a durable long term protective surface.

The powered modules are fired by transversely mounted silicon carbide resistance heating elements, which are insulated from the steel shell with ceramic sleeves. Electrical connections to the heating elements are made by attaching braided steel straps to their aluminized ends with spring tensioned C-clamps. These electrical connections are protected by ventilated wire ways.

The feed material enters the primary chamber through the top of the feed module for processing. It drops onto a high-temperature metal alloy belt, which transports the material through the chamber. The belt is supported in the chamber by high-temperature alloy top and bottom rollers and shafts which penetrate the chamber shell, and by externally mounted bearings. The belt is pulled through the chamber by a chain-driven roller system mounted in the feed module.

Chain-driven cakebreakers are mounted in the powered modules to stir the feed material periodically and increase process efficiency. The cakebreakers are rollers with an array of high-temperature alloy "fingers" attached which slowly rake through the material layer on the belt as it moves through the fired zones in the chamber.

The processed feed material drops off the end of belt in the time discharge module, where it is quenched with water sprays prior to being discharged by the screw conveyor system. The floor of the primary chamber consists of hoppers and screw conveyors, to collect and remove residual ash from the system. Each hopper has a small vibrator attached to the bottom clean-out tube to assist in the removal of ash into the center-mounted collector screw.

The combustion air for the primary chamber is provided by a blower mounted on the skid underneath the last powered module at the discharge end. The combustion air is carried by ducting up both sides of the chamber, along the top edges of the chamber, and into the chamber at strategically located ports.

Manually operated gate valves are used to control the flow of combustion air into the primary chamber. The combustion air flow within the chamber itself is from the discharge end toward the feed module, which is opposite to feed material movement (counter-current flow).

The exhaust gases from the primary chamber exit through the top of the chamber just prior to reaching the feed material inlet chute. They are directed to the adjacent secondary chamber by insulated crossover ducting.

Secondary Combustion Chamber

The secondary combustion chamber shell is constructed and insulated in a manner very similar to the primary chamber. Its purpose is to thermally destroy the combustible off-gas compounds carried in the exhaust gases, from the primary chamber. The chamber is fired by four propane or natural gas burners mounted on the forward end of the chamber.

Combustion air is provided by a blower, duct work, and plenum chamber mounted on the inlet end of the chamber. The four burner blocks also have an independently mounted combustion air blower and duct work for controlling burner flame patterns.

The primary chamber exhaust gases enter the secondary chamber from the top just above the burner flame patterns. The resultant flow turns 90° and passes through a series of unpowered silicon carbide rods, which increases combustion efficiency by creating turbulence. The waste gases then continue to the downstream end of the chamber, where they exit through the top of the chamber into a crossover duct leading to the base of the emergency bypass stack.

Emergency Bypass Stack

An emergency bypass stack is mounted in the system between the secondary chamber and the scrubber. It is designed to divert the high-temperature gases well above operational elevation during an emergency shutdown. The emergency bypass stack is a vertically mounted, rectangular cross section carbon steel shell, which is insulated similarly to the primary and secondary chamber. It is sealed at the top with counter-weighted doors, which will be opened by a compressed air cylinder during an emergency.

Scrubber

The normal flow of exhaust gases from the secondary chamber will be through the base of the emergency bypass stack, where the waste gases are split into two separate streams prior to entering the scrubber section. Both streams exit the emergency bypass stack into stainless steel tubes where the hot waste gases are cooled with quench water sprays prior to entering the dual fiberglass reinforced plastic (FRP) venturies.

Water injected into the venturi throats atomizes and increases particulate precipitation as the gases enter the front section of the FRP packed tower scrubber. The particulate entrapped in water droplets drains into an open blowdown holding area in the bottom of this section. The particulate-free waste gases continue into the aft section of the scrubber, where a caustic wash liquid is injected to neutralize acid vapor in the stream. The neutralized and cleaned gas stream exits the scrubber in a single duct leading to the induced draft blower.

Exhaust System

The scrubbed gases are drawn from the scrubber by the induced draft blower, which propels them up the FRP exhaust stack. The exhaust stack is mounted on a pad as a free-standing unit with sampling ladder, platform, and ports attached.

Control Van

The system control van is a specially designed unit built to house the primary controls required to start, run, and shutdown all subsystems. The control cabinet is located in the rear of the unit and contains all system alarms, annunciators, recorders, hand-off-auto (HOA) switches, process controllers, and process indicators. The stack gas analyzer and solid-state belt drive controller are located in the middle section, and the motor control center (MCC) is mounted in the front. All electrical power leads and control wiring exit the van through a conduit mounted in the flooring.

Full-scale Mobile Unit Process Description

The thermal treatment process system, just like the demonstration unit, consists of an electric-fired infrared continuous conveyor belt primary combustion chamber with associated material feed and discharge systems, process control and instrumentation equipment, gasfired secondary combustion chamber, and emission-control equipment. Waste material conveyed to the primary chamber must have a consistency that will assure adequate feeding to the furnace belt. The waste feed receives any required pre-treatment in an owner/operator's preparation unit. The prepared feed is put into the furnace live-bottom fee hopper, where it is fed out to the effective width of the furnace belt. The feed hopper discharge drive is manually synchronized with the furnace belt to control the furnace feed rate. The material then enters the furnace through a vertical drop chute.

The material moves through the primary chamber, where it is exposed to infrared radiation (electric heat) in multiple temperature-control zones. Zone temperatures are controlled by automatically varying the heating element input power to maintain preset zone setpoint temperatures. As the material moves through the chamber, it is exposed to the thermal environments necessary for drying, volatilization, and pyrolysis or incineration (depending on the mode selected) of volatiles and organics. Because of the density and thickness of the feed material, these steps may not occur in discrete phases. Particularly in the early stages, a given portion of the material may be undergoing any or all of these processes. In order to maximize process rates, the material layer is gently turned at several point in the furnace by means of cakebreakers. Upon completion of the primary chamber process, the material is discharged and spray cooled if necessary in a lateral screw conveyor. The material is discharged through an inclined screw conveyor and transported away from the furnace by whatever means site-specific conditions dictate.

As the material moves through the chamber on the conveyor belt, it can be exposed to combustion air (if that mode is selected) provided by means of a forced draft blower. The overall flow of the furnace gases is counter-current to the materal flow. This process feature allows for supplemental heating of the combustion air (if used) as it passes over the burnedout portions of the material layer; hence, a more energy-efficiency process is achieved. To assure proper turbulence of the gases during the incineration process, the combustion air is taken off of the main air ducting and injected at strategic points in the chamber through air jets.

The air jets are equipped with tight-closing valves for use in the pyrolysis mode. In order to remove the gases from the furnace, an induced draft blower is located at the scrubber exhaust. The blower maintains a sufficient draft on the system, assuring flow out of the primary chamber while preventing excess, unmetered in-leakage of air.

Once the combustive and/or vaporized volatiles and organics leave the furnace (and the air preheater), the gases pass through a gas-fired secondary combustion chamber and then a water and chemical scrubbing system, which both cools the exhaust gases and cleans the gas stream to acceptable air pollution limits. The scrubbed gas is exhausted to a stack through the induced draft blower.

Process Optimization

The basic objective of the process is to transform the feed material to another form, i.e., an ash that is acceptable for land disposal, while assuring that the exhaust gas products can likewise be dispersed without harm to the environment. Secondary objectives are to effect the process with minimum energy usage and minimum system maintenance costs. In order to achieve these objectives, the process is optimized by adjustment of a number of parameters. These process variables are classified as primary and secondary in terms of their effects and are discussed in the following paragraphs.

Primary Process Variables

The primary process variables in the infrared system are temperature, residence time, material layer thickness, and combustion air flow control. A material thickness of up to 2.00 in. is near-optimum in terms of throughput. During start up, then, the emphasis is primarily on residence time, temperature, and atmosphere control.

Termperature and residence time in the process are inversely related, i.e., the higher the temperature, the shorter the residence time required to achieve the desired discharge material and exhaust conditions and vice-versa.

During operation, these variables are optimized as follows: A residence time is selected (and set by means of the furnace conveyor belt variable-speed drive) such that the system throughput is slightly greater than that required to meet process requirements. This approach will assure that sufficient process capacity is available in the event of slight variations in the feed rate and in the properties of the feed. A recommended starting value for residence time is provided for each system, and normally will fall in the range of 5 to 50 minutes.

As an example, in the incineration process starting values of the primary chamber zone temperature control range from 1300°F to 1500°F in the initial (drying) zone to 1550°F to 1650°F in the final (combustion) zone. In the pyrolysis mode, zone temperatures can range down to 800°F.

The final primary variable is combustion air flow rate, if that mode has been selected. A balance between the amount of air entering the primary chamber from the combustion air duct at the discharge end of the furnace and the amount entering through the side jets will optimize combustion efficiency. This is optimized initial system start-up to increase combustion efficiency as well, by promoting proper combustion air mixing within the powered zones of the primary chamber.

Secondary Process Variables

One secondary process variable is chamber draft. This variable will exert some influence on the process, but once optimum values are determined for a constant set of feed conditions, it is either maintained automatically or requires infrequent adjustment.

Furnace draft is measured by means of a pressure transducer located in the materials discharge module of the furnace. The optimum draft setting is that which produces a neutral pressure at the discharge end of the furnace. Under these conditions, there is just sufficient draft at the furnace exhaust to remove the gases produced by the primary chamber process. Once established, the draft setting is maintained automatically be means of closed-loop control over the induced draft blower inlet vanes. If the draft setting is too low, pressurization of the furnace will result and hot gases may exit through openings in the furnace shell. If the draft setting is too high (pressure too negative), uncontrolled leakage of air into the furnace may result. The optimum draft setting is typically in the range from 0.0 to 0.1 in. WC.

Another secondary process variable is cooling of discharge material by liquid sprays. Depending on the heat content of the belt discharge material and the water spray rate, all or part of the spray or none can be allowed to vaporize. Then, depending on the vaporization-rate mode chosen, the primary chamber, downstream material, energy balance, and gas velocities are influenced. Therefore, the vaporization rate should be minimized when possible.

Summary of Performance Data

The Shirco infrared process has demonstrated its effectiveness on a variety of organic hazardous sludges and solid including dioxins, PNAs, PCP, and PCBs.

Times Beach Dioxin Research Facility

In July, 1985, the Shirco Pilot Unit was in operation at the Times Beach Dioxin Research Facility operated by the Missouri Department of Natural Resources (DNR). The purpose of the testing was to demonstrate Shirco's capability to successfully decontaminate soil laden with 2, 3, 7, 8, tetrachloro-dibenzo-p-dioxin (TCDD). The DNR Environmental Division coordinated the site preparation; emissions sampling and final analysis of the emissions and soil samples were performed by Roy F. Weston, Inc. Shirco prepared the test protocol and operated the furnace system.

Results of the testing have proven the infrared process to be a successful method of decontaminating the dioxin-laden soil present at Times Beach, Missouri, Presented in Table 25.1 is a summary of the results of the testing. An expanded report can be obtained from Shirco Infrared Systems, Inc.

International Paper Company Wood Preserving Plant

In November of 1985, the Shirco portable test unit was transported to a wood preserving plant in Joplin, Missouri. The plant was owned and operated by International Paper Company, the world's largest paper and forest products producer. The Region 7 EPA and the Missouri Department of Natural Resources authorized the tests under an amendment to International paper's Interim Status part A permit. The purpose of the tests was twofold: (1) to confirm the capability of the Shirco infrared process to effectively treat pentachlorophenol, dioxin, and creosote sludges and soils without the formation of adverse by-products in the stack gases and scrubber water; and (2) to establish the economic, operating, and throughput performance which would be experienced with a full-scale system.

Table 25.1

Test Results Summary

	EPA Standard	*15 Minute Residence*
Composite feed soil 2, 3, 7, 8 TCDD Concentration		156 ppb
Composite discharge soil 2, 3, 7, 8 TCDD Concentration	<1 ppb	Not deteced at 33 ppt
Particulate emissions at 7% O	.08 gr/dscf	0.0002 gr/dscf
Gas phase DRE of 2, 3, 7, 8 TCDD	>99.9999%	>99.999989%

Table 25. 2

On-Site Pilot Test Result: Wood-Preserving Waste Ash Quality

Compound	*Test 1*		*Test 2*	
	Primary Chamber Temperature:1650°F Solid Phase Residence Time: 15 min.		*Primary Chamber Temperature1650°F Solid Phase Residence Time: 25 min.*	
	Waste Concentration (ppm)	*Ash Concentration (ppb)*	*Waste Concentration (ppm)*	*Ash Concentration (ppb)*
Acenapthene	1,700	ND	4,200	ND
Anthracene	4,600	ND	11,200	ND
Benzo(a)anthracene	470	ND	1,300	ND
Carbazole	2,700	ND	5,400	ND
Chrysene	720	ND	2,200	ND
Fluoranthene	4,000	ND	14,000	ND
Fluorene	2,400	ND	4,600	ND
Penthachlorophenol	12,000	ND	11,000	ND
Phenanthrene	8,000	ND	22,000	ND
Pyrene	2,200	ND	7,400	sND

ND = Not detected at an average detection limit of 30 ppb.

Table 25.3

On-Site Pilot Test Result Wood-Preserving Waste POHC Destruction and Removal Efficiency

	Test 1		*Test 2*	
	Afterburner Temperature: 2200°F		*Afterburner Temperature: 1800°F*	
Compound	*Waste Concentration (ppm)*	*Flue Gas DRE (%)*	*Waste Concentration (ppm)*	*Flue Gas DRE (%)*
Acenapthene	1,700	99.99996	4,200	99.99996
Anthracene	4,600	99.99998	11,000	99.99998
Benzo (a) anthracene	470	99.99985	1,300	99.99998
Carbazole	2,700	99.99997	5,400	99.99997
Chrysene	720	99.99990	2,200	99.99993
Fluoranthene	4,000	99.99998	14,000	99.99997
Fluorene	2,400	99.99997	4,600	99.99997
Penthachlorophenol	12,000	99.99999	11,000	99.99998
Phenanthrene	8,000	99.99998	22,000	99.99996
Pyrene	2,200	99.99997	7,400	99.99998

Table 25.4

PCB-Contaminated Soil Demonstration Test Summary

Test Operating Parameter	*1*	*2*	*3*
Average Process Temperature (°F)	1603	1573	1471
Average Oxygen (%)	12,20	8.58	9.49
Particulate Concentration @ 7% 0 (gr/dscf)*	0.055	0.023	0.017
Waste Feed Rate (lb/hr)	61.5	61.5	79.8
PCB Concentration (ppm) Feed	2790	2560	2840
Ash**	ND	ND	ND
Destruction and Removal (%)	>99.9999	>99.99992	>99.9999

* *Federal Standard .08 (gr/dscf)*
** *Not Detectable at 5 ppb*

Emissions sampling and laboratory analysis were performed by ERT and final analysis was conducted by ENSECO laboratories, Shirco prepared the test protocols and operated the demonstration unit. Post-test analyses showed that the Shirco infrared process was highly effective in destroying heavy organic compounds such as pentachlorophenol and others. Summaries of the tests are shown in Tables 25.2 and 25.3.

Florida Steel Corporation Steel Mill Site

An on-site demonstration test was conducted May 13-15, 1986 for Florida Steel Corporation at an inactive specialty steel mill near Indiantown, Florida. The test was performed under authorization of an R&D permit issued by U.S. EPA TSCA headquarters office in Washington, D.C. The Florida Department of Environmental Regulation (FDER) approved of the temporary air emissions. State FDER and Region 4 EPA personnel observed the test, which was also widely publicized in local/regional TV and newspapers. The purpose of the test was to determine if the Shirco infrared unit could treat the PCB-contaminated soil to the required level of less than 2 ppm and evaluate the economic feasibility of on-site treatment with a full-scale Shirco unit.

Florida Steel provided support for site preparation, site utilities, and personnel to aid in waste material mixing. Sampling and process-stream monitoring were performed by TRC Corporation. ENSECO laboratories performed analysis of all ash, gaseous emissions, and scrubber water blowdown.

Table 25.4 illustrates test results for those tests performed in accordance with permit specifications.

Analysis of the scrubber effluent indicated that no PCBs were detected at a detection limit of 0.34 parts per billion. PCBs were not detected in the stack gases for any of the tests run in conformance with permit conditions. One test, due to a low secondary combustion oxygen level (outside the permit limit of 3%), did result in a PCB concentration of 2.4 mg/m^3.

26

Incineration for Pathological Waste Disposal

INTRODUCTION

The EPA's (Environmental Protection Agency) major concern in the 1980s is effective hazardous and toxic waste disposal. The managing of infectious/pathological wastes is considered to be critical, and the means of their disposal is being studied carefully since the impact of these wastes on the environment can be disastrous. This chapter examines incineration as a means of disposal for pathological wastes. Specifically, the design and control of the starved-air incinerator is discussed. Compliance to air regulations is a prime concern involving test methodology, calculations, permitting, and other environmental concerns. Incineration is recognized as one of the best techniques that applies to the treatment of all types of infectious wastes.

New regulations by federal and local authorities continue to strive for environmental improvement. The effective, efficient managing of hazardous wastes poses several considerations. The disposal of types 3 and 4 wastes, including infectious waste as characterized in the September 1982 *Draft Manual for Infectious Waste Management*, poses as potentially dangerous a problem as any. In order to prevent the danger of transporting infectious waste, more and more states now require on-site destruction.

Chereymisinoff cites incineration as "a most effective form of managing many wastes". Incineration is the only recommended technique that applies to the treatment of all six types of infectious wastes.

The scope of this chapter is to examine the controlled air incineration system which utilizes an oxygen-starved atmosphere concept in order to prevent the entrainment of particulates from the refuse bed out of the incinerator combustion chamber. This two-stage (double-chamber) unit will be discussed later in more detail.

Like any other vessel or reactor that is subject to air emissions, incinerators are regulated by federal as well as local laws. The environmental impact of these combustion units, as well as current design and air emission controlling criteria, are discussed.

WASTE CLASSIFICATION

The following is a general description of waste classification as defined in the *Interim Guide of Good Practice for Incineration at Federal Facilities*, by HEW.

Type 0

Type 0 waste is a mixture of highly combustible waste such as paper, cardboard cartons, wood boxes, and floor sweepings from commerical and industrial activities. The mixture contains up to 10% by weight of plastic bags, coated paper, laminated paper, treated corrugated cardboard, oily rags, and plastic or rubber scraps.

This type of waste contains 10% moisture and 5% noncombustible solids, and has a heating value of 8,500 Btu per pound as fired.

Type 1

Type 1 waste is also a mixture of combustible waste such as paper, cardboard cartons, wood scrap, foliage, and floor sweepings from domestic, commerical and industrial activities. The mixture contains up to 20% by weight of restaurant waste, but contains little or no treated paper, plastic, or rubber wastes.

This type of waste contains 25% moisture and 10% incombustible solids, and has a heating value of 6,500 Btu as fired.

Type 2

Type 2 waste is an approximately even mixture of rubbish and garbage by weight.

This type of waste, common to apartment and residential occupancy, consists of up to 50% moisture and 7% incombustible solids, and has a heating value of 4,300 Btu per pound as fired.

Type 3

Garbage such as animal and vegetable waste from restaurants, hotels, hospitals, markets, and similar installations, falls into the category of Type 3 waste.

This type of waste contains up to 70% moisture and up to 5% incombustible solids, and has a heating value of 2,500 Btu per pound as fired.

Type 4

Human and animal remains, such as organs, carcasses, and solid organic wastes from hospitals, laboratories, slaughterhouses, animal ponds, and similar sources, consisting of up to 85% moisture and 5% incombustible solids, and having a heating value as low as 1,000 Btu per pound as fired, fall under Type 4 waste.

Type 5

Type 5 waste includes gaseous, liquid, or semiliquid by-product waste, such as tar, paint, solvent, sludge, and fumes from industrial operations. Btu values must be determined by the individual materials to be destroyed.

Type 6

Solid by-product waste, such as rubber, plastic, and wood waste from industrial operations, is Type 6 waste. Btu values must be determined by the individual materials to be destroyed.

INCINERATION OF TYPES 3 AND 4 WASTES

Pathogens—medical and institutional solid waste—are potentially disease-producing. Organisms that can transmit disease are organic and will live on organic refuse as hosts; or, organic material necessary to maintain a culture may be an invisible coating on a plate or a container, or the culture might be maintained at the top of a hypodermic needle in the waste basket. This has frequently resulted in janitors becoming infected when they use their hands to clean out residue from waste containers. Pathological waste is mainly composed of carbon, hydrogen, and oxygen, with small amounts of minerals and a trace of nitrogen. The air emissions resulting from incinerating such wastes are constituted generally by smoke, gases, fly ash, and odor, which is usually regarded as "highly objectionable" and is regulated. Pathological wastes are not as completely destroyed as has been previously thought. In fact, emissions resulting from burning of such wastes could be a new source of infectious thought as can be the residue (non) burning material and ashes resulting from the burning process). Incomplete combustion of pathogens is a result of the following five parameters:

- Waste composition
- Waste feed rate
- Combustion temperature
- Air and fuel feed rates
- Internal flow paths

A typical research facility with an on-site pathological incinerator can include in its burnables mainly animal carcasses and low-level radioactive fecal material.

More specifically, animal carcasses may include:

- Horses
- Sheep
- Cattle
- Swine
- Chickens
- Dogs

The fecal material may contain no more than 0.05 microcuries of carbon-14 or tritium per gram of waste in accordance with Nuclear Regulatory Commission regulations.

Some paper and bedding material (ground-up corncobs) also may be incinerated.

Radioactive wastes, as mentioned above, are generally not accepted for incineration. There are cases, however, where wastes with small amounts of radioactivity may be accepted for burning. Etiological agents, additionally, are considered unacceptable as radioactive waste material.

AIR EMISSIONS

Types 3 and 4 Wastes

Basic design of a pathological incinerator, its operating conditions, and the refuse compositions have a considerable effect on the emissions released to the atmosphere. The greatest effect on the amount of particulate emissions comes from the way in which air is supplied to the combustion chamber. As indicated, air may be introduced in several locations, i.e., below the chamber, above the combustion area, or from the side. An increase of the underfire air may result in an increase in fly ash emissions.

Emissions of sulfur oxides depend on the sulfur content of the refuse. Poor combustion has as a result large amounts of carbon monoxide and unburned hydrocarbon emissions. Other types of emissions resulting from waste 3 and 4 incineration are nitrogen oxides, which may increase at higher temperatures of the combustion zone, higher residence time in the combustion zone before quenching, and an increase in the excess air rates to the point where dilution cooling overcomes the effect of increased oxygen concentrations. This is in general agreement with nitrogen fixation, usually encountered in the combustion process.

Following are criteria concerning the types of emissions mentioned above. These criteria are minimum requirements which typically must be met in order to secure permit approvals from regulatory agencies:

1. N.J.A.C. 7:27-11 limits particulate emissions to 0.1 grains per day standard cubic feet (gr/dscf) of stack gas corrected to 12% CO_2 from any auxiliary fluid. Emissions of particulate matter must be reduced to a maximum of 0.02 gr/dscf of exhaust gas corrected to 7% oxygen (7% oxygen is approximately equivalent to 12% CO_2). Maximum hourly emissions must not exceed 0.03 gr/dscf. The design of controls must be shown to achieve these emissions levels.

2. Hydrochloric acid (HCl) emissions: Outlet emission concentration of less than 50 parts per million by volume hourly average on a dry basis corrected to 7% O_2.

3. Sulfur dioxide (SO_2) emissions: Outlet hourly average emission concentration of less than 100 ppmv of SO_2 on a dry basis corrected to 7% O_2.

4. Carbon monoxide (CO) emissions: CO emissions less than 400 ppmv hourly average on a dry basis corrected to 7% O_2.

5. Nitrogen oxides (NO_x) emissions: NO_x emissions less than 300 ppmv hourly average on a dry basis corrected to 7% O_2.

6. Hydrocarbon emissions: Since New Jersey is a non-attainment area for ozone, if nonmethane hydrocarbon emissions exceed 50 tons per year, these emissions must be controlled to a level which represents the lowest achievable emission rate (LAER) as required by N.J.A.C. 7:27-8 (Emission Offset Rule). This rule also requires the applicant to secure emission offsets.

Radioactive Wastes

In the case where radioactive material is included in the waste feed, the resulting emissions must be monitored and the appropriate regulatory requirements met.

Waste radioactive carcasses may be disposed of by means of pathological incineration, provided they contain up to 0.05 μCi/gm of 3H or 14C. According to 10 CFR 20.306, materials containing this amount of radioactive material may be disposed of without regard to their radioactivity levels.

Records must be maintained of the materials to be incinerated to verify these levels. Emissions to the atmosphere of 3H and 14C must not exceed the Maximum Permissible Concentration (MPC) specified in both 10 CFR 20 ''Standards for Protection Against Radiation,'' and Bureau of Radiation Protection, N.J.A.C. 7:28-6.5.

Any waste that cannot be disposed of within the regulations cited will be removed from the industrial facilty to an approved burial site by an approved commerical carrier or contractor.

The emission concentration limit for 3H is 2E(-7) μCi/ml and that for 14C is 10E(-7) μCi/ml.

Visible Emissions

Sight is the most widely used and generally accepted measurement of air pollution internsity. Subchapter 11 of the New Jersey Air Pollution Control Code requires the use of a smoke chart, preferably the Ringelmann, published and described by the U.S. Bureau of Mines Information circular 7718, or a commissioner-approved equal. This technique will be discussed in more detail in the following section.

TEST REQUIREMENT AND METHODOLOGY

The purpose of emission testing is to quantify specific source emissions resulting from the operation of a multiple-chamber pathological incinerator. The data generated are then utilized by regulatory agencies such as the New Jersey Department of Environmental Protection (NJDEP) to make an equitable decision on granting an operating air permit for the particular incineration unit.

The emission testing program is performed for the determination of the following parameters, utilizing the test methods specified below:

Parameter	Test Methodology
Particulates	New Jersey Air Test Method 1, "Sampling and Analytical Procedures for determining Emissions of Particulates from Manufacturing Process and from Combustion of Fuels—Incineration Test Procedures."
Sulfur Dioxide	EPA Reference Method 6, "Determination of Sulfur Dioxide Emissions from Stationary Sources."
Nitrogen Oxides	EPA Reference Method 7, "Determination of Nitrogen Oxides Emissions from Stationary Sources."
Carbon Monoxide	EPA Reference Method 10, "Determination of Carbon Monoxide Emissions from Stationary Sources."
Total Hydrocarbons	New Jersey Air Test Method 3, "Sampling and Analytical Procedures for the Determination of Volatile Organic substances from Source Operations."
Hydrogen Chloride	API Method 767-54, "Determination of Hydrogen Chloride (Turbidimetric Method)."

In addition to these specified procedures, EPA Reference Methods I through 4 are performed for the determination of sampling points, temperature, moisture content, velocity, volumetric flow rate, and molecular weight. Here is a list of these additional requirements of emission performance tests:

- EPA Methods 1, "Sample and Velocity Traverses for stationary sources", to be conducted in conjunction with New Jersey Air Test Method 1 and API Method 767-54.

- EPa Method 2A, "Determination of Stack Gas Velocity and Volumetric Flow Rate (Type 5 Pitot Tube)."
- EPa Method 3, "Gas Analysis for CO_2, O_2, Excess Air and Dry Molecular Weight ," is to be conducted in conjunction with EPA Method 7.
- EPA Method 4, " Determination of Moisture Content in Stack Gases."

Regarding "Visible Emissions", the Ringelmann chart (smoke chart) is used. The Ringelmann technique develops varying shades of gray from a series of charts presenting graduated grids printed in varying widths of black. The charts, when compared with stack emissions, categorize the emission density into a Ringelmann number. The numbers vary from number 0 for a clear stack, to number 5 for a complete black emission. With a scrubber installation, the flue gases are saturated with water. Upon exiting the stack, a large vapor plume (steam) is generated. As the Ringelmann numbers are shades of gray, the numbers reported are not those at the stack exit but those sighted at the dissipation of the vapor plume.

CALCULATIONS

These example calculations are based on typical and maximum emission concentrations from carcasses to be disposed of in a pharmaceutical facility.

Calculation Method for Particulate Emissions Corrected to 12% CO_2

Various particulate emission regulations are based on a concentration corrected to 12% CO_2. This most often occurs with incinerator regulations

Case I

Regulation based on 12% CO_2 with no adjustment for auxiliary fuel contribution to CO_2. grains/dscf @ 12% CO_2 = (measured gr/dscf) × (12% CO_2 measured)

Case II

Regulations based on 12% CO_2 with CO_2 from auxiliary fuel excluded.

$$\text{grains/dscf @ 12\% } CO_2 = (\text{measured gr/dscf})\left(\frac{12}{\text{measured \% } CO_2 - \text{calc. fuel \% } CO_2}\right)$$

where calculated fuel % CO_2 =

(A) For typical fuel oil = X (gpm)(100)/total dscfm flue gas

where X = 193 for # 2 oil

= 204 for # 4 oil

= 221 for # 6 oil

(B) For natural gas = CFH (100)/60 (total dscfm flue gas)

(C) For propane = CFH (300)/60 (total dscfm flue gas)

(D) For butane = CFH (400)/60 (total dscfm flue gas)

DESIGN AND OPERATION

The technology of modern air incineration is a result of combined laboratory and field experience. In a laboratory setting when a constant waste stream is used, the characteristics of combustion follow certain predictable chemical reactions. Situations in actual field conditions differ. Waste streams are seldom consistent, as wastes themselves are heterogeneous. During the design, therefore, of any kind of incinerator it is important to consider the theoretical chemistry as well as specific field applications. This consideration plays a significant role in the development and successful implementation of an efficient controlled-air incineration system.

Controlled-air incineration, which is also called starved-air, pyrolytic, or hearth incineration, is a relatively recent and rapidly growing technology. On-site disposal of solid wastes from many industries utilize this method. With the proper design of an incineration system, a conversion of organic solid wastes to an inert sterile ash with a weight and volume reduction of as much as 95 to 98% can be achieved.

Controlled-air incineration is basically a two-stage combustion process. Figure 1 shows a typical controlled-air incineration. In the first chamber (primary chamber), the waste is fed and then heated (burned) with air that is less than stoichiometric. This condition (also called starved-air or oxygen-starved atmosphere) causes most of the volatiles to be destroyed pyrolytically with the required endothermic heat which is provided by the oxidation of the fixed carbon fraction.

The waste is heated by controlled introduction of air, thereby controlling the temperature. This results in very low velocities, so that ash particles are not entrained and therefore carried to the upper chamber. The absence of these particulates results in a clean stack exhaust. Temperatures in the chamber are relatively low, and remain below the melting and fusion temperature of most metals, glass, and other noncombustibles. All noncombustibles discharge with the residual ash.

In the upper (second) chamber, the smoke is reheated and additional air is introduced so that the very fine and hot smoke particles are oxidized rapidly and produce essentially harmless carbon dioxide and water vapor. After this reaction, additional air is introduced through the upper air inductor to reduce the temperature of stack gases before they exit into the atmosphere. These smoke and pyrolytic products consist primarily of volatile hydrocarbons and carbon monoxide, as well as other products of combustion. Figure 26.2 shows a schematic of control air incineration, while Figures 26.3 a and b illustrate the incineration system's theory and operation.

The ash removal of the incinerator is usually done manually, since auto ash-removal mechanisms are not feasible due to the fat and grease renderings that can be developed during incineration of pathological wastes. The transporting of these wastes is usually done in carts by special skids, similar to the set-up that is used for transporting biohazardous wastes.

Assistance in the design of incinerators may be obtained by the Federal Facilities branch Division of Abatement National Air Pollution Control Administration.

STATE-OF-THE-ART CONTROLS

Incinerators that burn more than 800 pounds per hour need to be equipped with a control device such as a scrubbing system. In addition, a controlling device is needed for a proper operating temperature

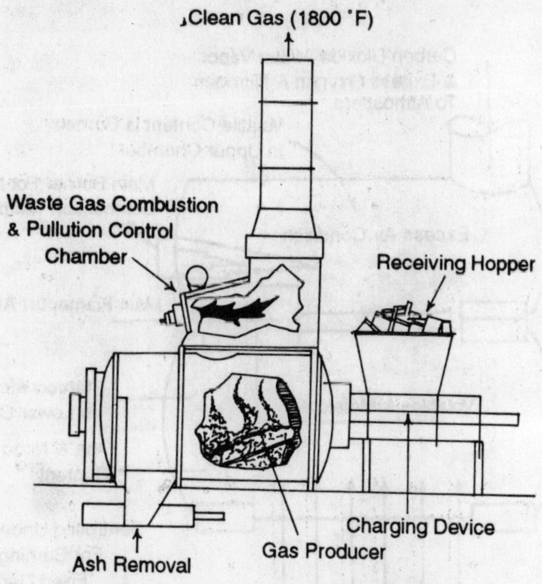

Figure 26.1. Controleed air incinerator.

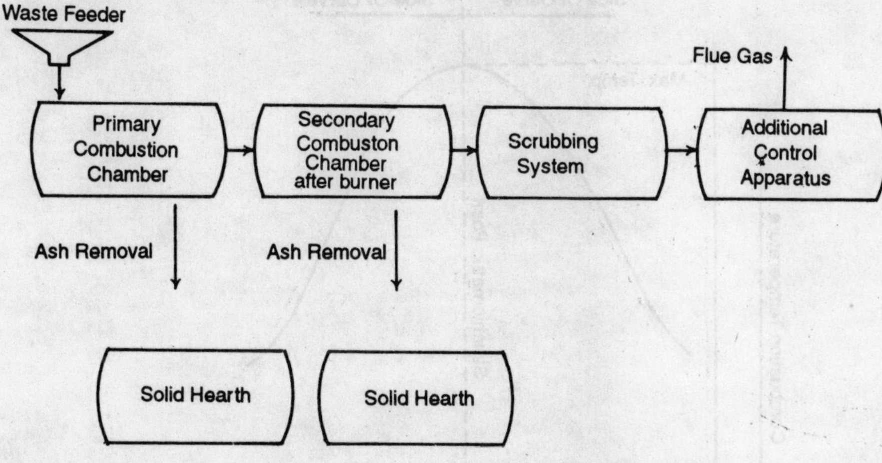

Figure 26.2. Controlled air incinerator (schematic).

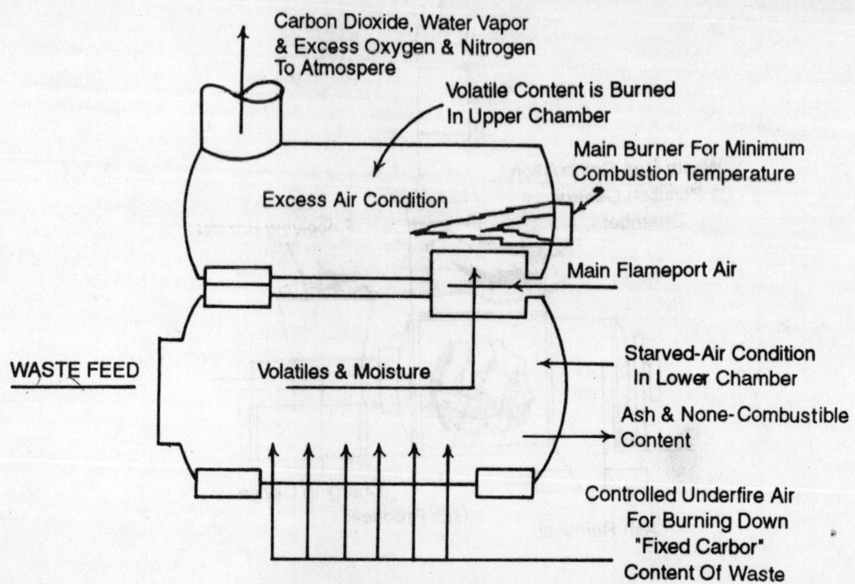

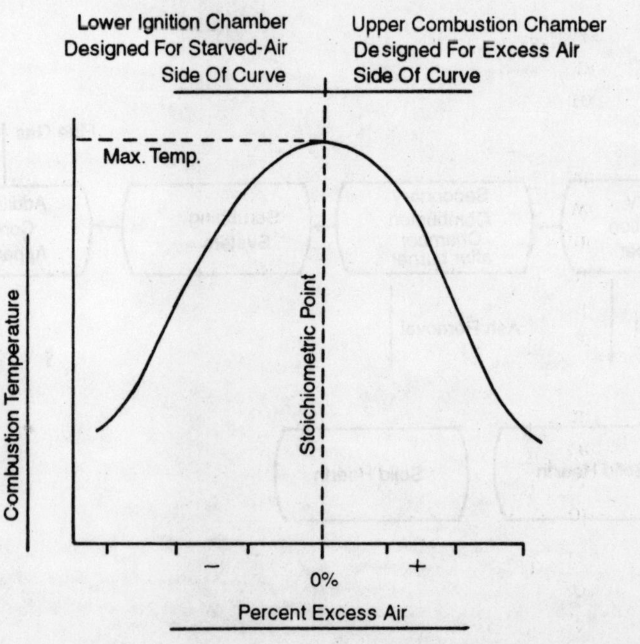

Figure 26.3. Incineration system—theory and operation.

in both chambers. A solid hearth is required by law for pathological destruction, to prevent discharge of noncombustibles into the sewer.

State-of-the-art controls include wet and dry scrubbers, auxiliary burners, bag houses, and electrostatic precipitators. Wet scrubbers produce a salty effluent,which in some cases may be discharged to a large publicly owned treatment plant. In other cases, the effluent must be discharged salt-free. The removal of the salt requires settling treatment, which furthermore may require first and second dewatering.

Dry scrubbers result in dry salt, which along with the fly ash, is collected in particulate control devices. Lime injection is recommended to the scrubber, as it provides the advantage of reducing the leachability of metals in the fly ash, and therefore makes land disposal of fly ash more environmentally acceptable.

Prevailing policy treats a manufacturing facility as a single facility, which due to existing sources of emissions might be beyond the 800 ib/hr threshold requiring a scrubber for a new incinerator of any size.

Should there be radioactive materials in addition to types 3 and 4 wastes to be incinerated, special on-line monitoring of emissions is required. The equipment should be designed for preheat and 2-second retention time at 1800°F, with continuous temperature recording to comply with applicable environmental regulations.

Venturi and packed-bed scrubbers are control devices that are very widely used in order to control particulate and HCI emissions. Venturi scrubbers use a scrubbing liquid, usually water or water-caustic solution, to neutralize hydrochloric acid gas. In the case of packed-bed scrubbers, the contaminants are removed by a gas absorption process that also depends on gas-liquid contact. Packed-bed columns are vessels that are filled with packing scrubbing material. The aforementioned scrubbing systems are widely used and accepted state-of-the-art control devices.

REGULATORY REQUIREMENTS AND OTHER ENVIRONMENTAL CONCERNS

On August 15, 1968, Subchapter 11, " Incinerators—Control and Prohibition of Air Pollution from Incinerators," was incorporated into a group of administrative regulations of the Air Pollution Control Code by the New Jersey State Department of Health. The following definition of a Special Incinerator is taken from Subchapter 11 and should be of interest:

"Special Incinerator: Municipal, pathological waste, or trade waste incinerator of any burning capacity, or any incinerator with a burning capacity in excess of 2,000 pounds per hour."

Section 3-"Emission Standards' states" "3.1 (d) Two years from the effective date of this chapter, no person shall use or cause to be used any existing special incinerator which shall emit more than 0.1 grains of particles, including the contribution of auxiliary fuel."

"3.2 (b) No person shall cause, suffer, allow or permit smoke from any incinerator the shade or appearance of which is darker than No.1 of the Ringleman Smoke Chart to be emitted into the open air; or emissions of such capacity with a stack or chimney, or exclusive of water vapor of such capacity leaving a stack or chimney to a degree greater than the emission designated as No.1 of the Ringlemann Smoke Chart."

Pathological incinerators may be regulated under hazardous waste regulations such as the Federal

Clean Air Act or the Resource Conservation Act, depending on the types of wastes. States and localities in many cases have applied emission standards, especially for particulate emissions and carbon monoxide, within their jurisdiction for all types of incinerators.

The EPA hazardous waste regulations address the management of toxic ignitable, corrosive, and reactive wastes where the Nuclear Regulatory Commission regulations address the management of radioactive wastes. Many states have passed hazardous-waste legislation to control the treatment, storage, and disposal of infectious waste since RCRA passed in 1976 and have already promulgated regulations conrolling infectious wastes, while some others are now preparing such regulations.

According to federal and state regulations, the incinerator of any facility requires an air permit for installation and operation. The application for such a permit must contain information regarding air contaminants (emissions) with calculated rates of emission in pounds per hour and equipment characteristics.

CONCLUSION

The intent of this chapter was to point out the potential hazards that may result from improper handling and disposal of pathological wastes, and to discuss the importance of incineration as the best available, demonstrated technology for such wastes' destruction. Even though pathological incineration is not a perfect technology due to the resulting emissions of unwanted products of incomplete combustion, considerable amounts of research and controls technology is currently in progress to ensure adequate and up-to-date incineration systems. With properly designed temperature interlocks controlling the critical parameters of waste feed-rate and combustion temperatures, compensating for various waste compositions, it is possible to safely, effectively, and efficiently manage and dispose of infectious pathological wastes within compliance of regulatory requirements.